Lecture Notes in Applied and Computational Mechanics

Volume 20

Series Editors

Prof. Dr.-Ing. Friedrich Pfeiffer
Prof. Dr.-Ing. Peter Wriggers

Lecture Notes in Applied and Computational Mechanics

Edited by F. Pfeiffer and P. Wriggers

Further volumes of this series found on our homepage: springer.com

Vol. 20: Zohdi T.I., Wriggers P.
Introduction to Computational Micromechanics
196 p. 2005 [978-3-540-22820-2]

Vol. 19: McCallen R., Browand F., Ross J. (Eds.)
The Aerodynamics of Heavy Vehicles:
Trucks, Buses, and Trains
567 p. 2004 [3-540-22088-7]

Vol. 18: Leine, R.I., Nijmeijer, H.
Dynamics and Bifurcations
of Non-Smooth Mechanical Systems
236 p. 2004 [3-540-21987-0]

Vol. 17: Hurtado, J.E.
Structural Reliability: Statistical Learning Perspectives
257 p. 2004 [3-540-21963-3]

Vol. 16: Kienzler R., Altenbach H., Ott I. (Eds.)
Theories of Plates and Shells:
Critical Review and New Applications
238 p. 2004 [3-540-20997-2]

Vol. 15: Dyszlewicz, J.
Micropolar Theory of Elasticity
356 p. 2004 [3-540-41835-0]

Vol. 14: Frémond M., Maceri F. (Eds.)
Novel Approaches in Civil Engineering
400 p. 2003 [3-540-41836-9]

Vol. 13: Kolymbas D. (Eds.)
Advanced Mathematical and Computational
Geomechanics
315 p. 2003 [3-540-40547-X]

Vol. 12: Wendland W., Efendiev M. (Eds.)
Analysis and Simulation of Multiїeld Problems
381 p. 2003 [3-540-00696-6]

Vol. 11: Hutter K., Kirchner N. (Eds.)
Dynamic Response of Granular and Porous Materials
under Large and Catastrophic Deformations
426 p. 2003 [3-540-00849-7]

Vol. 10: Hutter K., Baaser H. (Eds.)
Deformation and Failure in Metallic Materials
409 p. 2003 [3-540-00848-9]

Vol. 9: Skrzypek J., Ganczarski A.W. (Eds.)
Anisotropic Behaviour of Damaged Materials
366 p. 2003 [3-540-00437-8]

Vol. 8: Kowalski, S.J.
Thermomechanics of Drying Processes
365 p. 2003 [3-540-00412-2]

Vol. 7: Shlyannikov, V.N.
Elastic-Plastic Mixed-Mode Fracture Criteria and Parameters
246 p. 2002 [3-540-44316-9]

Vol. 6: Popp K., Schiehlen W. (Eds.)
System Dynamics and Long-Term Behaviour
of Railway Vehicles, Track and Subgrade
488 p. 2002 [3-540-43892-0]

Vol. 5: Duddeck, F.M.E.
Fourier BEM: Generalization
of Boundary Element Method by Fourier Transform
181 p. 2002 [3-540-43138-1]

Vol. 4: Yuan, H.
Numerical Assessments of Cracks in Elastic-Plastic Materials
311 p. 2002 [3-540-43336-8]

Vol. 3: Sextro, W.
Dynamical Contact Problems with Friction:
Models, Experiments and Applications
159 p. 2002 [3-540-43023-7]

Vol. 2: Schanz, M.
Wave Propagation in Viscoelastic
and Poroelastic Continua
170 p. 2001 [3-540-41632-3]

Vol. 1: Glocker, C.
Set-Valued Force Laws:
Dynamics of Non-Smooth Systems
222 p. 2001 [3-540-41436-3]

An Introduction to Computational Micromechanics

Corrected Second Printing

Tarek I. Zohdi · Peter Wriggers

With 66 Figures and 9 Tables

Springer

Prof. Tarek I. Zohdi
University of California
Department of Mechanical Engineering
Etcheverry Hall 6195
Berkeley, CA 94720-1740
USA
zohdi@me.berkeley.edu

Prof. Dr. Peter Wriggers
Universität Hannover
Fachbereich Bauingenieur- und Vermessungswesen
Institut für Baumechanik und Numerische Mechanik
Appelstr. 9A
30167 Hannover
Germany
wriggers@ibnm.uni-hannover.de

ISBN: 978-3-540-77482-2 e-ISBN: 978-3-540-32360-0

Lecture Notes in Applied and Computational Mechanics ISSN 1613-7736

Library of Congress Control Number: 2007942179

Cover design: WMXDesign GmbH, Heidelberg

Printed on acid-free paper

9 8 7 8 6 5 4 3 2 1 0

springer.com

Preface

Ideally, in an attempt to reduce laboratory costs, one would like to make a prediction of a new material's behavior by numerical simulation, with the primary goal being to accelerate trial and error experimental testing. The recent dramatic increase in computational power available for mathematical modeling and simulation raises the possibility that modern numerical methods can play a significant role in the analysis of heterogeneous microstructures. This fact has motivated the work that will be presented in this monograph, which contains basic homogenization theory, as well as introductions to topics such as microstructural optimization and multifield analysis of heterogeneous materials. The text can be viewed as a research monograph suitable for use in a first year graduate course for students in the applied sciences, mechanics and mathematics that have an interest in the computational micromechanical analysis of new materials.

Berkeley, USA, June 2004 — *Tarek I. Zohdi*
Hannover, Germany, June 2004 — *Peter Wriggers*

Contents

Chapter 1
Introduction

A key to the success of many modern structural components is the tailored behavior of the material. A relatively inexpensive way to obtain macroscopically desired responses is to enhance a base material's properties by the addition of microscopic matter, i.e. *to manipulate the microstructure.* Accordingly, in many modern engineering designs, materials with highly complex microstructures are now in use. The macroscopic characteristics of modified base materials are the aggregate response of an assemblage of different "pure" components, for example several particles or fibers suspended in a binding matrix material (Fig. 1.1). Thus, microscale inhomogeneities are encountered in metal matrix composites, concrete, etc. In the construction of such materials, the basic philosophy is to select material combinations to produce desired aggregate responses. For example, in structural engineering applications, the classical choice is a harder particulate phase that serves as a stiffening agent for a ductile, easy to form, base matrix material.

If one were to attempt to perform a direct numerical simulation, for example of the mechanical response of a macroscopic engineering structure composed of a microheterogeneous material, incorporating all of the microscale details, an extremely fine spatial discretization mesh, for example that of a finite element mesh, would be needed to capture the effects of the microscale heterogeneities. The resulting system of equations would contain literally billions of numerical unknowns. Such problems are beyond the capacity of computing machines for the foreseeable future. Furthermore, the exact subsurface geometry is virtually impossible to ascertain throughout the structure. In addition, even if one could solve such a system, the amount of information to process would be of such complexity that it would be difficult to extract any useful information on the desired macroscopic behavior. It is important to realize that solutions to partial differential equations, of even linear material models, at infinitesimal strains, describing the response of small bodies containing a few heterogeneities are still open problems. *In short, complete solutions are virtually impossible.*

Because of these facts, the use of *regularized* or *homogenized* material models (resulting in smooth coefficients in the partial differential equations) are commonplace in practically all branches of the physical sciences. The usual approach is

Fig. 1.1 Doping a base material with particulate additives

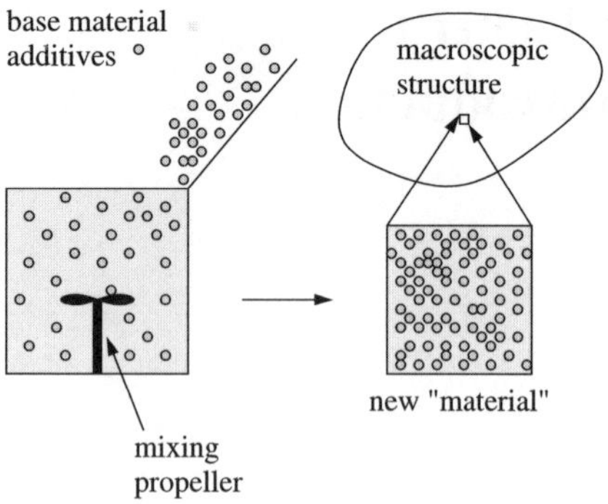

to compute a constitutive "relation between averages", relating volume averaged field variables. Thereafter, the regularized properties can be used in a macroscopic analysis (Fig. 1.2). The volume averaging takes place over a statistically representative sample of material, referred to in the literature as a representative volume element (RVE). The internal fields to be volumetrically averaged must be computed by solving a series boundary value problems with test loadings. Such regularization processes are referred to as "homogenization", "mean field theories", "theories of effective properties", etc. For overviews, we refer the interested reader to Jikov et al. [103] for mathematical aspects or to Aboudi [1], Hashin [76], Mura [150] and Nemat-Nasser and Hori [151] for mechanically inclined accounts of the subject.

For a sample to be statistically representative it must usually contain a large number of heterogeneities (Fig. 1.3) and, therefore, the computations over the RVE are still extremely large, but are of reduced computational effort in comparison with a direct attack on the "real" problem. Historically, most classical analytical methods for estimating the macroscopic response of such engineering materials have

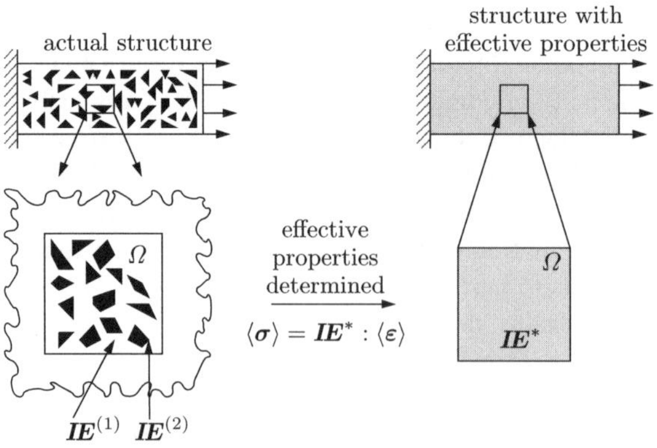

Fig. 1.2 The use of effective properties in structural engineering

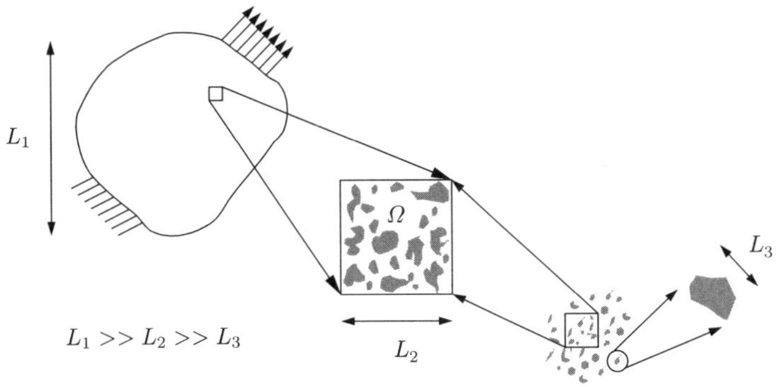

Fig. 1.3 The size requirements of a representative volume element

a strongly phenomenological basis and are, in reality, non-predictive of material responses that are unknown a-priori. This is true even in the linearly elastic, infinitesimal strain, range. In plain words, such models require extensive experimental data to "tune" parameters that have little or no physical significance. Criticisms, such as the one stated, have led to computational approaches which require relatively simple descriptions on the microscale, containing parameters that are physically meaningful. In other words, the phenomenological aspects of the material modeling are reduced, with the burden of the work being shifted to high performance computational methods. *Stated clearly, the "mission" of computational micro-macro mechanics is to determine relationships between the microstructure and the macroscopic response or "structural property" of a material, using models on the microscale that are as simple as possible.*[1]

1.1 Basic Concepts in Micro–Macro Modeling

Initially, for illustration purposes, we consider the relatively simple case of linear elasticity. In this context, the mechanical properties of microheterogeneous materials are characterized by a spatially variable elasticity tensor $\mathbb{E}$. Typically, in order to characterize the (homogenized) effective macroscopic response of such materials, a relation between averages

$$\langle \sigma \rangle_\Omega = \mathbb{E}^* : \langle \varepsilon \rangle_\Omega \tag{1.1}$$

is sought, where

$$\langle \cdot \rangle_\Omega \stackrel{\text{def}}{=} \frac{1}{|\Omega|} \int_\Omega \cdot \, d\Omega \,, \tag{1.2}$$

[1] Appropriately, the following quote, attributed to Albert Einstein: "*A model should be as simple as possible, but not simpler*", is a guiding principle.

and where σ and ε are the stress and strain tensor fields within a statistically representative volume element (RVE) of volume $|\Omega|$. The quantity, $\mathbb{E}^*$, is known as the effective property and is the elasticity tensor used in usual structural scale analyses. Similarly, one can describe other effective quantities such as conductivity or diffusivity, in virtually the same manner, relating other volumetrically averaged field variables. However, for the sake of brevity, we initially treat linear elastostatics problems.[2] *It is emphasized that effective quantities such as* $\mathbb{E}^*$ *are not material properties, but relations between averages.* A more appropriate term might be "apparent property", which is elaborated upon in Huet [88, 89]. However, to be consistent with the literature, we continue to refer to $\mathbb{E}^*$ by the somewhat inaccurate term effective "property".

1.2 Historical Overview

The analysis of microheterogeneous materials is not a recent development. Within the last 150 years, estimates of effective responses have been made under a variety of assumptions on the internal fields within the microstructure. Works dating back at least to Maxwell [138, 139] and Lord Rayleigh [172] have dealt with determining overall macroscopic *transport* phenomena of materials consisting of a matrix and distributions of spherical particles. Voigt [212] is usually credited with the first analysis of the effective *mechanical* properties of the microheterogeneous solids, with a complementary contribution given later by Reuss [173]. Voigt assumed that the strain field within an aggregate sample of heterogeneous material was uniform, leading to $\langle \mathbb{E} \rangle_\Omega$ as an approximation of the effective property, while Reuss approximated the stress fields within the aggregate of polycrystalline material as uniform. If the Reuss field is assumed within the RVE, then an approximation for the effective property is $\langle \mathbb{E}^{-1} \rangle_\Omega^{-1}$. A fundamental result (Hill [79]) is

$$\langle \mathbb{E}^{-1} \rangle_\Omega^{-1} \leq \mathbb{E}^* \leq \langle \mathbb{E} \rangle_\Omega . \tag{1.3}$$

These inequalities mean that the eigenvalues of the tensors $\mathbb{E}^* - \langle \mathbb{E}^{-1} \rangle_\Omega^{-1}$ and $\langle \mathbb{E} \rangle_\Omega - \mathbb{E}^*$ are non-negative. Therefore, one can interpret the Voigt and Reuss fields as providing two microfield extremes, since the Voigt stress field is one where the tractions at the phase boundaries cannot be in equilibrium, i.e. statically inadmissible, while the implied Reuss strains are such that the heterogeneities and the matrix could not be perfectly bonded, i.e. the field is kinematically inadmissible. Typically, the bounds are quite wide and provide only rough qualitative information.

Within the last 50 years improved estimates have been pursued. For example, the *Dilute* family of methods assume that there is no particle-particle interaction. With this assumption one requires only the solution to a single ellipsoidal particle

[2] By direct extension, in the geometrically nonlinear regime, such as $\langle \mathbf{S} \rangle_{\Omega_0}$ vs $\langle \mathbf{E} \rangle_{\Omega_0}$ or $\langle \mathbf{P} \rangle_{\Omega_0}$ vs $\langle \mathbf{F} \rangle_{\Omega_0}$, etc ..., where $\mathbf{P}$ and $\mathbf{S}$ are the First and Second Piola Kirchhoff stresses, $\mathbf{E}$ is the Green-Lagrange strain, $\mathbf{F}$ is the deformation gradient and where Ω_0 is the initial reference domain.

in an unbounded domain. This is the primary use of the elegant Eshelby [36] formalism, based on eigenstrain concepts, which is used to determine the solution to the problem of a single inclusion embedded in an infinite matrix of material under uniform exterior loading. By itself, this result is of little practical interest, however the solution is relatively compact and easy to use and, thus, has been the foundation for the development of many approximation methods based on non-interacting and weakly interacting (particle–particle) assumptions. *Non-interaction of particles is an unrealistic assumption for materials with randomly dispersed particulate microstructure, even at a few percent volume fraction.*

One classical attempt to improve the Dilute family of approximations is through ideas of self-consistency (Budiansky [19], Hill [80]). For example, in the standard *Self-Consistent* method, the idea is simply that the particle "sees" the effective medium instead of the matrix in the calculations. In other words the "matrix material" in the Eshelby analysis is simply the effective medium. Unfortunately, the Self-Consistent method can produce negative effective bulk and shear responses, for voids, with pore volume fractions of 50% and higher. For rigid inclusions it produces infinite effective bulk responses for any volume fraction and infinite effective shear responses above 40%. For details, see Aboudi [1]. Attempts have also been made to improve these approaches. For example the *Generalized Self-Consistent* method encases the particles in a shell of matrix material surrounded by the effective medium (see Christensen [27]). However, such methods also exhibit problems, primarily due to mixing scales of information in a phenomenological manner, which are critically discussed at length in Hashin [76]. For a relatively recent and thorough analysis of a variety of classical approaches, such as the ones briefly mentioned here, see Torquato [201, 202, 203, 204].

In addition to ad-hoc assumptions and estimates on the interaction between microscale constituents, many classical methods of analysis treat the microstructure as being a regular, periodic, infinite array of identical cells. Such an assumption is not justifiable for virtually any real material. *In this monograph we do not consider the highly specialized case of periodic microstructures.* There exists a large body of literature for periodic idealizations, and we refer the reader, for example, to Aboudi [1] for references.

1.3 Objectives of this Monograph

It is now commonly accepted that some type of numerical simulation is necessary in order to determine more accurate micro–macro structural responses. *Ideally, in an attempt to reduce laboratory expense, one would like to make predictions of a new material's behavior by numerical simulations, with the primary goal being to accelerate the trial and error laboratory development of new high performance materials.* Within the last 20 years, many numerical analyses have been performed two-dimensionally, which, in contrast to many structural problems, is relatively meaningless in micro–macro mechanics. For a statistically representative sample of microstructural material, three-dimensional numerical simulations are unavoidable for reliable results. As we have indicated, the analysis of even *linearly-elastic* materials, at infinitesimal strains, composed of a homogeneous matrix filled with microscale heterogeneities, is still extremely difficult. However, the recent dramatic increase in computational power available for mathematical modeling and simulation raises the possibility that modern numerical methods can play a significant role in the analysis of heterogeneous structures, even in the nonlinear range. This fact, among others, has motivated the work that will be presented in this monograph.

Chapter 2
Some Basics of the Mechanics of Solid Continua

Throughout this work, boldface symbols imply vectors or tensors. For the inner product of two vectors (first order tensors) $\mathbf{u}$ and $\mathbf{v}$ we have in three dimensions, $\mathbf{u} \cdot \mathbf{v} = v_i u_i = u_1 v_1 + u_2 v_2 + u_3 v_3$, where Cartesian bases and Einstein index summation notation are used. At the risk of over simplification, *we will ignore the difference between second order tensors and matrices.* Furthermore, we exclusively employ a Cartesian bases. Readers that feel uncomfortable with this approach should consult the wide range of texts which point out the subtle differences, for example the texts of Malvern [133], Marsden and Hughes [137] and others listed in the references. Accordingly, if we consider the second order tensor $\mathbf{A} = A_{ik} \mathbf{e}_i \otimes \mathbf{e}_k$ with its matrix representation

$$[\mathbf{A}] \stackrel{\text{def}}{=} \begin{bmatrix} A_{11} & A_{12} & A_{13} \\ A_{21} & A_{22} & A_{23} \\ A_{31} & A_{32} & A_{33} \end{bmatrix}, \tag{2.1}$$

then a first order contraction (inner product) of two second order tensors $\mathbf{A} \cdot \mathbf{B}$ is defined by the matrix product $[\mathbf{A}][\mathbf{B}]$, with components of $A_{ij} B_{jk} = C_{ik}$. It is clear that the range of the inner index j must be the same for $[\mathbf{A}]$ and $[\mathbf{B}]$. For three dimensions we have $i, j = 1, 2, 3$. The second order inner product of two tensors or matrices is $\mathbf{A} : \mathbf{B} = A_{ij} B_{ij} = tr([\mathbf{A}]^T [\mathbf{B}])$. The divergence of a vector $\mathbf{u}$, which results in a contraction to a scalar, is defined by $\nabla_x \cdot \mathbf{u} = u_{i,i}$, whereas for a second order tensor, $\mathbf{A}$, $\nabla_x \cdot \mathbf{A}$ describes a contraction to a vector with the components $A_{ij,j}$. The gradient of a vector $\mathbf{u}$(a dilation to a second order tensor) is given by $\nabla_x \mathbf{u}$ and has the components $u_{i,j}$, whereas for a second order tensor (a dilation to a third order tensor) $\nabla_x \mathbf{A}$ has components of $A_{ij,k}$. The gradient of a scalar ϕ (a dilation to a vector) is defined by $\nabla_x \phi$ and has the components $\phi_{,i}$. The scalar product of two second order tensors, for example the gradients of first order vectors, is defined as $\nabla_x \mathbf{v} : \nabla_x \mathbf{u} = \frac{\partial v_i}{\partial x_j} \frac{\partial u_i}{\partial x_j} \stackrel{\text{def}}{=} v_{i,j} u_{i,j}$, where $\partial u_i / \partial x_j, \partial v_i / \partial x_j$ are partial derivatives of u_i and v_i, and where u_i, v_i are the Cartesian components of $\mathbf{u}$ and $\mathbf{v}$. An example for the product of a tensor with a vector is $\nabla_x \mathbf{u} \cdot \mathbf{n}$ which has components of $u_{i,j} n_j$ in a Cartesian bases.

2.1 Kinematics of Deformations

The term deformation refers to a change in the shape of the continuum between a reference configuration and current configuration. In the reference configuration, a representative particle of the continuum occupies a point **p** in space and has the position vector (Fig. 2.1)

$$\mathbf{X} = X_1\mathbf{e}_1 + X_2\mathbf{e}_2 + X_3\mathbf{e}_3\,, \tag{2.2}$$

where $\mathbf{e}_1, \mathbf{e}_2, \mathbf{e}_3$ is a Cartesian reference triad, and X_1, X_2, X_3 (with center **O**) can be thought of as labels for a point. Sometimes the coordinates or labels (X_1, X_2, X_3, t) are called the referential coordinates. In the current configuration the particle originally located at point **p** is located at point $\mathbf{p}'$ and can be expressed also in terms of another position vector **x**, with the coordinates (x_1, x_2, x_3, t). These are called the current coordinates. It is obvious with this arrangement, that the displacement is $\mathbf{u} = \mathbf{x} - \mathbf{X}$ for a point originally at **X** and with final coordinates **x**.

When a continuum undergoes deformation (or flow), its points move along various paths in space. This motion may be expressed by

$$\mathbf{x}(X_1, X_2, X_3, t) = \mathbf{u}(X_1, X_2, X_3, t) + \mathbf{X}(X_1, X_2, X_3, t)\,, \tag{2.3}$$

which gives the present location of a point at time t, written in terms of the labels X_1, X_2, X_3. The previous position vector may be interpreted as a mapping of the initial configuration onto the current configuration. In classical approaches, it is assumed that such a mapping is a one-to-one and continuous, with continuous partial derivatives to whatever order is required. The description of motion or deformation expressed previously is known as the Lagrangian formulation. Alternatively, if the independent variables are the coordinates **x** and t, then $\mathbf{x}(x_1, x_2, x_3, t) = \mathbf{u}(x_1, x_2, x_3, t) + \mathbf{X}(x_1, x_2, x_3, t)$, and the formulation is denoted as Eulerian (Fig. 2.1).

2.1.1 Deformation of Line Elements

Partial differentiation of the displacement vector $\mathbf{u} = \mathbf{x} - \mathbf{X}$, with respect to **x** and **X**, produces the following displacement gradients:

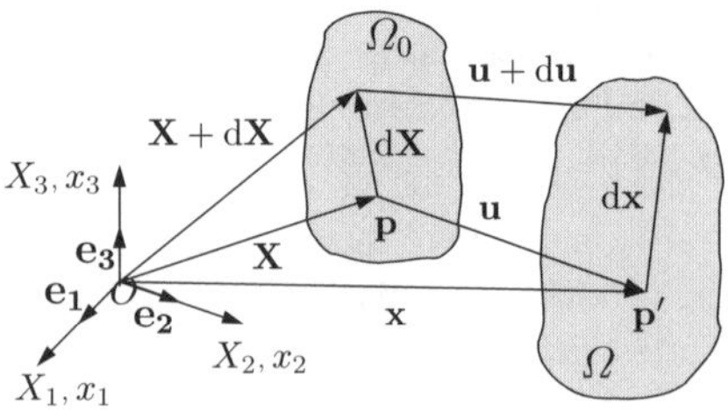

Fig. 2.1 Different descriptions of a deforming body

$$\nabla_X \mathbf{u} = \mathbf{F} - \mathbf{1} \qquad \text{and} \qquad \nabla_x \mathbf{u} = \mathbf{1} - \overline{\mathbf{F}} \tag{2.4}$$

where

$$\nabla_X \mathbf{x} \stackrel{\text{def}}{=} \frac{\partial \mathbf{x}}{\partial \mathbf{X}} = \mathbf{F} \stackrel{\text{def}}{=} \begin{bmatrix} \frac{\partial x_1}{\partial X_1} & \frac{\partial x_1}{\partial X_2} & \frac{\partial x_1}{\partial X_3} \\ \frac{\partial x_2}{\partial X_1} & \frac{\partial x_2}{\partial X_2} & \frac{\partial x_2}{\partial X_3} \\ \frac{\partial x_3}{\partial X_1} & \frac{\partial x_3}{\partial X_2} & \frac{\partial x_3}{\partial X_3} \end{bmatrix}, \tag{2.5}$$

and

$$\nabla_x \mathbf{X} \stackrel{\text{def}}{=} \frac{\partial \mathbf{X}}{\partial \mathbf{x}} = \overline{\mathbf{F}} \tag{2.6}$$

with the components $F_{ik} = x_{i,k}$ and $\overline{F}_{ik} = X_{i,k}$. $\mathbf{F}$ is known as the material deformation gradient, and $\overline{\mathbf{F}}$ is known as the spatial deformation gradient.

Now, consider the length of a differential element in the reference configuration $d\mathbf{X}$ and $d\mathbf{x}$ in the current configuration, $d\mathbf{x} = \nabla_X \mathbf{x} \cdot d\mathbf{X} = \mathbf{F} \cdot d\mathbf{X}$. Taking the difference in the magnitudes of these elements yields

$$\begin{aligned} d\mathbf{x} \cdot d\mathbf{x} - d\mathbf{X} \cdot d\mathbf{X} &= (\nabla_X \mathbf{x} \cdot d\mathbf{X}) \cdot (\nabla_X \mathbf{x} \cdot d\mathbf{X}) - d\mathbf{X} \cdot d\mathbf{X} \\ &= d\mathbf{X} \cdot (\mathbf{F}^T \cdot \mathbf{F} - \mathbf{1}) \cdot d\mathbf{X} \stackrel{\text{def}}{=} 2 d\mathbf{X} \cdot \mathbf{E} \cdot d\mathbf{X}\,. \end{aligned} \tag{2.7}$$

Alternatively, we have with $d\mathbf{X} = \nabla_x \mathbf{X} \cdot d\mathbf{x} = \overline{\mathbf{F}} \cdot d\mathbf{x}$ and

$$\begin{aligned} d\mathbf{x} \cdot d\mathbf{x} - d\mathbf{X} \cdot d\mathbf{X} &= d\mathbf{x} \cdot d\mathbf{x} - (\nabla_x \mathbf{X} \cdot d\mathbf{x}) \cdot (\nabla_x \mathbf{X} \cdot d\mathbf{x}) \\ &= d\mathbf{x} \cdot (\mathbf{1} - \overline{\mathbf{F}}^T \cdot \overline{\mathbf{F}}) \cdot d\mathbf{x} \stackrel{\text{def}}{=} 2 d\mathbf{x} \cdot \mathbf{e} \cdot d\mathbf{x}\,. \end{aligned} \tag{2.8}$$

Equation (2.7) defines the so-called *Lagrangian* strain tensor

$$\mathbf{E} \stackrel{\text{def}}{=} \tfrac{1}{2}(\mathbf{F}^T \cdot \mathbf{F} - \mathbf{1}) = \tfrac{1}{2}[\nabla_X \mathbf{u} + (\nabla_X \mathbf{u})^T + (\nabla_X \mathbf{u})^T \cdot \nabla_X \mathbf{u}]\,. \tag{2.9}$$

Frequently, the Lagrangian strain tensor is defined in terms of the so-called right Cauchy-Green strain, $\mathbf{C} = \mathbf{F}^T \cdot \mathbf{F}$ leading to $\mathbf{E} \stackrel{\text{def}}{=} \frac{1}{2}(\mathbf{C} - \mathbf{1})$. The Eulerian strain tensor is defined by (2.8) as

$$\mathbf{e} \stackrel{\text{def}}{=} \tfrac{1}{2}(\mathbf{1} - \overline{\mathbf{F}}^T \cdot \overline{\mathbf{F}}) = \tfrac{1}{2}(\nabla_x \mathbf{u} + (\nabla_x \mathbf{u})^T - (\nabla_x \mathbf{u})^T \cdot \nabla_x \mathbf{u})\,. \tag{2.10}$$

In a similar manner as for the Lagrangian strain tensor, the Eulerian strain tensor can be defined in terms of the so-called left Cauchy-Green strain $\mathbf{b} = \mathbf{F} \cdot \mathbf{F}^T$ which then yields $\mathbf{e} \stackrel{\text{def}}{=} \frac{1}{2}(\mathbf{1} - \mathbf{b}^{-1})$.

Remark. It should be clear that $d\mathbf{x}$ can be reinterpreted as the result of a mapping $\mathbf{F} \cdot d\mathbf{X} \longrightarrow d\mathbf{x}$, or a change in configuration (reference to current, Fig. 2.2) while $\overline{\mathbf{F}} \cdot d\mathbf{x} \longrightarrow d\mathbf{X}$, maps the current to the reference system. For the deformations to be invertible, and physically realizable, $\overline{\mathbf{F}} \cdot (\mathbf{F} \cdot d\mathbf{X}) = d\mathbf{X}$ and $\mathbf{F} \cdot (\overline{\mathbf{F}} \cdot d\mathbf{x}) = d\mathbf{x}$. We note

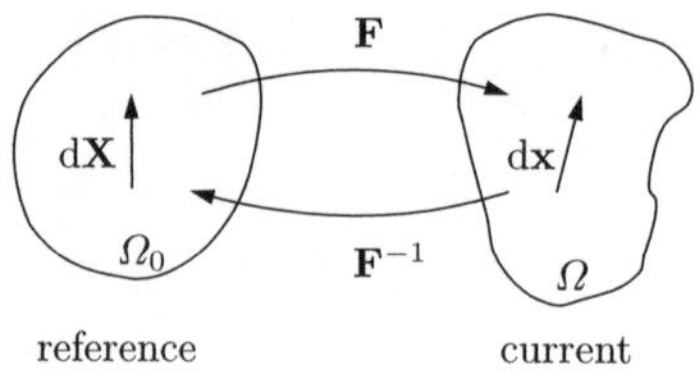

Fig. 2.2 Mapping back and forth with the deformation gradient

that $(det\overline{\mathbf{F}})(det\mathbf{F}) = 1$ and have the following obvious relation $\overline{\mathbf{F}} \cdot \mathbf{F} = \mathbf{1}$. It should be clear that $\overline{\mathbf{F}} = \mathbf{F}^{-1}$.

2.1.2 Infinitesimal Strain Measures

In infinitesimal deformation theory, the displacement gradient components being "small" implies that higher order terms like $(\nabla_X \mathbf{u})^T \cdot \nabla_X \mathbf{u}$ and $(\nabla_x \mathbf{u})^T \cdot \nabla_x \mathbf{u}$ can be neglected in the strain measures $\mathbf{e}$ and $\mathbf{E}$ leading to $\mathbf{e} \approx \varepsilon^E \stackrel{\text{def}}{=} \frac{1}{2}[\nabla_x \mathbf{u} + (\nabla_x \mathbf{u})^T]$ and $\mathbf{E} \approx \varepsilon^L \stackrel{\text{def}}{=} \frac{1}{2}[\nabla_X \mathbf{u} + (\nabla_X \mathbf{u})^T]$. If the displacement gradients are small compared with unity, ε^E and ε^L coincide closely to $\mathbf{e}$ and $\mathbf{E}$, respectively. If we assume that, $\frac{\partial}{\partial \mathbf{X}} \approx \frac{\partial}{\partial \mathbf{x}}$, we may use ε^E or ε^L interchangeably. Usually ε is the symbol used for infinitesimal strains. Furthermore, to avoid confusion, when using models employing the geometrically linear infinitesimal strain assumption we use the symbol of ∇ with no $\mathbf{X}$ or $\mathbf{x}$ subscript. Hence the infinitesimal strains are defined by

$$\varepsilon = \frac{1}{2}(\nabla \mathbf{u} + (\nabla \mathbf{u})^T)\,. \tag{2.11}$$

2.1.3 The Jacobian of the Deformation Gradient

The Jacobian of the deformation gradient, $\mathbf{F}$, is defined as

$$J \stackrel{\text{def}}{=} det\mathbf{F} = \begin{vmatrix} \frac{\partial x_1}{\partial X_1} & \frac{\partial x_1}{\partial X_2} & \frac{\partial x_1}{\partial X_3} \\ \frac{\partial x_2}{\partial X_1} & \frac{\partial x_2}{\partial X_2} & \frac{\partial x_2}{\partial X_3} \\ \frac{\partial x_3}{\partial X_1} & \frac{\partial x_3}{\partial X_2} & \frac{\partial x_3}{\partial X_3} \end{vmatrix}\,. \tag{2.12}$$

To interpret the Jacobian in a physical way, consider a reference differential volume in Fig. 2.3 which is given by $dS^3 = d\omega$, where $d\mathbf{X}^{(1)} = dS\mathbf{e}_1$, $d\mathbf{X}^{(2)} = dS\mathbf{e}_2$ and

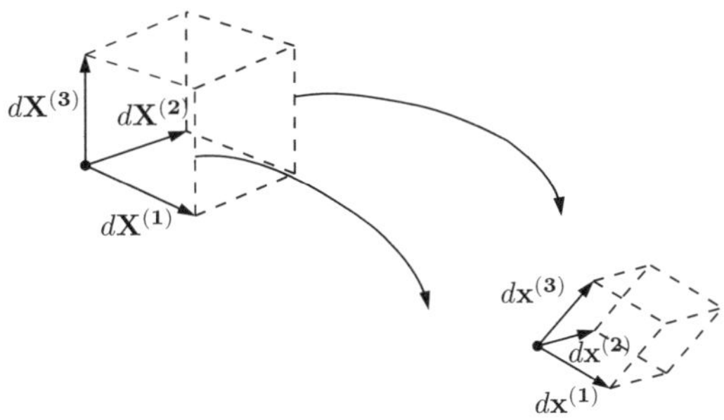

Fig. 2.3 A differential volume element

$d\mathbf{X}^{(3)} = dS\,\mathbf{e}_3$. The current differential element is described by $d\mathbf{x}^{(1)} = \frac{\partial \mathbf{x}_k}{\partial X_1} dS\,\mathbf{e}_k$, $d\mathbf{x}^{(2)} = \frac{\partial \mathbf{x}_k}{\partial X_2} dS\,\mathbf{e}_k$ and $d\mathbf{x}^{(3)} = \frac{\partial \mathbf{x}_k}{\partial X_3} dS\,\mathbf{e}_k$, where $\mathbf{e}_k$ is a unit vector, and

$$\underbrace{d\mathbf{x}^{(1)} \cdot (d\mathbf{x}^{(2)} \times d\mathbf{x}^{(3)})}_{\overset{\text{def}}{=} d\omega} = \begin{vmatrix} dx_1^{(1)} & dx_2^{(1)} & dx_3^{(1)} \\ dx_1^{(2)} & dx_2^{(2)} & dx_3^{(2)} \\ dx_1^{(3)} & dx_2^{(3)} & dx_3^{(3)} \end{vmatrix} = \begin{vmatrix} \frac{\partial x_1}{\partial X_1} & \frac{\partial x_2}{\partial X_1} & \frac{\partial x_3}{\partial X_1} \\ \frac{\partial x_1}{\partial X_2} & \frac{\partial x_2}{\partial X_2} & \frac{\partial x_3}{\partial X_2} \\ \frac{\partial x_1}{\partial X_3} & \frac{\partial x_2}{\partial X_3} & \frac{\partial x_3}{\partial X_3} \end{vmatrix} dS^3. \quad (2.13)$$

Therefore, $d\omega = J\,d\omega_0$. Thus, the Jacobian of the deformation gradient must remain positive definite, otherwise we obtain physically impossible "negative" volumes.

2.2 Equilibrium / Kinetics of Solid Continua

We start with the following postulated balance law for an arbitrary part ω around a point P with boundary $\partial\omega$ of a body Ω, see Fig. 2.4,

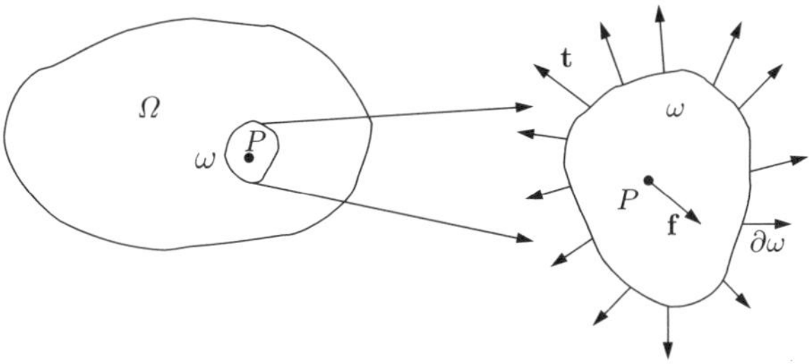

Fig. 2.4 Newton's laws applied to a continuum

$$\underbrace{\int_{\partial\omega} \mathbf{t} da}_{\text{surface forces}} + \underbrace{\int_{\omega} \mathbf{f} d\omega}_{\text{body forces}} = \underbrace{\frac{d}{dt}\int_{\omega} \rho \dot{\mathbf{u}} d\omega}_{\text{inertial forces}}, \tag{2.14}$$

where ρ is the material density, $\mathbf{b}$ is the body force per unit mass ($\mathbf{f} = \rho\mathbf{b}$) and $\dot{\mathbf{u}}$ is the time derivative of the displacement. When the actual molecular structure is considered on a sub-microscopic scale, the force densities, $\mathbf{t}$, which we commonly refer to as "surface forces", are taken to involve short-range intermolecular forces. Tacitly we assume that the effects of radiative forces, and others which do not require momentum transfer through a continuum, are negligible. This is a so-called local action postulate. As long as the volume element is large, our resultant body and surface forces may be interpreted as sums of these intermolecular forces. When we pass to larger scales, we can justifiably use the continuum concept.

2.2.1 Postulates on Volume and Surface Quantities

Now, consider a tetrahedron in equilibrium, as shown in Fig. 2.5. From Newton's laws,

$$\mathbf{t}^{(n)}\Delta A^{(n)} + \mathbf{t}^{(-1)}\Delta A^{(1)} + \mathbf{t}^{(-2)}\Delta A^{(2)} + \mathbf{t}^{(-3)}\Delta A^{(3)} + \mathbf{f}\Delta V = \rho \Delta V \ddot{\mathbf{u}}, \tag{2.15}$$

where $\Delta A^{(n)}$ is the surface area of the face of the tetrahedron with normal $\mathbf{n}$, and ΔV is the tetrahedron volume. Clearly, as the distance between the tetrahedron base

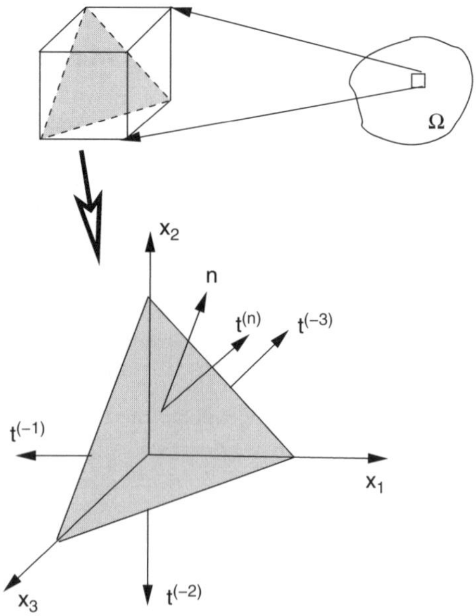

Fig. 2.5 Cauchy tetrahedron: a "sectioned material point"

(located at (0,0,0)) and the surface center, denoted h, goes to zero ($h \to 0$) we have $\Delta A^{(n)} \to 0 \Rightarrow \frac{\Delta V}{\Delta A^{(n)}} \to 0$. Geometrically, we have $\frac{\Delta A^{(i)}}{\Delta A^{(n)}} = \cos(x_i, x_n) \stackrel{\text{def}}{=} n_i$, and therefore $\mathbf{t}^{(n)} + \mathbf{t}^{(-1)} \cos(x_1, x_n) + \mathbf{t}^{(-2)} \cos(x_2, x_n) + \mathbf{t}^{(-3)} \cos(x_3, x_n) = \mathbf{0}$. It is clear that forces on the surface areas could be decomposed into three linearly independent components. It is convenient to introduce the concept of stress at a point, representing the surface forces there, pictorially represented by a cube surrounding a point. The fundamental issue that must be resolved is the characterization of these surface forces. We can represent the force density vector, the so-called "traction", on a surface by the component representation:

$$\mathbf{t} \stackrel{\text{def}}{=} \left\{ \begin{array}{c} \sigma_{i1} \\ \sigma_{i2} \\ \sigma_{i3} \end{array} \right\}, \tag{2.16}$$

where the second index represents the direction of the component and the first index represents the normal to corresponding coordinate plane. From this point forth, we will drop the superscript notation of $\mathbf{t}^{(n)}$, where it is implicit that $\mathbf{t} \stackrel{\text{def}}{=} \mathbf{t}^{(n)} = \sigma^T \cdot \mathbf{n}$, where

$$\sigma \stackrel{\text{def}}{=} \begin{bmatrix} \sigma_{11} & \sigma_{12} & \sigma_{13} \\ \sigma_{21} & \sigma_{22} & \sigma_{23} \\ \sigma_{31} & \sigma_{32} & \sigma_{33} \end{bmatrix}, \tag{2.17}$$

or explicitly ($\mathbf{t}^{(1)} = -\mathbf{t}^{(-1)}$, $\mathbf{t}^{(2)} = -\mathbf{t}^{(-2)}$, $\mathbf{t}^{(3)} = -\mathbf{t}^{(-3)}$)

$$\mathbf{t} = \mathbf{t}^{(1)} n_1 + \mathbf{t}^{(2)} n_2 + \mathbf{t}^{(3)} n_3 = \sigma^T \cdot \mathbf{n} = \begin{bmatrix} \sigma_{11} & \sigma_{12} & \sigma_{13} \\ \sigma_{21} & \sigma_{22} & \sigma_{23} \\ \sigma_{31} & \sigma_{32} & \sigma_{33} \end{bmatrix}^T \left\{ \begin{array}{c} n_1 \\ n_2 \\ n_3 \end{array} \right\}, \tag{2.18}$$

where σ is the so-called Cauchy stress tensor.[1]

2.2.2 Balance Law Formulations

Substitution of (2.18) into (2.14) yields ($\omega \subset \Omega$)

$$\underbrace{\int_{\partial\omega} \sigma \cdot \mathbf{n} \, da}_{\text{surface forces}} + \underbrace{\int_{\omega} \mathbf{f} \, d\omega}_{\text{body forces}} = \underbrace{\frac{d}{dt} \int_{\omega} \rho \dot{\mathbf{u}} \, d\omega}_{\text{inertial forces}}. \tag{2.19}$$

A relationship can be determined between the densities in the current and reference configurations, $\int_{\omega} \rho \, d\omega = \int_{\omega_0} \rho J \, d\omega_0 = \int_{\omega_0} \rho_0 \, d\omega_0$. Therefore, the Jacobian

[1] Some authors follow the notation with the first index represents the direction of the component and the second index represents the normal to corresponding coordinate plane. This leads to $\mathbf{t} \stackrel{\text{def}}{=} \mathbf{t}^{(n)} = \sigma \cdot \mathbf{n}$. In the absence of couple stresses, a balance of angular momentum implies a symmetry of stress, $\sigma = \sigma^T$, and thus the difference in notations become immaterial.

Fig. 2.6 Stress at a point

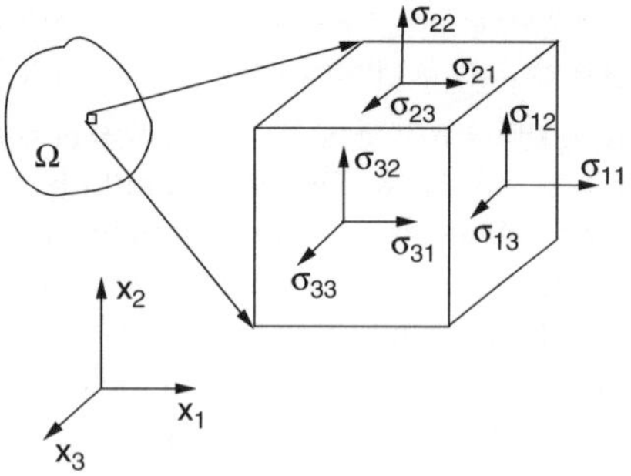

can also be interpreted as the ratio of material densities at a point. Since the volume is arbitrary, we can assume that $\rho J = \rho_0$ holds at every point in the body. Therefore, we may write $\frac{d}{dt}(\rho_0) = \frac{d}{dt}(\rho J) = 0$, when the system is mass conservative over time. This leads to writing the last term in (2.19) as $\frac{d}{dt}\int_{\omega} \rho \dot{\mathbf{u}} d\omega = \int_{\omega_0} \frac{d(\rho J)}{dt} \dot{\mathbf{u}} d\omega_0 + \int_{\omega_0} \rho \ddot{\mathbf{u}} J d\omega_0 = \int_{\omega} \rho \ddot{\mathbf{u}} d\omega$. From Gauss's Divergence theorem, and an implicit assumption that σ is differentiable, we have $\int_{\omega} (\nabla_x \cdot \sigma + \mathbf{f} - \rho \ddot{\mathbf{u}})\, d\omega = \mathbf{0}$. If the volume is argued as being arbitrary, then the relation in the integral must hold pointwise, yielding

$$\nabla_x \cdot \sigma + \mathbf{f} = \rho \ddot{\mathbf{u}}. \tag{2.20}$$

Symmetry of the Stress Tensor

Starting with an angular momentum balance, under the assumptions that no infinitesimal "micro-moments" or so-called "couple-stresses" exist, then it can be shown that the stress tensor must be symmetric (Malvern [133]), i.e. $\int_{\partial\omega} \mathbf{x} \times \mathbf{t} da + \int_{\omega} \mathbf{x} \times \mathbf{f} d\omega = \frac{d}{dt}\int_{\omega} \mathbf{x} \times \rho \dot{\mathbf{u}} d\omega$, which implies $\sigma^T = \sigma$. It is somewhat easier to consider a differential element, such as in Fig. 2.6 and to simply sum moments about the center. Doing this one immediately obtains $\sigma_{12} = \sigma_{21}, \sigma_{23} = \sigma_{32}$ and $\sigma_{13} = \sigma_{31}$. Therefore

$$\mathbf{t} = \mathbf{t}^{(1)} n_1 + \mathbf{t}^{(2)} n_2 + \mathbf{t}^{(3)} n_3 = \sigma \cdot \mathbf{n} = \sigma^T \cdot \mathbf{n}. \tag{2.21}$$

2.3 Referential Descriptions of Balance Laws

In some cases it is quite difficult to perform a stress analysis for finite deformation solid mechanics problem, in the current configuration, primarily because it is unknown a priori. Therefore, frequently, all quantities are usually transformed ("pulled") back to the original coordinates, the referential frame. Therefore, it is preferable to think of a formulation in terms of the referential fixed coordinates $\mathbf{X}$, a so called *Lagrangian* formulation. With this in mind, there are two commonly used referential measures of stresses. We start by developing a purely mathematical result, leading to the so-called "Nanson" formula for transformation of surface elements (Fig. 2.7). Consider the cross product of two differential line elements in a current configuration, $d\mathbf{x}^{(1)} \times d\mathbf{x}^{(2)} = (\mathbf{F} \cdot d\mathbf{X}^{(1)}) \times (\mathbf{F} \cdot d\mathbf{X}^{(2)})$. The right hand side of this relation can be reformulated[2] $(\mathbf{F} \cdot d\mathbf{X}^{(1)}) \times (\mathbf{F} \cdot d\mathbf{X}^{(2)}) = (det\,\mathbf{F})\,\mathbf{F}^{-T} \cdot (d\mathbf{X}^{(1)} \times d\mathbf{X}^{(2)})$. This leads to the so-called Nanson formula

$$d\mathbf{a} = \mathbf{n}\,da = (det\mathbf{F})\,\mathbf{F}^{-T} \cdot \mathbf{n}_0\,da_0 = (det\mathbf{F})\,\mathbf{F}^{-T} \cdot d\mathbf{a}_0. \tag{2.22}$$

Knowing this, we now formulate the equations of equilibrium in the current or a reference configuration. Consider two surface elements, one on the current configuration, and one on a reference configuration. Let us form a new kind of stress tensor, call it $\mathbf{P}$, such that the amount of force is the same: $\mathbf{P} \cdot \mathbf{n}_0\,da_0 = \sigma \cdot \mathbf{n}\,da = \sigma \cdot \mathbf{F}^{-T}\,(det\mathbf{F}) \cdot \mathbf{n}_0\,da_0$. This implies

$$\mathbf{P} = (det\mathbf{F})\,\sigma \cdot \mathbf{F}^{-T}. \tag{2.23}$$

The tensor $\mathbf{P}$ is called the first Piola-Kirchhoff stress, and gives the actual force on the current area, but calculated per unit area of reference area. However, it is is not symmetric, and this sometimes causes difficulties in an analysis. Therefore, we symmetrize it by $\mathbf{F}^{-1} \cdot \mathbf{P}$. This yields the so-called second Piola-Kirchhoff stress

$$\mathbf{S} = \mathbf{S}^T = (det\mathbf{F})\,\mathbf{F}^{-1} \cdot \sigma \cdot \mathbf{F}^{-T}. \tag{2.24}$$

By definition we have $\int_{\partial\omega_0} \mathbf{n}_0 \cdot \mathbf{P}\,da_0 = \int_{\partial\omega} \mathbf{n} \cdot \sigma\,da$, and thus

$$\underbrace{\int_{\partial\omega_0} \mathbf{n}_0 \cdot \mathbf{P}\,da_0}_{\text{surface forces}} + \underbrace{\int_{\omega_0} \mathbf{f}J\,d\omega_0}_{\text{body forces}} = \int_{\omega_0} \rho_0 \frac{d\dot{\mathbf{u}}}{dt}\,d\omega_0\,, \tag{2.25}$$

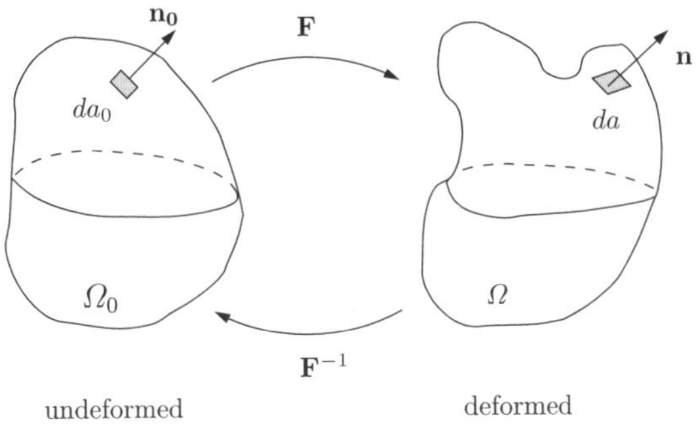

Fig. 2.7 A current and reference surface element

[2] An important vector identity (see Chandriashakiah and Debnath [24]) for a tensor $\mathbf{T}$ and two first order vectors $\mathbf{a}$ and $\mathbf{b}$ is $(\mathbf{T} \cdot \mathbf{a}) \times (\mathbf{T} \cdot \mathbf{b}) = \mathbf{T}^* \cdot (\mathbf{a} \times \mathbf{b})$, where the $\mathbf{T}^*$ is the transpose of the adjoint defined by $\mathbf{T}^* \stackrel{\text{def}}{=} (det\mathbf{T})\mathbf{T}^{-T}$.

and therefore

$$\underbrace{\int_{\omega_0} \nabla_X \cdot \mathbf{P}\, d\omega_0}_{\text{surface forces}} + \underbrace{\int_{\omega_0} \mathbf{f} J\, d\omega_0}_{\text{body forces}} = \int_{\omega_0} \rho_0 \frac{d\dot{\mathbf{u}}}{dt}\, d\omega_0 \,. \tag{2.26}$$

With $\mathbf{P} = \mathbf{F} \cdot \mathbf{S}$ the last equation yields $\int_{\omega_0} \nabla_X \cdot (\mathbf{F} \cdot \mathbf{S})\, d\omega_0 + \int_{\omega_0} \mathbf{f} J\, d\omega_0 = \int_{\omega_0} \rho_0 \frac{d\dot{\mathbf{u}}}{dt}$ $d\omega_0$. Since the control volume is arbitrary, we have

$$\nabla_X \cdot \mathbf{P} + \mathbf{f} J = \rho_0 \frac{d\dot{\mathbf{u}}}{dt} \quad \text{or} \quad \nabla_X \cdot (\mathbf{F} \cdot \mathbf{S}) + \mathbf{f} J = \rho_0 \frac{d\dot{\mathbf{u}}}{dt}. \tag{2.27}$$

2.4 The First Law of Thermodynamics/An Energy Balance

The interconversions of mechanical, thermal and chemical energy in a system are governed by the first law of thermodynamics. It states that the time rate of change of the total energy, $\mathscr{K} + \mathscr{I}$, is equal to the work rate, $\mathscr{P}$ and the net heat supplied, $\mathscr{H} + \mathscr{Q}$,

$$\frac{d}{dt}(\mathscr{K} + \mathscr{I}) = \mathscr{P} + \mathscr{H} + \mathscr{Q}. \tag{2.28}$$

Here the kinetic energy of a subvolume of material contained in Ω, denoted ω, is $\mathscr{K} \stackrel{\text{def}}{=} \int_\omega \frac{1}{2} \rho \dot{\mathbf{u}} \cdot \dot{\mathbf{u}}\, d\omega$, the rate of work or power of external forces acting on ω is given by $\mathscr{P} \stackrel{\text{def}}{=} \int_\omega \rho \mathbf{b} \cdot \dot{\mathbf{u}}\, d\omega + \int_{\partial\omega} \boldsymbol{\sigma} \cdot \mathbf{n} \cdot \dot{\mathbf{u}}\, da$, the heat flow into the volume by conduction is $\mathscr{Q} \stackrel{\text{def}}{=} -\int_{\partial\omega} \mathbf{q} \cdot \mathbf{n}\, da = -\int_\omega \nabla_x \cdot \mathbf{q}\, d\omega$, the heat generated due to sources, *such as chemical reactions*, is $\mathscr{H} \stackrel{\text{def}}{=} \int_\omega \rho z\, d\omega$ and the stored energy is $\mathscr{I} \stackrel{\text{def}}{=} \int_\omega \rho w\, d\omega$. If we make the assumption that the mass in the system is constant, one has,

$$\text{current mass} = \int_\omega \rho\, d\omega = \int_{\omega_0} \rho J\, d\omega_0 \approx \int_{\omega_0} \rho_0\, d\omega_0 = \text{original mass}, \tag{2.29}$$

which implies $\rho J = \rho_0 \Rightarrow \dot{\rho} J + \rho \dot{J} = 0$. Using this and the energy balance leads to

$$\begin{aligned}
\frac{d}{dt} \int_\omega \frac{1}{2} \rho \dot{\mathbf{u}} \cdot \dot{\mathbf{u}}\, d\omega &= \int_{\omega_0} \frac{d}{dt} \frac{1}{2} (\rho J \dot{\mathbf{u}} \cdot \dot{\mathbf{u}})\, d\omega_0 \\
&= \int_{\omega_0} (\frac{d}{dt} \rho_0) \frac{1}{2} \dot{\mathbf{u}} \cdot \dot{\mathbf{u}}\, d\omega_0 + \int_\omega \rho \frac{d}{dt} \frac{1}{2} (\dot{\mathbf{u}} \cdot \dot{\mathbf{u}})\, d\omega \\
&= \int_\omega \rho \dot{\mathbf{u}} \cdot \ddot{\mathbf{u}}\, d\omega.
\end{aligned} \tag{2.30}$$

We also have

$$\frac{d}{dt} \int_\omega \rho w\, d\omega = \frac{d}{dt} \int_{\omega_0} \rho J w\, d\omega_0 = \int_{\omega_0} \frac{d}{dt} (\rho_0) w\, d\omega_0 + \int_\omega \rho \dot{w}\, d\omega. \tag{2.31}$$

By using the divergence theorem, we obtain

$$\int_{\partial\omega} \sigma \cdot \mathbf{n} \cdot \dot{\mathbf{u}}\, da = \int_{\omega} \nabla_x \cdot (\sigma \cdot \dot{\mathbf{u}})\, d\omega = \int_{\omega} (\nabla_x \cdot \sigma) \cdot \dot{\mathbf{u}}\, d\omega + \int_{\omega} \sigma : \nabla_x \dot{\mathbf{u}}\, d\omega. \quad (2.32)$$

Combining the results, and enforcing balance of momentum, leads to

$$\int_{\omega} (\rho \dot{w} + \dot{\mathbf{u}} \cdot (\rho \ddot{\mathbf{u}} - \nabla_x \cdot \sigma - \rho \mathbf{b}) - \sigma : \nabla_x \dot{\mathbf{u}} + \nabla_x \cdot \mathbf{q} - \rho z)\, d\omega$$
$$= \int_{\omega} (\rho \dot{w} - \sigma : \nabla_x \dot{\mathbf{u}} + \nabla_x \cdot \mathbf{q} - \rho z)\, d\omega = 0. \quad (2.33)$$

Since the volume ω is arbitrary, the integrand must hold locally and we have

$$\rho \dot{w} - \sigma : \nabla_x \dot{\mathbf{u}} + \nabla_x \cdot \mathbf{q} - \rho z = 0. \quad (2.34)$$

In later chapters dealing with multifield problems, this equation will be investigated more extensively.

Remark. Through similar arguments as above, one can reformulate the First Law of Thermodynamics in the reference configuration as

$$\rho_0 \dot{w} - \mathbf{S} : \dot{\mathbf{E}} + \nabla_X \cdot \mathbf{q}_0 - \rho_0 z = 0, \quad (2.35)$$

where $\mathbf{q}_0 = \mathbf{q} J \cdot \mathbf{F}^{-T}$.

2.5 The Second Law of Thermodynamics/A Restriction

Consider the quantity commonly known as the "entropy" of a volume $H \stackrel{\text{def}}{=} \int_{\omega} \rho \eta\, d\omega$, where η is the entropy per unit mass, or "specific" entropy.[3] The rate of entropy input is $Q \stackrel{\text{def}}{=} \int_{\omega} \rho \frac{z}{\theta}\, d\omega - \int_{\partial\omega} \frac{\mathbf{q}}{\theta} \cdot \mathbf{n}\, da$, where $\rho \frac{z}{\theta}$ is the entropy source and $\frac{\mathbf{q}}{\theta}$ is the entropy "influx". *The basic postulate of Clausius (1854)–Duhem (1901), is*

$$\dot{H} \geq Q \Rightarrow \frac{d}{dt} \int_{\omega} \rho \eta\, d\omega - \int_{\omega} \rho \frac{z}{\theta}\, d\omega + \int_{\partial\omega} \frac{\mathbf{q}}{\theta} \cdot \mathbf{n}\, da \geq 0. \quad (2.36)$$

In other words, the time rate of change of entropy in the body is no less than the entropy input into the body. The last expression can be recast in various useful forms. We have $\frac{d}{dt} \int_{\omega} \rho \eta\, d\omega = \frac{d}{dt} \int_{\omega_0} \rho J \eta\, d\omega_0 = \int_{\omega_0} \frac{d}{dt}(\rho_0) \eta\, d\omega_0 + \int_{\omega} \rho \dot{\eta}\, d\omega$. Furthermore, by the divergence theorem we have $\int_{\partial\omega} \frac{\mathbf{q}}{\theta} \cdot \mathbf{n}\, da = \int_{\omega} \nabla_x \cdot \frac{\mathbf{q}}{\theta}\, d\omega$. Inserting these relations, and since the volume ω is arbitrary, the integrand must hold locally and we have

$$\rho \dot{\eta} - \rho \frac{z}{\theta} + \nabla_x \cdot \frac{\mathbf{q}}{\theta} \geq 0. \quad (2.37)$$

[3] Loosely speaking, the entropy is a measure of the disorder in a system.

By expanding $\nabla_x \cdot \frac{\mathbf{q}}{\theta} = \frac{1}{\theta}\nabla_x \cdot \mathbf{q} - \frac{1}{\theta^2}(\nabla_x\theta)\cdot\mathbf{q}$, we obtain $\rho\dot{\eta} - \rho\frac{z}{\theta} + (\frac{1}{\theta}\nabla_x \cdot \mathbf{q} - \frac{1}{\theta^2}(\nabla_x\theta)) \geq 0$. Employing the First Law of Thermodynamics, we obtain $\rho\theta\dot{\eta} - \rho\dot{w} + \sigma : \nabla_x\dot{\mathbf{u}} - \rho z - \frac{1}{\theta}(\nabla_x\theta)\cdot\mathbf{q} \geq 0$. A consistent physical observation is that, in the absence of external input, heat flows from a region of higher temperature to that of a lower temperature (hot to cold), i.e. heat never flows against the temperature gradient, $\nabla_x\theta\cdot\mathbf{q} \leq 0$. Thus, if Fourier's law ($\mathbf{q} = -\mathbf{K}\cdot\nabla_x\theta$) is used, $-\nabla_x\theta\cdot\mathbf{K}\cdot\nabla_x\theta \leq 0$ implies that $\mathbf{K}$ must be positive definite. Since $\nabla_x\theta\cdot\mathbf{q} \leq 0$, we have $\rho\theta\dot{\eta} - \rho\dot{w} + \sigma : \nabla_x\dot{\mathbf{u}} \geq \frac{1}{\theta}(\nabla_x\theta)\cdot\mathbf{q} \leq 0$ or $\rho\theta\dot{\eta} - \rho z + \nabla_x\cdot\mathbf{q} \geq \frac{1}{\theta}(\nabla_x\theta)\cdot\mathbf{q} \leq 0$. When $\nabla_x\theta = \mathbf{0}$ then we have $\rho\theta\dot{\eta} - \rho\dot{w} + \sigma : \nabla_x\dot{\mathbf{u}} \geq 0$ or $\rho\theta\dot{\eta} - \rho z + \nabla_x\cdot\mathbf{q} \geq 0$, either of which is referred to as the Clausius (1854)–Planck (1887) inequality. Such inequalities can be used to invalidate or construct (complicated) constitutive laws, although throughout this monograph, standard models, where the Second Law of Thermodynamics is satisfied automatically, are employed.

2.6 Linearly Elastic Constitutive Equations

The fundamental mechanism that produces forces in elastic deformation is the stretching of atomic bonds. In Fig. 2.8, a force/atom-separation curve is depicted. If the deformations are small, then one can argue that we are dealing with only the linear portion of the curve. Ultimately, this allows us to usually use linear relationships for models relating forces to deformations. Atoms in materials are held together by electronic forces, or bonds, whose behavior resembles that of little springs. The stiffness of an atomic bond force between two atoms is shown in Fig. 2.8. If we compute simple expressions for the force versus strain, we have $F = \frac{K(1.25-1)r}{r} = 0.25K \Rightarrow F = \frac{K}{4}$. Some materials, mainly ceramics, truly exhibit this theoretical stiffness. Most materials do not exactly exhibit this stiffness,

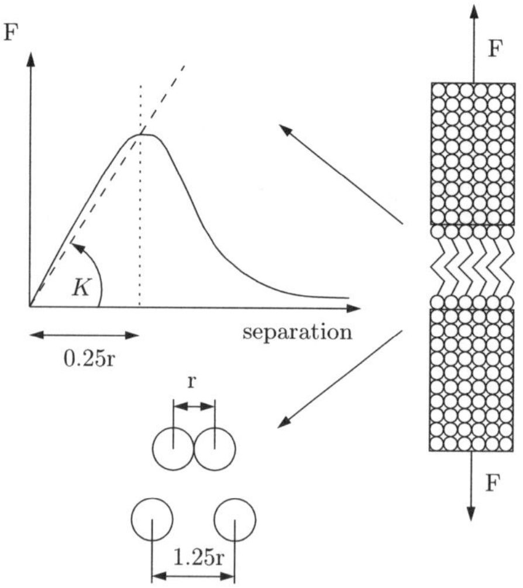

Fig. 2.8 The atomic bond force as a function of atom separation

however, the relationships are approximately *linear* between the atomic force and atom separation, for small deformations.

In order to determine the material properties of structural materials, engineers usually resort to a macroscopic tension test for material data. The usual procedure to determine tensile properties of materials is to place samples of material in testing machines, apply the loads, and then to measure the resulting deformations, such as lengths and changes in diameter in a portion of the specimen, of circular cross-section, called the gage length. The location of the gage length is away from the attachments to the testing machine (Fig. 2.9). The ends where the samples are attached to the machine are larger so that failure will not occur there first, which would ruin the experimental measurements. Slow rates of deformation are applied, and the response is usually measured with strain gauges or extensometers. For a metal, samples are usually 1.25 cm in diameter and 5 cm in length. Compression tests are usually on cubes (5 cm × 5 cm × 5 cm) or cylinders (2.5 cm in diameter and 2.5 cm in length). Here both the load applied by the machine, and the shortening of the specimen, should be measured over the gage length. For concrete, a material that is expected to carry compressive loads, the specimens are usually on the order of 15 cm in diameter and 30 cm long.[4]

The immediate result of a tension test is the axial force (F) divided by the original area (A_0), denoted, loosely speaking, as the "stress", and change in length ($\Delta L \stackrel{\text{def}}{=} L - L_0$) per unit length ($\varepsilon_0$) or "engineering strain", which is typically measured by strain gauges. As a first approximation we define the tensile stiffness of the material, known as Young's Modulus, denoted E, by $\sigma_0 \stackrel{\text{def}}{=} \frac{F}{A_0} = E\frac{\Delta L}{L} \stackrel{\text{def}}{=} E\varepsilon_0$. As we know, the terms σ_0 and ε_0 are, strictly speaking, not the true stress and strain, but simply serve our presentation purposes. Clearly, what we have presented is somewhat ad hoc, therefore, next we present the classical theory of linearly elastic material responses for three dimensional states of stress and strain.

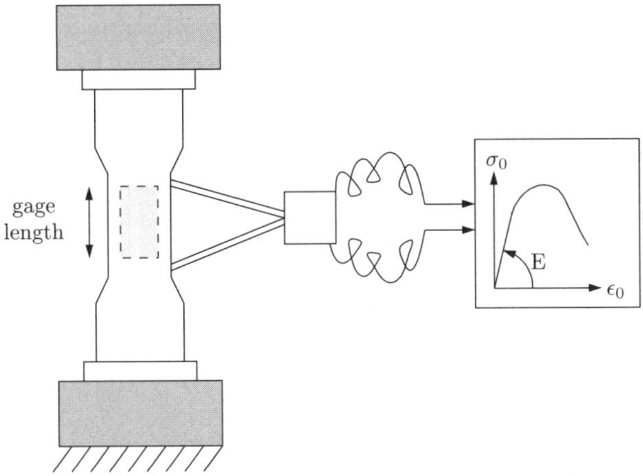

Fig. 2.9 A standard metals testing machine

[4] Concrete is usually aged 28 days before testing.

2.6.1 The Infinitesimal Strain Case

Initially we discuss relationships between the stress and strain, so-called "material laws" or "constitutive relations" for geometrically linear problems (infinitesimal deformations). Accordingly, if we neglect thermal effects, (2.34) implies $\rho\dot{w} = \sigma : \nabla_x \dot{\mathbf{u}}$, which in the infinitesimal strain linearly elastic case is $\rho\dot{w} = \sigma : \dot{\varepsilon}$. From the chain rule of differentiation we have

$$\rho\dot{w} = \rho\frac{\partial w}{\partial \varepsilon} : \frac{d\varepsilon}{dt} = \sigma : \dot{\varepsilon} \Rightarrow \sigma = \rho\frac{\partial w}{\partial \varepsilon} . \tag{2.38}$$

The starting point to develop a constitutive theory is to assume a stored elastic energy function exists, a function denoted $W \stackrel{\mathrm{def}}{=} \rho w$, which depends only on the mechanical deformation. The simplest function that fulfills $\sigma = \rho\frac{\partial w}{\partial \varepsilon}$ is $W = \frac{1}{2}\varepsilon : \mathbb{E} : \varepsilon$. Such a function satisfies the intuitive physical requirement that, for any small strain from an undeformed state, energy must be stored in the material. Alternatively, a small strain material law can be derived from $\sigma = \frac{\partial W}{\partial \varepsilon}$ and $W \approx c_0 + \mathbf{c}_1 : \varepsilon + \frac{1}{2}\varepsilon : \mathbb{E} : \varepsilon + \ldots$ which implies $\sigma \approx \mathbf{c}_1 + \mathbb{E} : \varepsilon + \ldots$. We are free to set $c_0 = 0$ (it is arbitrary) in order to have zero strain energy at zero strain and, furthermore, we assume that no stresses exist in the reference state ($\mathbf{c}_1 = \mathbf{0}$). With these assumptions, we obtain the familiar relation

$$\sigma \approx \mathbb{E} : \varepsilon . \tag{2.39}$$

This is a linear (tensorial) relation between stresses and strains. The existence of a strictly positive stored energy function in the reference configuration implies that the linear elasticity tensor must have positive eigenvalues at every point in the body. Typically, different materials are classified according to the number of independent constants in $\mathbb{E}$. A general material has 81 independent constants, since it is a fourth order tensor relating 9 components of stress to strain. However, the number of constants can be reduced to 36 since the stress and strain tensors are symmetric. This is easily seen from the matrix representation[5] of $\mathbb{E}$:

$$\underbrace{\begin{Bmatrix} \sigma_{11} \\ \sigma_{22} \\ \sigma_{33} \\ \sigma_{12} \\ \sigma_{23} \\ \sigma_{31} \end{Bmatrix}}_{\stackrel{\mathrm{def}}{=}\{\sigma\}} = \underbrace{\begin{bmatrix} E_{1111} & E_{1122} & E_{1133} & E_{1112} & E_{1123} & E_{1113} \\ E_{2211} & E_{2222} & E_{2233} & E_{2212} & E_{2223} & E_{2213} \\ E_{3311} & E_{3322} & E_{3333} & E_{3312} & E_{3323} & E_{3313} \\ E_{1211} & E_{1222} & E_{1233} & E_{1212} & E_{1223} & E_{1213} \\ E_{2311} & E_{2322} & E_{2333} & E_{2312} & E_{2323} & E_{2313} \\ E_{1311} & E_{1322} & E_{1333} & E_{1312} & E_{1323} & E_{1313} \end{bmatrix}}_{\stackrel{\mathrm{def}}{=}[\mathbb{E}]} \underbrace{\begin{Bmatrix} \varepsilon_{11} \\ \varepsilon_{22} \\ \varepsilon_{33} \\ 2\varepsilon_{12} \\ 2\varepsilon_{23} \\ 2\varepsilon_{31} \end{Bmatrix}}_{\stackrel{\mathrm{def}}{=}\{\varepsilon\}} . \tag{2.40}$$

[5] The symbol $[\cdot]$ is used to indicate the matrix notation equivalent to a tensor form, while $\{\cdot\}$ is used to indicate the vector representation.

The existence of a scalar energy function forces $\mathbb{E}$ to be symmetric since the strains are symmetric, in other words $W = \frac{1}{2}\varepsilon : \mathbb{E} : \varepsilon = \frac{1}{2}(\varepsilon : \mathbb{E} : \varepsilon)^T = \frac{1}{2}\varepsilon^T : \mathbb{E}^T : \varepsilon^T = \frac{1}{2}\varepsilon : \mathbb{E}^T : \varepsilon$ which implies $\mathbb{E}^T = \mathbb{E}$. Consequently, $\mathbb{E}$ has only 21 free constants. The nonnegativity of W imposes the restriction that $\mathbb{E}$ remains positive definite. At this point, based on many factors that depend on the material microstructure, it can be shown that the components of $\mathbb{E}$ may be written in terms of anywhere between 21 and 2 independent parameters. We explore such concepts further via the ideas of elastic symmetry.

2.6.2 Material Symmetry

Transformation matrices are used in determining the elastic symmetries. Consider a plane of symmetry, the $x_2 - x_3$ plane (Fig. 2.10). A plane of symmetry implies that the material has the same properties with respect to that plane. Therefore, we should be able to flip the axes with respect to that plane, and have no change in the constitutive law. By definition, a plane of elastic symmetry exists at a point where the elastic constants have the same value for a pair of coordinate systems. The axes are referred to as "equivalent elastic directions". Also by definition, an axis of symmetry of order K exists at a point when there are sets of equivalent elastic directions which can be superposed by a rotation through an angle $K/2\pi$ about an axis. The way to determine elastic symmetry is as follows: first one forms $[\hat{\sigma}] = [\mathbf{Q}]^{-1}[\sigma][\mathbf{Q}]$ implying $\{\hat{\sigma}\} = [\mathbf{T}]\{\sigma\} = [\hat{\mathbb{E}}][\mathbf{T}]\{\varepsilon\} = [\hat{\mathbb{E}}]\{\hat{\varepsilon}\}$ which implies $\{\sigma\} = [\mathbf{T}^{-1}][\hat{\mathbb{E}}][\mathbf{T}]\{\varepsilon\}$, where $[\mathbf{Q}]$ is a rotational transformation matrix. Imposing elastic symmetry means the components are invariant with respect to the transformation, hence $[\mathbb{E}] = [\mathbf{T}^{-1}][\hat{\mathbb{E}}][\mathbf{T}]$. Therefore, all components that are not identical must be zero if the material has the assumed elastic symmetry. In this fashion one can "carve" away components from a general anisotropic material tensor. The central point of such symmetries is that in a new transformed state, $\mathbb{E}$ should not change.

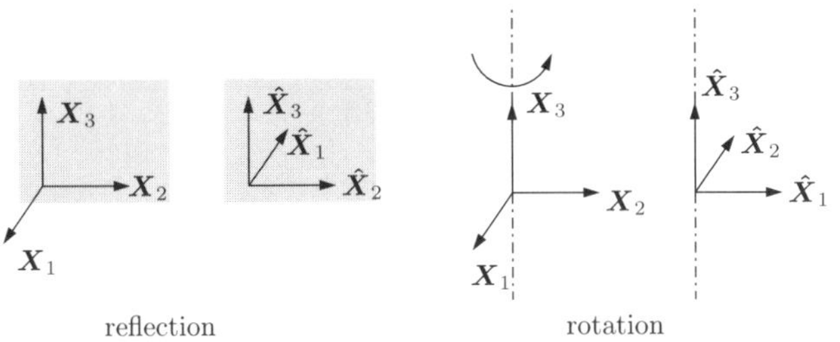

Fig. 2.10 Left: reflection with respect to the $x_2 - x_3$ plane. **Right**: rotation with respect to the x_3 axis

Examples of Elastic Symmetry

To make things clear, consider the following steps for one plane of symmetry starting with an originally Triclinic material with 21 free constants (2.40), defined by $\mathbb{E}$:

- STEP 1: Reflect the x_1 axis

$$R(x_1) \stackrel{\text{def}}{=} \begin{bmatrix} -1 & 0 & 0 \\ 0 & 1 & 0 \\ 0 & 0 & 1 \end{bmatrix} \begin{Bmatrix} x_1 \\ x_2 \\ x_3 \end{Bmatrix} = \begin{Bmatrix} \hat{x}_1 \\ \hat{x}_2 \\ \hat{x}_3 \end{Bmatrix} = \begin{Bmatrix} -x_1 \\ x_2 \\ \hat{x}_3 \end{Bmatrix}. \tag{2.41}$$

- STEP 2: Transform the stress and strain tensors with the same transformation, but for second order tensor rules:

$$\begin{bmatrix} -1 & 0 & 0 \\ 0 & 1 & 0 \\ 0 & 0 & 1 \end{bmatrix} \begin{bmatrix} \sigma_{11} & \sigma_{12} & \sigma_{13} \\ \sigma_{21} & \sigma_{22} & \sigma_{23} \\ \sigma_{31} & \sigma_{32} & \sigma_{33} \end{bmatrix} \begin{bmatrix} -1 & 0 & 0 \\ 0 & 1 & 0 \\ 0 & 0 & 1 \end{bmatrix} = \begin{bmatrix} \hat{\sigma}_{11} & \hat{\sigma}_{12} & \hat{\sigma}_{13} \\ \hat{\sigma}_{21} & \hat{\sigma}_{22} & \hat{\sigma}_{23} \\ \hat{\sigma}_{31} & \hat{\sigma}_{32} & \hat{\sigma}_{33} \end{bmatrix}, \tag{2.42}$$

and thus

$$\begin{bmatrix} \hat{\sigma}_{11} & \hat{\sigma}_{12} & \hat{\sigma}_{13} \\ \hat{\sigma}_{21} & \hat{\sigma}_{22} & \hat{\sigma}_{23} \\ \hat{\sigma}_{31} & \hat{\sigma}_{32} & \hat{\sigma}_{33} \end{bmatrix} = \begin{bmatrix} \sigma_{11} & -\sigma_{12} & -\sigma_{13} \\ -\sigma_{21} & \sigma_{22} & \sigma_{23} \\ -\sigma_{31} & \sigma_{32} & \sigma_{33} \end{bmatrix}. \tag{2.43}$$

Also,

$$\begin{bmatrix} -1 & 0 & 0 \\ 0 & 1 & 0 \\ 0 & 0 & 1 \end{bmatrix} \begin{bmatrix} \varepsilon_{11} & \varepsilon_{12} & \varepsilon_{13} \\ \varepsilon_{21} & \varepsilon_{22} & \varepsilon_{23} \\ \varepsilon_{31} & \varepsilon_{32} & \varepsilon_{33} \end{bmatrix} \begin{bmatrix} -1 & 0 & 0 \\ 0 & 1 & 0 \\ 0 & 0 & 1 \end{bmatrix}, \tag{2.44}$$

and also

$$\begin{bmatrix} \hat{\varepsilon}_{11} & \hat{\varepsilon}_{12} & \hat{\varepsilon}_{13} \\ \hat{\varepsilon}_{21} & \hat{\varepsilon}_{22} & \hat{\varepsilon}_{23} \\ \hat{\varepsilon}_{31} & \hat{\varepsilon}_{32} & \hat{\varepsilon}_{33} \end{bmatrix} = \begin{bmatrix} \varepsilon_{11} & -\varepsilon_{12} & -\varepsilon_{13} \\ -\varepsilon_{21} & \varepsilon_{22} & \varepsilon_{23} \\ -\varepsilon_{31} & \varepsilon_{32} & \varepsilon_{33} \end{bmatrix}. \tag{2.45}$$

- STEP 3: Form the constitutive law in the primed frame:

$$\begin{Bmatrix} \sigma_{11} \\ \sigma_{22} \\ \sigma_{33} \\ -\sigma_{12} \\ \sigma_{23} \\ -\sigma_{31} \end{Bmatrix} = \begin{Bmatrix} \hat{\sigma}_{11} \\ \hat{\sigma}_{22} \\ \hat{\sigma}_{33} \\ \hat{\sigma}_{12} \\ \hat{\sigma}_{23} \\ \hat{\sigma}_{31} \end{Bmatrix}, \tag{2.46}$$

and

$$\begin{Bmatrix} \hat{\sigma}_{11} \\ \hat{\sigma}_{22} \\ \hat{\sigma}_{33} \\ \hat{\sigma}_{12} \\ \hat{\sigma}_{23} \\ \hat{\sigma}_{31} \end{Bmatrix} = \begin{bmatrix} E_{1111} & E_{1122} & E_{1133} & E_{1112} & E_{1123} & E_{1113} \\ E_{2211} & E_{2222} & E_{2233} & E_{2212} & E_{2223} & E_{2213} \\ E_{3311} & E_{3322} & E_{3333} & E_{3312} & E_{3323} & E_{3313} \\ E_{1211} & E_{1222} & E_{1233} & E_{1212} & E_{1223} & E_{1213} \\ E_{2311} & E_{2322} & E_{2333} & E_{2312} & E_{2323} & E_{2313} \\ E_{1311} & E_{1322} & E_{1333} & E_{1312} & E_{1323} & E_{1313} \end{bmatrix} \begin{Bmatrix} \hat{\varepsilon}_{11} \\ \hat{\varepsilon}_{22} \\ \hat{\varepsilon}_{33} \\ 2\hat{\varepsilon}_{12} \\ 2\hat{\varepsilon}_{23} \\ 2\hat{\varepsilon}_{31} \end{Bmatrix}. \tag{2.47}$$

- STEP 4: Put everything in terms of the original variables, which implies that the constitutive law must be the same as before, if the plane was a plane of symmetry, and thus the tensor relating σ and ε is :

$$\begin{bmatrix} E_{1111} & E_{1122} & E_{1133} & -E_{1112} & E_{1123} & -E_{1113} \\ E_{2211} & E_{2222} & E_{2233} & -E_{2212} & E_{2223} & -E_{2213} \\ E_{3311} & E_{3322} & E_{3333} & -E_{3312} & E_{3323} & -E_{3313} \\ -E_{1211} & -E_{1222} & -E_{1233} & E_{1212} & -E_{1223} & E_{1213} \\ E_{2311} & E_{2322} & E_{2333} & -E_{2312} & E_{2323} & -E_{2313} \\ -E_{1311} & -E_{1322} & -E_{1333} & E_{1312} & -E_{1323} & E_{1313} \end{bmatrix}. \tag{2.48}$$

- STEP 5: All components that are not equal before and after the reflection of axes are zero:

$$\begin{Bmatrix} \sigma_{11} \\ \sigma_{22} \\ \sigma_{33} \\ \sigma_{12} \\ \sigma_{23} \\ \sigma_{31} \end{Bmatrix} = \begin{bmatrix} E_{1111} & E_{1122} & E_{1133} & 0 & E_{1123} & 0 \\ E_{2211} & E_{2222} & E_{2233} & 0 & E_{2223} & 0 \\ E_{3311} & E_{3322} & E_{3333} & 0 & E_{3323} & 0 \\ 0 & 0 & 0 & E_{1212} & 0 & E_{1213} \\ E_{2311} & E_{2322} & E_{2333} & 0 & E_{2323} & 0 \\ 0 & 0 & 0 & E_{1312} & 0 & E_{1313} \end{bmatrix} \begin{Bmatrix} \varepsilon_{11} \\ \varepsilon_{22} \\ \varepsilon_{33} \\ 2\varepsilon_{12} \\ 2\varepsilon_{23} \\ 2\varepsilon_{31} \end{Bmatrix}. \tag{2.49}$$

The end result is a Monoclinic material, i.e, one plane of elastic symmetry ($x_2 - x_3$ plane)(13 free constants)

$$\mathbb{E} \stackrel{\text{def}}{=} \begin{bmatrix} E_{1111} & E_{1122} & E_{1133} & 0 & E_{1123} & 0 \\ E_{2211} & E_{2222} & E_{2233} & 0 & E_{2223} & 0 \\ E_{3311} & E_{3322} & E_{3333} & 0 & E_{3323} & 0 \\ 0 & 0 & 0 & E_{1212} & 0 & E_{1213} \\ E_{2311} & E_{2322} & E_{2333} & 0 & E_{2323} & 0 \\ 0 & 0 & 0 & E_{1312} & 0 & E_{1313} \end{bmatrix}, \tag{2.50}$$

or ($x_1 - x_3$ plane)(13 free constants)

$$\mathbb{E} \stackrel{\text{def}}{=} \begin{bmatrix} E_{1111} & E_{1122} & E_{1133} & 0 & 0 & E_{1113} \\ E_{2211} & E_{2222} & E_{2233} & 0 & 0 & E_{2213} \\ E_{3311} & E_{3322} & E_{3333} & 0 & 0 & E_{3313} \\ 0 & 0 & 0 & E_{1212} & E_{1223} & 0 \\ 0 & 0 & 0 & E_{1223} & E_{2323} & 0 \\ E_{1311} & E_{1322} & E_{1333} & 0 & 0 & E_{1313} \end{bmatrix}. \tag{2.51}$$

The basic procedure is the same, for reflectional symmetry, rotational symmetries, etc. What follows is a catalog of commonly referred to materials of various symmetries.

- Two mutually perpendicular planes of symmetry reduce the material symmetry to nine free constants, known as an *orthotropic* material. It also has a third mutually perpendicular plane of symmetry, without changing the number of elastic constants. In other words, if one were to reflect a reference frame located at a material point (rotated by 180 degrees) the properties are the same. Accordingly, for an orthotropic material, there are two planes of symmetry (nine free constants)

$$\mathbb{E} \stackrel{\text{def}}{=} \begin{bmatrix} E_{1111} & E_{1122} & E_{1133} & 0 & 0 & 0 \\ E_{2211} & E_{2222} & E_{2233} & 0 & 0 & 0 \\ E_{3311} & E_{3322} & E_{3333} & 0 & 0 & 0 \\ 0 & 0 & 0 & E_{1212} & 0 & 0 \\ 0 & 0 & 0 & 0 & E_{2323} & 0 \\ 0 & 0 & 0 & 0 & 0 & E_{1313} \end{bmatrix}. \tag{2.52}$$

- In addition, if there is one plane in which the material properties are equal in all directions, there are only five free constants and the material is termed *transversely isotropic*. Accordingly, for transversely isotropic material: two planes of symmetry and one plane of directional independence (five free constants)

$$\mathbb{E} \stackrel{\text{def}}{=} \begin{bmatrix} E_{1111} & E_{1122} & E_{1133} & 0 & 0 & 0 \\ E_{2211} & E_{2222} & E_{2233} & 0 & 0 & 0 \\ E_{3311} & E_{3322} & E_{3333} & 0 & 0 & 0 \\ 0 & 0 & 0 & E_{1212} & 0 & 0 \\ 0 & 0 & 0 & 0 & E_{2323} & 0 \\ 0 & 0 & 0 & 0 & 0 & E_{1313} \end{bmatrix} \tag{2.53}$$

  and $E_{1111} = E_{2222}$, $E_{1133} = E_{2233}$, $E_{1313} = E_{2323}$, $E_{1212} = \frac{1}{2}(E_{1111} - E_{1122})$
- Cubic materials have two planes of symmetry and two planes of directional independence, and can be shown to have three free constants. Accordingly, for a cubic material: two planes of symmetry and two planes of directional independence (three free constants)

$$\mathbb{E} \stackrel{\text{def}}{=} \begin{bmatrix} E_{1111} & E_{1122} & E_{1133} & 0 & 0 & 0 \\ E_{2211} & E_{2222} & E_{2233} & 0 & 0 & 0 \\ E_{3311} & E_{3322} & E_{3333} & 0 & 0 & 0 \\ 0 & 0 & 0 & E_{1212} & 0 & 0 \\ 0 & 0 & 0 & 0 & E_{2323} & 0 \\ 0 & 0 & 0 & 0 & 0 & E_{1313} \end{bmatrix}, \tag{2.54}$$

and $E_{1111} = E_{2222} = E_{3333}$, $E_{1122} = E_{1133} = E_{2233}$, $E_{1313} = E_{3333} = E_{1212}$

- Finally, if there are an infinite number of planes where the material properties are equal (in all directions), there are two free constants, the Lamé parameters, and the material is of the familiar *isotropic* variety. An isotropic body has material properties that are the same in every direction at a point in the body, i.e., the properties are not a function of orientation at a point in a body. Accordingly, for isotropic materials: two planes of symmetry and an infinite number of planes of directional independence (two free constants),

$$\mathbb{E} \stackrel{\text{def}}{=} \begin{bmatrix} \kappa+\frac{4}{3}\mu & \kappa-\frac{2}{3}\mu & \kappa-\frac{2}{3}\mu & 0 & 0 & 0 \\ \kappa-\frac{2}{3}\mu & \kappa+\frac{4}{3}\mu & \kappa-\frac{2}{3}\mu & 0 & 0 & 0 \\ \kappa-\frac{2}{3}\mu & \kappa-\frac{2}{3}\mu & \kappa+\frac{4}{3}\mu & 0 & 0 & 0 \\ 0 & 0 & 0 & \mu & 0 & 0 \\ 0 & 0 & 0 & 0 & \mu & 0 \\ 0 & 0 & 0 & 0 & 0 & \mu \end{bmatrix}. \tag{2.55}$$

In this case we have

$$\mathbb{E} : \varepsilon = 3\kappa\frac{tr\varepsilon}{3}\mathbf{1} + 2\mu\varepsilon' \Rightarrow \varepsilon : \mathbb{E} : \varepsilon = 9\kappa(\frac{tr\varepsilon}{3})^2 + 2\mu\varepsilon' : \varepsilon', \tag{2.56}$$

where $tr\varepsilon = \varepsilon_{ii}$ and $\varepsilon' = \varepsilon - \frac{1}{3}(tr\varepsilon)\mathbf{1}$ is the deviatoric strain. The eigenvalues of an isotropic elasticity tensor are $(3\kappa, 2\mu, 2\mu, \mu, \mu, \mu)$. Therefore, we must have $\kappa > 0$ and $\mu > 0$ to retain positive definiteness of $\mathbb{E}$.

Remark. Obviously, the elasticity tensor, regardless of the degree of anisotropy, must have positive eigenvalues. By forming the similarity transform $[\mathbf{Z}^T][\mathbb{E}][\mathbf{Z}] = [\Lambda]$, where the 6×6 collection of 6×1 mutually orthonormal eigenvectors of the elasticity is denoted as $\mathbf{Z}$ and where $[\Lambda]$ is diagonalized, we have $\{\sigma\} = [\mathbf{Z}][\Lambda][\mathbf{Z}^T]\{\varepsilon\}$, which implies $[\mathbf{Z}^T]\{\sigma\} = \{\hat{\sigma}\} = [\Lambda][\mathbf{Z}^T]\{\varepsilon\} = [\Lambda]\{\hat{\varepsilon}\}$. Clearly, the constitutive law in the transformed basis has the form:

$$\begin{Bmatrix} \hat{\sigma}_{11} \\ \hat{\sigma}_{22} \\ \hat{\sigma}_{33} \\ \hat{\sigma}_{12} \\ \hat{\sigma}_{23} \\ \hat{\sigma}_{13} \end{Bmatrix} = \begin{bmatrix} \Lambda_1 & 0 & 0 & 0 & 0 & 0 \\ 0 & \Lambda_2 & 0 & 0 & 0 & 0 \\ 0 & 0 & \Lambda_3 & 0 & 0 & 0 \\ 0 & 0 & 0 & \Lambda_4 & 0 & 0 \\ 0 & 0 & 0 & 0 & \Lambda_5 & 0 \\ 0 & 0 & 0 & 0 & 0 & \Lambda_6 \end{bmatrix} \begin{Bmatrix} \hat{\varepsilon}_{11} \\ \hat{\varepsilon}_{22} \\ \hat{\varepsilon}_{33} \\ 2\hat{\varepsilon}_{12} \\ 2\hat{\varepsilon}_{23} \\ 2\hat{\varepsilon}_{13} \end{Bmatrix}. \tag{2.57}$$

The strain energy is simply

$$W = \frac{1}{2}\left(\Lambda_1(\hat{\varepsilon}_{11})^2 + \Lambda_2(\hat{\varepsilon}_{22})^2 + \Lambda_3(\hat{\varepsilon}_{33})^2 + 4\Lambda_4(\hat{\varepsilon}_{12})^2 + 4\Lambda_5(\hat{\varepsilon}_{23})^2 + 4\Lambda_6(\hat{\varepsilon}_{13})^2\right). \tag{2.58}$$

One interpretation is that of a generalized decomposition of the energy into pieces associated with the principle directions of the the elasticity tensor in matrix form. Clearly, each eigenvalue, $\Lambda_i, i = 1,...,6$, must be positive.

2.6.3 *Material Constant Interpretation*

There are a variety of ways to write isotropic constitutive laws, each time with a physically meaningful pair of material constants.

Splitting the Strain

It is sometimes important to split infinitesimal strains into two physically meaningful parts

$$\varepsilon = \frac{tr\varepsilon}{3}\mathbf{1} + (\varepsilon - \frac{tr\varepsilon}{3}\mathbf{1}). \tag{2.59}$$

The Jacobian, J, of the deformation gradient $\mathbf{F}$ is $det(\mathbf{1}+\nabla_X\mathbf{u})$, and can be written as

$$J = \det\mathbf{F} = \det(\mathbf{1} + \nabla_X\mathbf{u}). \tag{2.60}$$

Its expansion yields $J = det(\mathbf{1}+\nabla_X\mathbf{u}) \approx 1 + tr\nabla_X\mathbf{u} + \mathcal{O}(\nabla_X\mathbf{u}) = 1 + tr\varepsilon + ...$ Therefore, with infinitesimal strains $(1 + tr\varepsilon)d\omega_0 = d\omega$ and we can write $tr\varepsilon = \frac{d\omega - d\omega_0}{d\omega_0}$. Hence, $tr\varepsilon$ is associated with the *volumetric part of the deformation*. Furthermore, since $\varepsilon' \stackrel{\text{def}}{=} \varepsilon - \frac{tr\varepsilon}{3}\mathbf{1} = \mathbf{0}$, the so-called "strain deviator" describes distortion in the material.

Infinitesimal Strain Material Laws

The stress σ can be split into two parts (dilatational and a deviatoric):

$$\sigma = \frac{tr\sigma}{3}\mathbf{1} + (\sigma - \frac{tr\sigma}{3}\mathbf{1}) \stackrel{\text{def}}{=} -p\mathbf{1} + \sigma', \tag{2.61}$$

where we call the symbol p the hydrostatic pressure and σ' the stress deviator. With (2.56) we write

$$p = -3\kappa\left(\frac{tr\varepsilon}{3}\right) \quad \text{and} \quad \sigma' = 2\mu\,\varepsilon'. \tag{2.62}$$

This is one form of Hooke's Law. The resistance to change in the volume is measured by κ. We note that $(\frac{tr\sigma}{3}\mathbf{1})' = \mathbf{0}$, which indicates that this part of the stress produces no distortion.

Another fundamental form of Hooke's law is

$$\sigma = \frac{E}{1+\nu}\left(\varepsilon + \frac{\nu}{1-2\nu}tr\varepsilon\mathbf{1}\right), \tag{2.63}$$

which implies the inverse form

$$\varepsilon = \frac{1+\nu}{E}\sigma - \frac{\nu}{E}tr\sigma\mathbf{1}. \tag{2.64}$$

To interpret the constants, consider a uniaxial tension test (pulled in the x_1 direction) where $\sigma_{12} = \sigma_{13} = \sigma_{23} = 0$, which implies $\varepsilon_{12} = \varepsilon_{13} = \varepsilon_{23} = 0$. Also, we have $\sigma_{22} = \sigma_{33} = 0$. Under these conditions we have $\sigma_{11} = E\varepsilon_{11}$ and $\varepsilon_{22} = \varepsilon_{33} = -\nu\varepsilon_{11}$. Therefore, E, the so-called "Young's" modulus, is the ratio of the uniaxial stress to the corresponding strain component. The Poisson ratio, ν, is the ratio of the transverse strains to the uniaxial strain.

Another commonly used set of stress-strain forms are the Lamé relations,

$$\sigma = \lambda tr\varepsilon\mathbf{1} + 2\mu\varepsilon \quad \text{or} \quad \varepsilon = -\frac{\lambda}{2\mu(3\lambda+2\mu)}tr\sigma\mathbf{1} + \frac{\sigma}{2\mu}. \tag{2.65}$$

To interpret the constants, consider a pressure test where $\sigma_{12} = \sigma_{13} = \sigma_{23} = 0$, and where $\sigma_{11} = \sigma_{22} = \sigma_{33}$. Under these conditions we have $\kappa = \lambda + \frac{2}{3}\mu = \frac{E}{3(1-2\nu)}$, $\mu = \frac{E}{2(1+\nu)}$ and $\frac{\kappa}{\mu} = \frac{2(1+\nu)}{3(1-2\nu)}$. We observe that $\frac{\kappa}{\mu} \to \infty$ implies $\nu \to \frac{1}{2}$, and $\frac{\kappa}{\mu} \to 0$ implies $\Rightarrow \nu \to -1$. Therefore, since both κ and μ must be positive and finite, this implies $-1 < \nu < 1/2$ and $0 < E < \infty$. For example, some polymeric foams exhibit $\nu < 0$, steels $\nu \approx 0.3$, and some forms of rubber have $\nu \to 1/2$. We note that λ *can be positive or negative.*

2.6.4 Consequences of Positive-Definiteness

Under general conditions, with a (positive-definite) linear elastic material law, at infinitesimal strains, the solution is unique, in other words, there exists only one solution. To see this, let us first suppose that there were two solutions that satisfy equilibrium

$$\nabla \cdot (\mathbb{E} : \nabla\mathbf{u}^{(1)}) + \mathbf{f} = \rho\ddot{\mathbf{u}}^{(1)} \text{ and } \nabla \cdot (\mathbb{E} : \nabla\mathbf{u}^{(2)}) + \mathbf{f} = \rho\ddot{\mathbf{u}}^{(2)}, \tag{2.66}$$

with specified traction boundary conditions on Γ_t and with specified displacement boundary conditions on Γ_u ($\overline{\Gamma_t \cup \Gamma_u} = \overline{\partial\Omega}$).

Elastostatic Case

Multiplying each equation by $(\mathbf{u}^{(1)} - \mathbf{u}^{(2)})$, integrating over the volume, using the divergence theorem, and subtracting each equation from one another we obtain

$$\int_{\Omega} \nabla(\mathbf{u}^{(1)} - \mathbf{u}^{(2)}) : \mathbb{E} : \nabla(\mathbf{u}^{(1)} - \mathbf{u}^{(2)})\, \mathrm{d}\Omega = \int_{\partial\Omega} \mathbf{n} \cdot (\sigma^{(1)} - \sigma^{(2)}) \cdot (\mathbf{u}^{(1)} - \mathbf{u}^{(2)})\, \mathrm{d}A. \tag{2.67}$$

The integral on the right vanishes since the tractions from both solution must be the same value on the traction specified boundary, and zero on the displacement part of the boundary since the solutions are the same on the displacement part of the boundary. Since the domain is arbitrary, the integrand must vanish. Since $\mathbb{E}$ is positive definite the integrand can only vanish if the solutions are the same, i.e. $\mathbf{u}^{(1)} - \mathbf{u}^{(2)} = \mathbf{0}$. For completeness, we discuss the elastodynamic case.

Elastodynamic Case

As in the static case, multiplying each equation by $(\dot{\mathbf{u}}^{(1)} - \dot{\mathbf{u}}^{(2)})$, integrating over the volume, using the divergence theorem and subtracting each equation from one another, we obtain

$$\int_{\Omega} \nabla(\mathbf{u}^{(1)} - \mathbf{u}^{(2)}) : \mathbb{E} : \nabla(\dot{\mathbf{u}}^{(1)} - \dot{\mathbf{u}}^{(2)})\, \mathrm{d}\Omega + \int_{\Omega} \rho(\ddot{\mathbf{u}}^{(1)} - \ddot{\mathbf{u}}^{(2)}) \cdot (\dot{\mathbf{u}}^{(1)} - \dot{\mathbf{u}}^{(2)})\, \mathrm{d}\Omega = 0. \tag{2.68}$$

Let us define a function

$$\begin{aligned} \frac{d\Psi}{dt} \stackrel{\text{def}}{=} & 2\int_{\Omega} \nabla(\mathbf{u}^{(1)} - \mathbf{u}^{(2)}) : \mathbb{E} : \nabla(\dot{\mathbf{u}}^{(1)} - \dot{\mathbf{u}}^{(2)})\, \mathrm{d}\Omega \\ & +2\int_{\Omega} \rho(\ddot{\mathbf{u}}^{(1)} - \ddot{\mathbf{u}}^{(2)}) \cdot (\dot{\mathbf{u}}^{(1)} - \dot{\mathbf{u}}^{(2)})\, \mathrm{d}\Omega = 0. \end{aligned} \tag{2.69}$$

We have that

$$\begin{aligned} \Psi \stackrel{\text{def}}{=} & \int_{\Omega} \nabla(\mathbf{u}^{(1)} - \mathbf{u}^{(2)}) : \mathbb{E} : \nabla(\mathbf{u}^{(1)} - \mathbf{u}^{(2)})\, \mathrm{d}\Omega \\ & + \int_{\Omega} \rho(\dot{\mathbf{u}}^{(1)} - \dot{\mathbf{u}}^{(2)}) \cdot (\dot{\mathbf{u}}^{(1)} - \dot{\mathbf{u}}^{(2)})\, \mathrm{d}\Omega = C, \end{aligned} \tag{2.70}$$

where C is a constant. Because the initial and boundary conditions are the same for both solutions, the constant $C = 0$ and, thus, the integral is equal to zero. Since the domain is arbitrary, the integrand must vanish. Since $\mathbb{E}$ is positive definite, ρ is positive, and the fact that both are independent of one another, this implies that the integrand can only vanish if the solutions are the same, i.e. $\mathbf{u}^{(1)} - \mathbf{u}^{(2)} = \mathbf{0}$ and $\dot{\mathbf{u}}^{(1)} - \dot{\mathbf{u}}^{(2)} = \mathbf{0}$.

2.7 Hyperelastic Finite Strain Material Laws

A commonly used model for purely elastic materials, such as some polymers, which are expected to experience finite deformations, is hyperelasticity. Analogous to the infinitesimal strain case, an elastic material is called hyperelastic if there exists a stored energy function, W, that is only a function of the mechanical deformation, and that $\mathbf{S} = 2\frac{\partial W}{\partial \mathbf{C}} = \frac{\partial W}{\partial \mathbf{E}}$ or $\mathbf{P} = \frac{\partial W}{\partial \mathbf{F}}$. These relations stem from the fact that the work, generally, in a closed process (cycle), must be non-negative, and exactly zero for a hyperelastic material. Various definitions for a nonnegative stored energy function, W, in the initial configuration, follow from

$$\begin{aligned}\int_{t_0}^{t_1}\left(\int_{\omega_0}\dot{W}\,d\omega_0\right)dt = \text{work} &= \int_{t_0}^{t_1}\left(\int_{\omega_0}\mathbf{P}:\dot{\mathbf{F}}\,d\omega_0\right)dt\\ &= \frac{1}{2}\int_{t_0}^{t_1}\left(\int_{\omega_0}\mathbf{S}:\dot{\mathbf{C}}\,d\omega_0\right)dt\\ &= \int_{t_0}^{t_1}\left(\int_{\omega_0}\mathbf{S}:\dot{\mathbf{E}}\,d\omega_0\right)dt \geq 0.\end{aligned} \tag{2.71}$$

Since the integrals are arbitrary, we have $\dot{W} = \mathbf{P} : \dot{\mathbf{F}}$, $2\dot{W} = \mathbf{S} : \dot{\mathbf{C}}$ and $\dot{W} = \mathbf{S} : \dot{\mathbf{E}}$. Alternatively, these relations could have been directly obtained from the First Law of Thermodynamics, with no dissipation $\rho_0\dot{w} = \dot{W} = \mathbf{S} : \dot{\mathbf{E}}$ (2.35). Analogously to the infinitesimal case, the results are

$$\begin{aligned}\frac{\partial W}{\partial \mathbf{F}} : \dot{\mathbf{F}} = \mathbf{P} : \dot{\mathbf{F}} &\Rightarrow \frac{\partial W}{\partial \mathbf{F}} = \mathbf{P},\\ 2\frac{\partial W}{\partial \mathbf{C}} : \dot{\mathbf{C}} = \mathbf{S} : \dot{\mathbf{C}} &\Rightarrow 2\frac{\partial W}{\partial \mathbf{C}} = \mathbf{S},\\ \frac{\partial W}{\partial \mathbf{E}} : \dot{\mathbf{E}} = \mathbf{S} : \dot{\mathbf{E}} &\Rightarrow \frac{\partial W}{\partial \mathbf{E}} = \mathbf{S}.\end{aligned} \tag{2.72}$$

2.7.1 Basic Requirements for Finite Strain Laws

In addition to being material frame invariant, any stored energy function must obey four criteria:

1. $\mathbf{C} = \mathbf{1} \Leftrightarrow (I_C = II_C = 3, III_C = 1) \Leftrightarrow W = 0$, where (I_C, II_C, III_C) are the principal invariants of $\mathbf{C}$,
2. $W \geq 0$,
3. $\mathbf{S} = \mathbf{0}$ for $\mathbf{C} = \mathbf{1}$ and
4. the material constants in a finite deformation material law must give responses with known material constants, for example in the isotropic case, $\lambda\,(= \kappa - \frac{2\mu}{3})$ and μ, when perturbed around the undeformed configuration (see Ciarlet [28] for more details).

Condition (**4**) implies that the material constants in a finite deformation material law must be such that they match hyperelastic responses with known Lame constants, λ and μ, when perturbed around the undeformed configuration, thus matching the simplest (moderate strain) hyperelastic law, the Kirchhoff-St. Venant[6]

$$\mathbf{S} = \underbrace{\lambda tr(\mathbf{E})\mathbf{1} + 2\mu\mathbf{E}}_{\text{Kirchhoff–St.Venant law}} + \mathcal{O}(\mathbf{E}). \tag{2.73}$$

For illustration purposes, consider a somewhat more general standard form for a stored energy function, W, as follows (compressible Mooney-Rivlin):

$$W = \underbrace{K_1(\overline{I}_C - 3) + K_2(\overline{II}_C - 3)}_{\text{incompressible part}} + \underbrace{\frac{\kappa}{2}(\sqrt{III_C} - 1)^2}_{\text{compressible part}}. \tag{2.74}$$

This is one possible form of a compressible material response function. In this model, the first and second invariants of $\mathbf{C}$, I_C and II_C, have been scaled by the square-root of the third invariant, i.e. $\overline{I}_C = I_C III_C^{-\frac{1}{3}} = I_C J^{-\frac{2}{3}}$ and $\overline{II}_C = II_C III_C^{-\frac{2}{3}} = II_C J^{-\frac{4}{3}}$, to ensure that they contribute nothing to the compressible part of the response. The factors stem from the eigenvalues of $\mathbf{C} = \{\Lambda_1^2, \Lambda_2^2, \Lambda_3^2\}$, where $I_C = \Lambda_1^2 + \Lambda_2^2 + \Lambda_3^2$, $II_C = \Lambda_1^2\Lambda_2^2 + \Lambda_2^2\Lambda_3^2 + \Lambda_3^2\Lambda_1^2$ and $III_C = \Lambda_1^2\Lambda_2^2\Lambda_3^2$, and where $\overline{\Lambda}_1 = J^{-\frac{1}{3}}\Lambda_1$, $\overline{\Lambda}_2 = J^{-\frac{1}{3}}\Lambda_2$, $\overline{\Lambda}_3 = J^{-\frac{1}{3}}\Lambda_3$, which imply $\overline{\Lambda_1\Lambda_2\Lambda_3} = \frac{\Lambda_1\Lambda_2\Lambda_3}{J} = 1$. We further define an incompressible deformation gradient, $\overline{\mathbf{F}} \stackrel{\text{def}}{=} J^{-\frac{1}{3}}\mathbf{F} = III_C^{-\frac{1}{6}}\mathbf{F}$ and its associated Cauchy-Green strain tensor $\overline{\mathbf{C}} \stackrel{\text{def}}{=} \overline{\mathbf{F}}^T \cdot \overline{\mathbf{F}} = J^{-\frac{2}{3}}\mathbf{C} = III_C^{-\frac{1}{3}}\mathbf{C}$. Note that because of the fact that $\overline{J} = 1$, the first terms form a so-called incompressible function. In other words, the corresponding scaled third invariant is always unity. We may write

$$\begin{aligned}
\mathbf{S} &= 2\left(\frac{\partial W}{\partial I_C}\frac{\partial I_C}{\partial \mathbf{C}} + \frac{\partial W}{\partial II_C}\frac{\partial II_C}{\partial \mathbf{C}} + \frac{\partial W}{\partial III_C}\frac{\partial III_C}{\partial \mathbf{C}}\right) \\
&= 2\left(\frac{\partial W}{\partial \overline{I}_C}\frac{\partial \overline{I}_C}{\partial I_C}\mathbf{1} + \frac{\partial W}{\partial \overline{II}_C}\frac{\partial \overline{II}_C}{\partial II_C}(I_C\mathbf{1} - \mathbf{C}) + \frac{\partial W}{\partial III_C} III_C \mathbf{C}^{-1}\right),
\end{aligned} \tag{2.75}$$

where the following relations hold for the derivative of the invariants with respect to the Cauchy-Green strain tensor

$$\frac{\partial I_C}{\partial \mathbf{C}} = \mathbf{1}, \quad \frac{\partial II_C}{\partial \mathbf{C}} = I_C\mathbf{1} - \mathbf{C} \quad \frac{\partial III_C}{\partial \mathbf{C}} = III_C\mathbf{C}^{-1} \tag{2.76}$$

and the derivatives of the scaled invariants with respect to the invariants

[6] One can alternatively write the equation in terms of the bulk $\kappa = \lambda + \frac{2\mu}{3}$ and shear moduli μ. In general a constitutive law of the form $\mathbf{S} = \mathbb{E} : \mathbf{E}$ is known as a Kirchhoff-St. Venant material law. It is the simplest possible finite strain law which is hyperelastic and frame indifferent.

$$\frac{\partial \bar{I}_C}{\partial I_C} = III_C^{-\frac{1}{3}}, \quad \frac{\partial \bar{II}_C}{\partial II_C} = III_C^{-\frac{2}{3}}. \tag{2.77}$$

For proofs see Ciarlet [28]. Furthermore, for the particular strain energy function in (2.74) we obtain

$$\begin{aligned}
\frac{\partial W}{\partial \bar{I}_C} &= K_1, \qquad \frac{\partial W}{\partial \bar{II}_C} = K_2, \\
\frac{\partial W}{\partial III_C} &= \frac{\kappa}{2}(1 - III_C^{-\frac{1}{2}}) - \frac{K_1}{3} I_C III_C^{-\frac{4}{3}} - \frac{2K_2}{3} II_C III_C^{-\frac{5}{3}}.
\end{aligned} \tag{2.78}$$

2.7.2 Determination of Material Constants

The first three conditions for an admissible energy function are satisfied by construction. The use of (2.74) in (2.75) leads to

$$\begin{aligned}
\mathbf{S} = 2\Big[& K_1 III_C^{-\frac{1}{3}} \mathbf{1} + K_2 III_C^{-\frac{2}{3}} (I_C \mathbf{1} - \mathbf{C}) + (\frac{\kappa}{2}(III_C - III_C^{\frac{1}{2}}) \\
& - \frac{K_1}{3} I_C III_C^{-\frac{1}{3}} - \frac{2K_2}{3} II_C III_C^{-\frac{2}{3}}) \mathbf{C}^{-1} \Big],
\end{aligned} \tag{2.79}$$

which, when evaluated at $\mathbf{C} = \mathbf{1}$, yields $\mathbf{S} = 2(K_1 + 2K_2 - K_1 - 2K_2)\mathbf{1} = \mathbf{0}$. Therefore, condition (**3**) from Sect. 2.7.1 is also automatically satisfied. What remains is to satisfy condition (**4**). We have a general tangent

$$\begin{aligned}
\tfrac{\partial \mathbf{S}}{\partial \mathbf{E}} : \delta \mathbf{E} = \tfrac{\partial \mathbf{S}}{\partial \mathbf{C}} : \delta \mathbf{C} & \\
= 2([K_1(-\tfrac{1}{3} III_C^{-\frac{4}{3}} & III_C \mathbf{C}^{-1} : \delta \mathbf{C}]\mathbf{1} \\
+[K_2(-\tfrac{2}{3} III_C^{-\frac{5}{3}} & III_C \mathbf{C}^{-1} : \delta \mathbf{C}](I_C \mathbf{1} - \mathbf{C}) \\
+[K_2(III_C^{-\frac{2}{3}} & (\mathbf{1} : \delta \mathbf{C})\mathbf{1} - \delta \mathbf{C})] \\
+[(\tfrac{\kappa}{2}(1 - III_C^{-\frac{1}{2}}) & III_C \mathbf{C}^{-1} : \delta \mathbf{C})\mathbf{1} - \tfrac{K_1}{3}((\mathbf{1} : \delta \mathbf{C})\mathbf{1} III_C^{-\frac{1}{3}} \\
+\tfrac{K_1}{9} I_C III_C^{-\frac{4}{3}} & III_C \mathbf{C}^{-1} : \delta \mathbf{C} \\
-\tfrac{2}{3} K_2((I_C \mathbf{1} - \mathbf{C}) & : \delta \mathbf{C}) III_C^{-\frac{2}{3}} + \tfrac{4}{9} K_2 II_C III_C^{-\frac{5}{3}} III_C \mathbf{C}^{-1} : \delta \mathbf{C}] \mathbf{C}^{-1} \\
-[\tfrac{\kappa}{2}(III_C - III_C^{\frac{1}{2}}) & - K_1 \tfrac{1}{3} I_C III_C^{-\frac{1}{3}} - K_2 \tfrac{2}{3} II_C III_C^{-\frac{2}{3}}] \mathbf{C}^{-2} \cdot \delta \mathbf{C}).
\end{aligned} \tag{2.80}$$

Allowing $\mathbf{C} \to \mathbf{1}$, and equating coefficients of the tangent relations implied in (2.80), we have from $(tr\delta\mathbf{C})\mathbf{1}$, $\frac{\lambda}{2} = -\frac{2}{3}K_2 - \frac{2}{3}K_1 + \frac{\kappa}{2}$ which implies $\mu = 2(K_1 + K_2)$, and from $\delta\mathbf{C}$, $\mu = 2(K_1 + K_2)$. Therefore, the coefficients must obey $\mu = 2(K_1 + K_2)$, and we have in the general case

$$W = \underbrace{K_1(\bar{I}_C - 3) + (\frac{\mu}{2} - K_1)(\overline{II}_C - 3)}_{\overline{W}} + \underbrace{\frac{\kappa}{2}(\sqrt{III_C} - 1)^2}_{U}. \tag{2.81}$$

We remark that when $K_1 = \frac{\mu}{2}$ and $K_2 = 0$, the material is called a Compressible Neo-Hookean material, with a strain energy function of

$$W = \frac{\mu}{2}(\bar{I}_C - 3) + \frac{\kappa}{2}(\sqrt{III_C} - 1)^2. \tag{2.82}$$

Remark. The Cauchy stress can be split in the following manner, $\sigma = \sigma' + p\mathbf{1}$, where $p \stackrel{\text{def}}{=} \frac{tr\sigma}{3}$, and thus

$$\mathbf{S} = J\mathbf{F}^{-1} \cdot (\sigma' + p\mathbf{1}) \cdot \mathbf{F}^{-T} = \underbrace{J\mathbf{F}^{-1} \cdot \sigma' \cdot \mathbf{F}^{-T}}_{\stackrel{\text{def}}{=}\mathbf{S}'} + Jp\mathbf{C}^{-1}. \tag{2.83}$$

We also have, by definition,

$$\mathbf{S} = 2\frac{\partial W}{\partial \mathbf{C}} = 2\frac{\partial \overline{W}}{\partial \mathbf{C}} + 2\frac{\partial U}{\partial \mathbf{C}} = 2\frac{\partial \overline{W}}{\partial \mathbf{C}} + 2\frac{\partial U}{\partial J}\frac{\partial J}{\partial \mathbf{C}} = \underbrace{2\frac{\partial \overline{W}}{\partial \mathbf{C}}}_{\mathbf{S}'} + \frac{\partial U}{\partial J}J\mathbf{C}^{-1}. \tag{2.84}$$

Since $\sigma' = \frac{1}{J}\mathbf{F} \cdot \left(2\frac{\partial \overline{W}}{\partial \mathbf{C}}\right) \cdot \mathbf{F}^T$, (2.84) implies $p = \frac{\partial U}{\partial J} = \kappa(J - 1)$.

2.8 Moderate Strain Constitutive Relations

Consider a rigid motion, a rotation and translation, of the form $\tilde{\mathbf{x}} = \mathbf{R} \cdot \mathbf{x} + \mathbf{c}$. An original differential element, $d\mathbf{x}$, will be oriented differently after the motion, $d\tilde{\mathbf{x}}$, however, its length will remain unchanged, i.e. $||d\tilde{\mathbf{x}}|| = ||d\mathbf{x}||$. A quantity such as distance is said to be objective under rigid body motions. The term "objective" is given to any vector that consistently transforms according to $\tilde{\mathbf{a}} = \mathbf{R} \cdot \mathbf{a}$. Analogously, for second order tensors, we require $\tilde{\mathbf{a}} = \mathbf{R} \cdot \mathbf{a} \cdot \mathbf{R}^T$. The velocity vector is an example of a non-objective vector, because differentiating the rotated mapping $\tilde{\mathbf{u}} = \mathbf{R} \cdot \mathbf{u}$, with respect to time yields

$$\tilde{\mathbf{v}} = \frac{d\tilde{\mathbf{u}}}{dt} = \dot{\mathbf{R}} \cdot \mathbf{u} + \mathbf{R} \cdot \dot{\mathbf{u}} = \dot{\mathbf{R}} \cdot \mathbf{u} + \mathbf{R} \cdot \mathbf{v}, \tag{2.85}$$

thus, $\tilde{\mathbf{v}} \neq \mathbf{R} \cdot \mathbf{v}$. The concept of objectivity under rigid body motions is central to the construction of reliable constitutive relations. In this context, objectivity can be used by studying the effect of rigid motions superposed on existing deformations. An important example of a nonobjective tensor is the temporal derivative of the Cauchy stress,

$$\dot{\tilde{\sigma}} = \frac{d(\mathbf{R} \cdot \sigma \cdot \mathbf{R}^T)}{dt} = \dot{\mathbf{R}} \cdot \sigma \cdot \mathbf{R}^T + \mathbf{R} \cdot \dot{\sigma} \cdot \mathbf{R}^T + \mathbf{R} \cdot \sigma \cdot \dot{\mathbf{R}^T} \neq \mathbf{R} \cdot \dot{\sigma} \cdot \mathbf{R}^T. \tag{2.86}$$

Thus, unless $\dot{\mathbf{R}} = \mathbf{0}$, it is not objective. The application of such concepts to constitutive relations is straightforward. Our goal is always to employ constitutive relations that transform consistently, i.e. their character does not change under superposed rigid body motions. Consider the definition of the deformation gradient, $d\mathbf{x} = \mathbf{F} \cdot d\mathbf{X}$. Superposed rigid motions leads to

$$\mathbf{R} \cdot d\mathbf{x} = \mathbf{R} \cdot (\mathbf{F} \cdot d\mathbf{X}) \Rightarrow \tilde{\mathbf{F}} = \mathbf{R} \cdot \mathbf{F}. \tag{2.87}$$

As a consequence, we have for the Green-Lagrange referential strain measure

$$\tilde{\mathbf{E}} = \frac{1}{2}(\tilde{\mathbf{F}}^T \cdot \tilde{\mathbf{F}} - \mathbf{1}) = \frac{1}{2}((\mathbf{R} \cdot \mathbf{F})^T \cdot \mathbf{R} \cdot \mathbf{F} - \mathbf{1}) = \frac{1}{2}(\mathbf{F}^T \cdot \mathbf{F} - \mathbf{1}) = \mathbf{E}, \tag{2.88}$$

since $\mathbf{R}^T \cdot \mathbf{R} = \mathbf{1}$. Also, for the second Piola-Kirchhoff strain measure,

$$\tilde{\mathbf{S}} = \tilde{J}\tilde{\mathbf{F}}^{-1} \cdot \tilde{\sigma} \cdot \tilde{\mathbf{F}}^{-T} = \tilde{J}(\mathbf{F}^{-1} \cdot \mathbf{R}^{-1}) \cdot (\mathbf{R} \cdot \tilde{\sigma} \cdot \mathbf{R}^T) \cdot (\mathbf{R}^{-T} \cdot \mathbf{F}^{-T}) = J\mathbf{F}^{-1} \cdot \sigma \cdot \mathbf{F}^{-T} = \mathbf{S}. \tag{2.89}$$

Therefore, a constitutive law such as the Kirchhoff-St. Venant law $\mathbf{S} = \mathbb{E} : \mathbf{E}$, is said to be frame-indifferent, since it employs objective stress and strain measures, i.e. $\tilde{\mathbf{S}} = \mathbb{E} : \tilde{\mathbf{E}} = \mathbb{E} : \mathbf{E} = \mathbf{S}$.

A necessary and sufficient condition for a constitutive relation to be frame indifferent is if $\sigma = \mathscr{F}(\mathbf{F})$, then $\mathbf{Q} \cdot \mathscr{F}(\mathbf{F}) \cdot \mathbf{Q}^T = \mathscr{F}(\mathbf{Q} \cdot \mathbf{F})$, for arbitrary orthogonal $\mathbf{Q}$. For our purposes, we can take $\mathbf{Q} = \mathbf{R}$. Clearly, the Kirchhoff-St. Venant law is frame indifferent, due to the objectivity of $\mathbf{S}$ and $\mathbf{E}$. Furthermore, it is intuitively obvious that a relation such as the Compressible-Mooney material model is also frame indifferent, due to the fact that it is constructed from the principle invariants of $\mathbf{C}$. However, for many seemingly general constitutive relations, there are restrictions. For example, consider the following law $\sigma = \mathbb{E} : \mathbf{e}$. Such a relation must have an isotropic modulus for it to be frame indifferent under all superposed rigid body rotations.[7] In order to see this, consider $\mathbf{R} \cdot \sigma \cdot \mathbf{R}^T = \mathbb{E} : (\mathbf{R} \cdot \mathbf{e} \cdot \mathbf{R}^T)$.

[7] We note that simply because a strain or stress measure employs quantities such as the deformation gradient in its definition, does not mean that it will remain unaltered under rigid motions, for example, take the Almansi (Eulerian) strain

$$\tilde{\mathbf{e}} = \frac{1}{2}(\mathbf{1} - \tilde{\mathbf{F}}^{-T} \cdot \tilde{\mathbf{F}}^{-1}) = \mathbf{R} \cdot \left(\frac{1}{2}(\mathbf{1} - \mathbf{F}^{-T} \cdot \mathbf{F}^{-1})\right) \cdot \mathbf{R}^T = \mathbf{R} \cdot \mathbf{e} \cdot \mathbf{R}^T. \tag{2.90}$$

However, such a measure will nonetheless prove to be useful for isotropic materials.

In order to make the process transparent, consider the matrix notation introduced earlier, $[\hat{\sigma}] = [\mathbf{R}][\sigma][\mathbf{R}]^T$ implying $\{\hat{\sigma}\} = [\mathbf{T}]\{\sigma\} = [\hat{\mathbb{E}}][\mathbf{T}]\{\mathbf{e}\} = [\hat{\mathbb{E}}]\{\hat{\mathbf{e}}\}$ which implies $\{\sigma\} = [\mathbf{T}^{-1}][\hat{\mathbb{E}}][\mathbf{T}]\{\mathbf{e}\}$, where $[\mathbf{R}]$ is a transformation matrix, and where $[\cdot]$ is used to indicate matrix notation equivalent to a tensor form, while $\{\cdot\}$ to indicate a vector representation. Frame indifference requires that $[\mathbb{E}] = [\mathbf{T}^{-1}][\hat{\mathbb{E}}][\mathbf{T}]$, $\forall[\mathbf{R}]$ and thus $\forall[\mathbf{T}]$. This relation holds only if $[\mathbb{E}]$ is isotropic.[8] However, such a relation is only elastic, i.e. the deformation is only a function of the deformed state, and not hyperelastic, i. e. it is not derivable from a differentiation of a potential energy function. It is said to be *Cauchy-elastic*. Hyperelastic constitutive relations, those derived from scalar energy functions employing objective strain measures, such as Kirchhoff St. Venant, $W = \frac{1}{2}\mathbf{E} : \mathbb{E} : \mathbf{E}$ or Compressible Mooney Rivlin materials, which employ the principle invariants of $\mathbf{C}$, are automatically frame indifferent. The simple Kirchhoff-St. Venant and Almansi/Eulerian

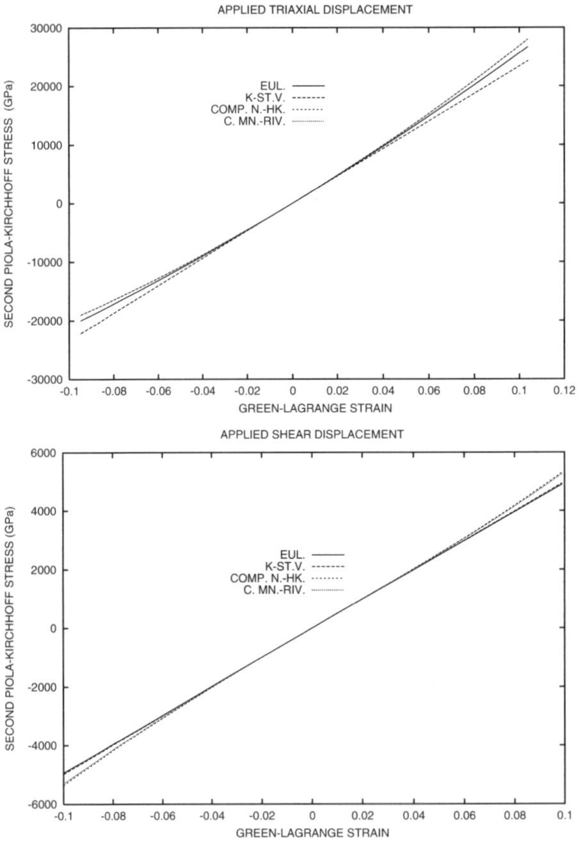

Fig. 2.11 A comparison of various finite deformation laws

[8] The Kirchhoff St. Venant relation can be anisotropic.

laws are extremely useful in applications where there are small or moderate elastic strains, and large inelastic strains. For finite deformations with moderate elastic strains ($\leq 3\%$), the constitutive laws discussed yield virtually identical responses. Consider

- Eulerian $\sigma = \mathbb{E} : \mathbf{e}, \Rightarrow \mathbf{S} = J\mathbf{F}^{-1} \cdot (\mathbb{E} : \mathbf{e}) \cdot \mathbf{F}^{-T}$
- Kirchhoff-St. Venant $\mathbf{S} = \mathbb{E} : \mathbf{E}$,
- Compressible Mooney-Rivlin $\mathbf{S} = 2\frac{\partial W}{\partial \mathbf{C}}, W = K_1(\bar{I}_C - 3) + K_2(\overline{II}_C - 3) + \frac{\kappa}{2}(\sqrt{III_C} - 1)^2$
- $2(K_1 + K_2) = \mu$, with a special case of a Neo-Hookean material being $K_2 = 0$.

We consider the following homogeneous loadings at a material point

$$\begin{bmatrix} u_1|_{\partial\Omega} \\ u_2|_{\partial\Omega} \\ u_3|_{\partial\Omega} \end{bmatrix} = \alpha \begin{bmatrix} 1\,0\,0 \\ 0\,1\,0 \\ 0\,0\,1 \end{bmatrix} \begin{bmatrix} X_1 \\ X_2 \\ X_3 \end{bmatrix} \quad \begin{bmatrix} u_1|_{\partial\Omega} \\ u_2|_{\partial\Omega} \\ u_3|_{\partial\Omega} \end{bmatrix} = \alpha \begin{bmatrix} 0\,1\,0 \\ 1\,0\,0 \\ 0\,0\,0 \end{bmatrix} \begin{bmatrix} X_1 \\ X_2 \\ X_3 \end{bmatrix}. \tag{2.91}$$

The responses are shown in Fig. 2.11. The relative proximity of these relation should not be a surprise for under 3% strains since, $(\nabla_X \mathbf{u})^T \cdot (\nabla_X \mathbf{u})$ and $(\nabla_x \mathbf{u})^T \cdot (\nabla_x \mathbf{u})$ are quite small.

Chapter 3
Fundamental Weak Formulations

In many problems of mathematical physics the true solutions are nonsmooth, i.e. the strains and the stresses are not differentiable in the classical sense. For example in the equation of static equilibrium $\nabla \cdot \sigma + \mathbf{f} = \mathbf{0}$, there is an implicit requirement that the stress was differentiable.[1] *In many applications, this is too strong of a requirement.* Therefore, when solving such problems we have two options: (1) enforcement of jump conditions at every interface where continuity is in question or (2) weak formulations (weakening the regularity requirements). Weak forms, which are designed to accommodate irregular data and solutions, are usually preferred. *Numerical techniques employing weak forms, such as the Finite Element Method, have been developed with the essential property that whenever a smooth classical solution exists, it is also a solution to the weak form problem.* Therefore, we lose nothing by reformulating a problem in a weaker way. However, an important feature of such formulations is the ability to allow natural and easy approximations to solutions in an energetic sense, which is desirable in the framework of mechanics.

3.1 Direct Weak Formulations

To derive a direct weak form for a body, we take the equilibrium equations (denoted the strong form) and form a scalar product with an arbitrary smooth vector valued function $\mathbf{v}$, and integrate over the body, $\int_\Omega (\nabla \cdot \sigma + \mathbf{f}) \cdot \mathbf{v}\, d\Omega = \int_\Omega \mathbf{r} \cdot \mathbf{v}\, d\Omega = 0$, where $\mathbf{r}$ is called the residual. We call $\mathbf{v}$ a "test" function. If we were to add a condition that we do this for all ($\stackrel{\text{def}}{=} \forall$) possible "test" functions then $\int_\Omega (\nabla \cdot \sigma + \mathbf{f}) \cdot \mathbf{v}\, d\Omega = \int_\Omega \mathbf{r} \cdot \mathbf{v}\, d\Omega = 0$, $\forall \mathbf{v}$, implies $\mathbf{r} = \mathbf{0}$. Therefore if every possible test function was considered, then $\mathbf{r} = \nabla \cdot \sigma + \mathbf{f} = \mathbf{0}$ on any finite region in Ω. Consequently, the weak and strong statements would be equivalent provided thetrue solution is smooth

[1] Throughout this chapter, we consider only static linear elasticity, at infinitesimal strains, and specialize approaches later for nonlinear and time dependent problems.

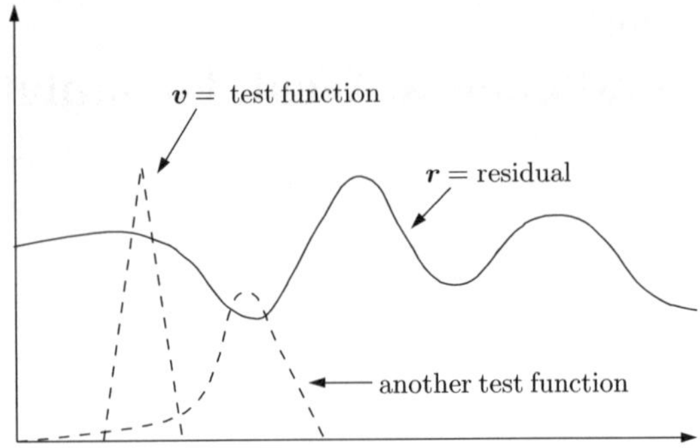

Fig. 3.1 The idea of a test function

enough to have a strong solution. Clearly, $\mathbf{r}$ can never be nonzero over any finite region in the body, because the test function will "find" them (Fig. 3.1).

Using the product rule of differentiation, $\nabla\cdot(\boldsymbol{\sigma}\cdot\mathbf{v}) = (\nabla\cdot\boldsymbol{\sigma})\cdot\mathbf{v} + \nabla\mathbf{v} : \boldsymbol{\sigma}$ leads to, $\forall\mathbf{v}$, $\int_\Omega(\nabla\cdot(\boldsymbol{\sigma}\cdot\mathbf{v}) - \nabla\mathbf{v}:\boldsymbol{\sigma})\,d\Omega + \int_\Omega \mathbf{f}\cdot\mathbf{v}\,d\Omega = 0$, where we choose the $\mathbf{v}$ from an admissible set, to be discussed momentarily. Using the divergence theorem leads to, $\forall\mathbf{v}$, $\int_\Omega \nabla\mathbf{v}:\boldsymbol{\sigma}\,d\Omega = \int_\Omega \mathbf{f}\cdot\mathbf{v}\,d\Omega + \int_{\partial\Omega}\boldsymbol{\sigma}\cdot\mathbf{n}\cdot\mathbf{v}\,dA$, which leads to $\int_\Omega \nabla\mathbf{v}:\boldsymbol{\sigma}\,d\Omega = \int_\Omega \mathbf{f}\cdot\mathbf{v}\,d\Omega + \int_{\partial\Omega}\mathbf{t}\cdot\mathbf{v}\,dA$. If we decide to restrict our choices of $\mathbf{v}$'s to those such that $\mathbf{v}|_{\Gamma_u} = \mathbf{0}$, we have, where $\mathbf{d}$ is the applied boundary displacement on Γ_u, for infinitesimal strain linear elasticity

$$\text{Find } \mathbf{u}, \mathbf{u}|_{\Gamma_u} = \mathbf{d}, \text{ such that } \forall\mathbf{v}, \mathbf{v}|_{\Gamma_u} = \mathbf{0}$$

$$\underbrace{\int_\Omega \nabla\mathbf{v} : \mathbb{E} : \nabla\mathbf{u}\,d\Omega}_{\stackrel{\text{def}}{=}\mathscr{B}(\mathbf{u},\mathbf{v})} = \underbrace{\int_\Omega \mathbf{f}\cdot\mathbf{v}\,d\Omega + \int_{\Gamma_t}\mathbf{t}\cdot\mathbf{v}\,dA}_{\stackrel{\text{def}}{=}\mathscr{F}(\mathbf{v})}. \tag{3.1}$$

This is called a "weak" form because it does not require the differentiability of the stress $\boldsymbol{\sigma}$. In other words, the differentiability requirements have been *weakened.* It is clear that we are able to consider problems with quite irregular solutions. We observe that if we test the solution with all possible test functions of sufficient smoothness, then the weak solution is equivalent to the strong solution. *We emphasize that provided the true solution is smooth enough, the weak and strong forms are equivalent, which can be seen by the above constructive derivation.* To see this a bit more clearly, we consider a simple one-dimensional example.

3.1.1 An Example

Let us define a one-dimensional continuous function $r \in C^0(\Omega)$, on a one-dimensional domain, $\Omega = (0, L)$. Our claim is that $\int_\Omega rv\,dx = 0$, $\forall v \in C^0(\Omega)$, which implies

Fig. 3.2 A residual function and a test function

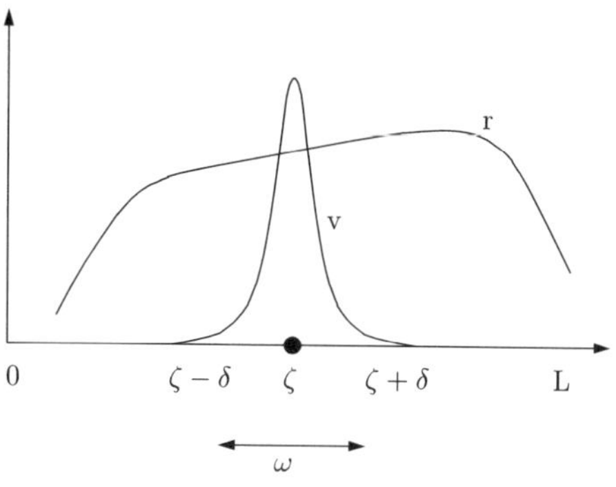

$r = 0$. This can be easily proven by contradiction. Suppose $r \neq 0$ at some point $\zeta \in \Omega$. Since $r \in C^0(\Omega)$, there must exist a subdomain (subinterval), $\omega \in \Omega$, defined through δ, $\omega \stackrel{\text{def}}{=} \zeta \pm \delta$ such that r has the same sign as at point ζ. Since v is arbitrary, we may choose v to be zero outside of this interval, and positive inside (Fig. 3.2). This would imply that $0 < \int_\Omega rv\ d\Omega = \int_\omega rv\ d\Omega = 0$ which is acontradiction. Now select $r = \frac{d\sigma}{dx} + f \in C^0(\Omega) \Rightarrow \frac{d(E\varepsilon)}{dx} + f \in C^0(\Omega) \Rightarrow u \in C^2(\Omega)$. Therefore, for example in one-dimensional infinitesimal strain linear elasticity, the equivalence of weak and strong forms occurs if $u \in C^2(\Omega)$.

Remark. For the general case of geometrical and materialnonlinearities, the equivalence of weak and strong forms occurs if $\sigma \in C^1(\Omega)$.

3.1.2 Some Restrictions

A key question is the selection of the sets of functions in the weak form. Somewhat naively, the answer is simple, the integrals must remain finite. Therefore, the following restrictions hold ($\forall \mathbf{v}$),

$$|\int_\Omega \mathbf{f} \cdot \mathbf{v} d\Omega| < \infty, \quad |\int_{\partial\Omega} \sigma \cdot \mathbf{n} \cdot \mathbf{v} d\Omega| < \infty \quad \text{and} \quad |\int_\Omega \nabla \mathbf{v} : \sigma d\Omega| < \infty, \tag{3.2}$$

and govern the selection of the approximation spaces. These relations simply mean that the functions must be square integrable. In order to make precise statements one must have a method of "book keeping". A commonly used framework is to employ so-called Hilbertian Sobolev spaces. We recall that a norm has three main characteristics for any vectors $\mathbf{u}$ and $\mathbf{v}$ such that $||\mathbf{u}|| < \infty$ and $||\mathbf{v}|| < \infty$ are

(1) $||\mathbf{u}|| > 0$, $||\mathbf{u}|| = 0$ if and only if $\mathbf{u} = \mathbf{0}$,
(2) $||\mathbf{u}+\mathbf{v}|| \leq ||\mathbf{u}|| + ||\mathbf{v}||$ and
(3) $||\alpha\mathbf{u}|| = |\alpha|||\mathbf{u}||$, where α is a scalar.

Certain types of norms, so-called Hilbert space norms, are frequently used in solid mechanics. Following standard notation, we denote $H^1(\Omega)$ as the usual space of scalar functions with generalized partial derivatives of order ≤ 1 in $L^2(\Omega)$, i.e. square integrable, in other words $u \in H^1(\Omega)$ if $||u||^2_{H^1(\Omega)} \stackrel{\text{def}}{=} \int_\Omega \frac{\partial u}{\partial x_j}\frac{\partial u}{\partial x_j}\,d\Omega + \int_\Omega uu\,d\Omega < \infty$. We define $\mathbf{H}^1(\Omega) \stackrel{\text{def}}{=} [H^1(\Omega)]^3$ as the space of vector-valued functions whose components are in $H^1(\Omega)$, i.e.

$$\mathbf{u} \in \mathbf{H}^1(\Omega) \text{ if } ||\mathbf{u}||^2_{\mathbf{H}^1(\Omega)} \stackrel{\text{def}}{=} \int_\Omega \frac{\partial u_i}{\partial x_j}\frac{\partial u_i}{\partial x_j}\,d\Omega + \int_\Omega u_i u_i\,d\Omega < \infty, \tag{3.3}$$

and we denote $\mathbf{L}^2(\Omega) \stackrel{\text{def}}{=} [L^2(\Omega)]^3$. Using these definitions, a complete weak boundary value problemcan be written as follows. The data (loads) are assumed to be such that $\mathbf{f} \in \mathbf{L}^2(\Omega)$ and $\mathbf{t} \in \mathbf{L}^2(\Gamma_t)$, but less smooth data can be considered without complications. Implicitly we require that $\mathbf{u} \in \mathbf{H}^1(\Omega)$ and $\sigma \in \mathbf{L}^2(\Omega)$ without continually making such references. Therefore in summary we assume that our solutions obey these restrictions, leading to the following infinitesimal strainlinear elasticity weak statement:

$$\begin{gathered}\text{Find } \mathbf{u} \in \mathbf{H}^1(\Omega), \mathbf{u}|_{\Gamma_u} = \mathbf{d}, \text{ such that } \forall \mathbf{v} \in \mathbf{H}^1(\Omega), \mathbf{v}|_{\Gamma_u} = \mathbf{0} \\ \int_\Omega \nabla \mathbf{v} : \mathbb{E} : \nabla \mathbf{u}\,d\Omega = \int_\Omega \mathbf{f} \cdot \mathbf{v}\,d\Omega + \int_{\Gamma_t} \mathbf{t} \cdot \mathbf{v}\,dA.\end{gathered} \tag{3.4}$$

We note that if the data in (3.4) are smooth and if (3.4) possesses a solution $\mathbf{u}$ that is sufficiently regular, then $\mathbf{u}$ is the solution of the classical linear elastostatics problem in strong form:

$$\begin{aligned} &\nabla \cdot (\mathbb{E} : \nabla \mathbf{u}) + \mathbf{f} = \mathbf{0}, && \mathbf{x} \in \Omega, \\ &\mathbf{u} = \mathbf{d}, && \mathbf{x} \in \Gamma_u, \\ &(\mathbb{E} : \nabla \mathbf{u}) \cdot \mathbf{n} = \mathbf{t}, && \mathbf{x} \in \Gamma_t. \end{aligned} \tag{3.5}$$

3.1.3 The Principle of Minimum Potential Energy

A useful concept in mechanics is that of minimum principles involving the $||\cdot||_{E(\Omega)}$ energy (semi) norm. By direct manipulation we have[2]

$$
\begin{aligned}
||\mathbf{u}-\mathbf{w}||^2_{E(\Omega)} &\stackrel{\text{def}}{=} \mathscr{B}(\mathbf{u}-\mathbf{w},\mathbf{u}-\mathbf{w})\\
&= \mathscr{B}(\mathbf{u},\mathbf{u})+\mathscr{B}(\mathbf{w},\mathbf{w})-2\mathscr{B}(\mathbf{u},\mathbf{w})\\
&= \mathscr{B}(\mathbf{w},\mathbf{w})-\mathscr{B}(\mathbf{u},\mathbf{u})-2\mathscr{B}(\mathbf{u},\mathbf{w})+2\mathscr{B}(\mathbf{u},\mathbf{u})\\
&= \mathscr{B}(\mathbf{w},\mathbf{w})-\mathscr{B}(\mathbf{u},\mathbf{u})-2\mathscr{B}(\mathbf{u},\mathbf{w}-\mathbf{u})\\
&= \mathscr{B}(\mathbf{w},\mathbf{w})-\mathscr{B}(\mathbf{u},\mathbf{u})-2\mathscr{F}(\mathbf{w}-\mathbf{u})\\
&= \mathscr{B}(\mathbf{w},\mathbf{w})-2\mathscr{F}(\mathbf{w})-(\mathscr{B}(\mathbf{u},\mathbf{u})-2\mathscr{F}(\mathbf{u}))\\
&= 2\mathscr{J}(\mathbf{w})-2\mathscr{J}(\mathbf{u}),
\end{aligned}
\tag{3.6}
$$

where we define the "elastic potential" $\mathscr{J}(\mathbf{w})\stackrel{\text{def}}{=}\frac{1}{2}\mathscr{B}(\mathbf{w},\mathbf{w})-\mathscr{F}(\mathbf{w})=\frac{1}{2}\int_\Omega \nabla\mathbf{w}:\mathbb{E}:\nabla\mathbf{w}\,d\Omega-\int_\Omega \mathbf{f}\cdot\mathbf{w}\,d\Omega-\int_{\Gamma_t}\mathbf{t}\cdot\mathbf{w}\,dA$. Equation 3.6 implies

$$
0\leq ||\mathbf{u}-\mathbf{w}||^2_{E(\Omega)}=2(\mathscr{J}(\mathbf{w})-\mathscr{J}(\mathbf{u}))\ \textit{or}\ \mathscr{J}(\mathbf{u})\leq\mathscr{J}(\mathbf{w}),
\tag{3.7}
$$

where (3.7) is known as the Principle of Minimum Potential Energy (PMPE). In other words, the true solution possesses the minimum potential.

The minimum property of the exact solution can be proven by an alternative technique. Let us construct a potential function, for a perturbation away from the exact solution $\mathbf{u}$, denoted $\mathbf{u}+\lambda\mathbf{v}$, where λ is a scalar and $\mathbf{v}$ is any admissible variation (test function),

$$
\begin{aligned}
\mathscr{J}(\mathbf{u}+\lambda\mathbf{v}) = &\int_\Omega \frac{1}{2}\nabla(\mathbf{u}+\lambda\mathbf{v}):\mathbb{E}:\nabla(\mathbf{u}+\lambda\mathbf{v})\,d\Omega-\int_\Omega \mathbf{f}\cdot(\mathbf{u}+\lambda\mathbf{v})\,d\Omega\\
&-\int_{\Gamma_t}\mathbf{t}\cdot(\mathbf{u}+\lambda\mathbf{v})\,dA.
\end{aligned}
\tag{3.8}
$$

If we differentiate with respect to λ,

$$
\frac{\partial\mathscr{J}(\mathbf{u}+\lambda\mathbf{v})}{\partial\lambda}=\int_\Omega \nabla\mathbf{v}:\mathbb{E}:\nabla(\mathbf{u}+\lambda\mathbf{v})\,d\Omega-\int_\Omega \mathbf{f}\cdot\mathbf{v}\,d\Omega-\int_{\Gamma_t}\mathbf{t}\cdot\mathbf{v}\,dA=0,
\tag{3.9}
$$

and set $\lambda=0$, because we know that the exact solution is at $\lambda=0$, we have

$$
\frac{\partial\mathscr{J}(\mathbf{u}+\lambda\mathbf{v})}{\partial\lambda}\Big|_{\lambda=0}=\int_\Omega \nabla\mathbf{v}:\mathbb{E}:\nabla\mathbf{u}\,d\Omega-\int_\Omega \mathbf{f}\cdot\mathbf{v}\,d\Omega-\int_{\Gamma_t}\mathbf{t}\cdot\mathbf{v}\,dA=0.
\tag{3.10}
$$

This is sometimes referred to as the *Euler-Lagrange* equation of the potential. Clearly, the minimizer of the potential is the solution to the field equations, since it

[2] $\mathbf{w}$ is kinematically admissible, i.e. it statisfies the specified displacement boundary conditions and it has a finite energy norm.

produces the weak formulation as a result. This is a minimum since $\frac{\partial^2 \mathscr{J}(\mathbf{u}+\lambda\mathbf{v})}{\partial\lambda^2}|_{\lambda=0} = \int_\Omega \nabla\mathbf{v} : \mathbb{E} : \nabla\mathbf{v}\, d\Omega \geq 0$.

It is important to note that the weak form, derived earlier, requires no such potential, and thus is a more general approach than a minimum principle. Thus, in the elastic case, the weak formulation can be considered as a minimization of an potential energy function. This is sometimes referred to as a Rayleigh-Ritz method.

3.1.4 Complementary Weak Forms

There exists another set of weak formulations and corresponding minimum principles called complementary principles. Starting with the equilibrium without body forces for the (tensor-type) test function γ, where $\nabla\cdot\gamma = \mathbf{0}, \gamma\cdot\mathbf{n}|_{\Gamma_n} = \mathbf{0}$ and multiplying by the solution $\mathbf{u}$ leads to $\int_\Omega \nabla\cdot\gamma\cdot\mathbf{u}\, d\Omega = 0 = \int_\Omega \nabla\cdot(\gamma\cdot\mathbf{u})\, d\Omega - \int_\Omega \gamma : \nabla\mathbf{u}\, d\Omega$. Using the divergence theorem yields

Find $\sigma, \nabla\cdot\sigma + \mathbf{f} = \mathbf{0}, \sigma\cdot\mathbf{n}|_{\Gamma_t} = \mathbf{t}$ such that

$$\underbrace{\int_\Omega \gamma : \mathbb{E}^{-1} : \sigma\, d\Omega}_{\stackrel{\text{def}}{=}\mathscr{A}(\sigma,\gamma)} = \underbrace{\int_{\Gamma_u} \gamma\cdot\mathbf{n}\cdot\mathbf{u}\, dA}_{\stackrel{\text{def}}{=}\mathscr{G}(\gamma)} \quad \forall\gamma, \nabla\cdot\gamma = \mathbf{0},\ \gamma\cdot\mathbf{n}|_{\Gamma_t} = \mathbf{0}. \quad (3.11)$$

This is called the complementary form of (3.1). Similar restrictions are placed on the trial and test fields to force the integrals to make sense, i.e. to be finite. Similar boundedness restrictions control the choice ofadmissible complementary functions. In other words we assume that the solutions produce finite energy.*Despite the apparent simplicity of suchprinciples they are rarely used in practical computations because of the fact that it is very hard to find even approximate functions, σ, that satisfy $\nabla\cdot\sigma + \mathbf{f} = \mathbf{0}$ a priori.*

The Principle of Minimum Complementary Potential Energy

As in the primal case, a similar process is repeated using the complementaryweak formulation. We define a complementary norm

$$0 \leq ||\sigma - \gamma||^2_{E^{-1}(\Omega)} \stackrel{\text{def}}{=} \int_\Omega (\sigma - \gamma) : \mathbb{E}^{-1} : (\sigma - \gamma)\, d\Omega = \mathscr{A}(\sigma - \gamma, \sigma - \gamma). \quad (3.12)$$

Again, by direct manipulation, we have

$$
\begin{aligned}
\|\sigma-\gamma\|^2_{E^{-1}(\Omega)} &= \mathscr{A}(\sigma-\gamma,\sigma-\gamma)\\
&= \mathscr{A}(\sigma,\sigma)+\mathscr{A}(\gamma,\gamma)-2\mathscr{A}(\sigma,\gamma)\\
&= \mathscr{A}(\gamma,\gamma)-\mathscr{A}(\sigma,\sigma)-2\mathscr{A}(\sigma,\gamma)+2\mathscr{A}(\sigma,\sigma)\\
&= \mathscr{A}(\gamma,\gamma)-\mathscr{A}(\sigma,\sigma)-2\mathscr{A}(\sigma,\gamma-\sigma)\\
&= \mathscr{A}(\gamma,\gamma)-\mathscr{A}(\sigma,\sigma)-2\mathscr{G}(\gamma-\sigma)\\
&= \mathscr{A}(\gamma,\gamma)-2\mathscr{G}(\gamma)-(\mathscr{A}(\sigma,\sigma)-2\mathscr{G}(\sigma))\\
&= 2\mathscr{K}(\gamma)-2\mathscr{K}(\sigma),
\end{aligned}
\tag{3.13}
$$

where we define $\mathscr{K}(\gamma)\stackrel{\text{def}}{=}\frac{1}{2}\mathscr{A}(\gamma,\gamma)-\mathscr{G}(\gamma)=\frac{1}{2}\int_\Omega \gamma:\mathbb{E}^{-1}:\gamma d\Omega-\int_{\Gamma_u}\gamma\cdot\mathbf{n}\cdot\mathbf{u}dA$. Therefore,

$$
\|\sigma-\gamma\|^2_{E^{-1}(\Omega)} = 2(\mathscr{K}(\gamma)-\mathscr{K}(\sigma)) \ or\ \mathscr{K}(\sigma)\leq\mathscr{K}(\gamma), \tag{3.14}
$$

which is the Principle of Minimum Complementary Potential Energy (PMCPE). By directly adding together the potential energy and the complementary energy we obtain an equation of energy balance:

$$
\begin{aligned}
\mathscr{J}(\mathbf{u})+\mathscr{K}(\sigma) &= \frac{1}{2}\int_\Omega \nabla\mathbf{u}:\mathbb{E}:\nabla\mathbf{u}d\Omega-\int_\Omega \mathbf{f}\cdot\mathbf{u}d\Omega-\int_{\Gamma_t}\mathbf{t}\cdot\mathbf{u}dA\\
&\quad+\frac{1}{2}\int_\Omega \sigma:\mathbb{E}^{-1}:\sigma d\Omega-\int_{\Gamma_u}\mathbf{t}\cdot\mathbf{u}dA\\
&= 0.
\end{aligned}
\tag{3.15}
$$

These relations will be important later.

Chapter 4
Fundamental Micro–Macro Concepts

As stated in the introduction, it is clear that for the relation between averages to be useful it must be computed over a sample containing a statistically representative amount of material. This requirement can be formulated in a very precise mathematical way, which has a clear physical meaning. A commonly accepted macro/micro criterion used in effective property calculations is the so-called Hill condition, $\langle \sigma : \varepsilon \rangle_\Omega = \langle \sigma \rangle_\Omega : \langle \varepsilon \rangle_\Omega$. Hill's condition (Hill [79]) dictates the size requirements on the RVE. For any perfectly bonded heterogeneous body, in the absence of body forces, two physically important loading states satisfy Hill's condition. They are

(1) pure linear displacements of the form:

$$\mathbf{u}|_{\partial\Omega} = \mathscr{E} \cdot \mathbf{x} \Rightarrow \langle \varepsilon \rangle_\Omega = \mathscr{E} \tag{4.1}$$

(2) pure tractions of the form:

$$\mathbf{t}|_{\partial\Omega} = \mathscr{L} \cdot \mathbf{n} \Rightarrow \langle \sigma \rangle_\Omega = \mathscr{L}; \tag{4.2}$$

where $\mathscr{E}$ and $\mathscr{L}$ are constant strain and stress tensors, respectively. Clearly, for Hill's conditions to be satisfied within a macroscopic body, under nonuniform external loading, the sample must be large enough to possess small boundary field fluctuations relative to its size. Therefore applying (1)- or (2)-type boundary conditions to a large sample is a way of reproducing approximately what may be occurring in a statistically representative microscopic sample of material in a macroscopic body. Thus, there is a clear interpretation to these test boundary conditions. *Our requirement that the sample must be large enough to have relatively small boundary field fluctuations relative to its size and small enough relative to the macroscopic engineering structure, forces us to choose boundary conditions that are uniform (Fig. 1.3). This is not optional.*

Next, we will derive a testing procedure for the computation of the effective constitutive tensor $\mathbb{E}^*$, which provides the structural scale constitutive properties of a microheterogeneous material, since it yields mapping between the average stress and strain measures

$$\langle \sigma \rangle_\Omega = \mathbb{E}^* : \langle \varepsilon \rangle_\Omega, \tag{4.3}$$

where $\langle\cdot\rangle_\Omega \stackrel{\text{def}}{=} \frac{1}{|\Omega|}\int_\Omega \cdot \, d\Omega$, and where σ and ε are the stress and strain tensor fields within a microscopic sample of material, with volume $|\Omega|$. Furthermore, bounds will be derived for $\mathbb{E}^*$ which are computed from the constitutive properties of the phases of the microheterogeneous material.

4.1 Testing Procedures

To determine $\mathbb{E}^*$, one specifies six linearly independent loadings of the form,

(1) $\mathbf{u}|_{\partial\Omega} = \mathscr{E}^{(I\rightarrow VI)} \cdot \mathbf{x}$ or

(2) $\mathbf{t}|_{\partial\Omega} = \mathscr{L}^{(I\rightarrow VI)} \cdot \mathbf{n}$

where $\mathscr{E}^{(I\rightarrow VI)}$ and $\mathscr{L}^{(I\rightarrow VI)}$ are symmetric second order strain and stress tensors, with spatially constant components. This loading is applied to the sample in Fig. 4.1 which depicts a microheterogeneous material. Each independent loading yields six different averaged stress components and hence provides six equations for the constitutive constants in $\mathbb{E}^*$.

The sample, shown in Fig. 4.1, is called a statistically representative volume element (RVE) in the mechanics literature. In order for such an analysis to be valid, i.e. to make the material data reliable, the sample must be small enough that it can be considered as a material point with respect to the size of the domain under analysis, but large enough to be a statistically representative sample of the microstructure, see also Fig. 1.3.

If the effective response is assumed to be isotropic, then only one test loading (instead of usually six), containing non-zero dilatational ($\frac{tr\sigma}{3}$ and $\frac{tr\varepsilon}{3}$) and deviatoric components ($\sigma' \stackrel{\text{def}}{=} \sigma - \frac{tr\sigma}{3}\mathbf{I}$ and $\varepsilon' \stackrel{\text{def}}{=} \varepsilon - \frac{tr\varepsilon}{3}\mathbf{I}$), is necessary to determine the effective bulk and shear moduli:

$$3\kappa^* \stackrel{\text{def}}{=} \frac{\langle \frac{tr\sigma}{3} \rangle_\Omega}{\langle \frac{tr\varepsilon}{3} \rangle_\Omega} \quad \text{and} \quad 2\mu^* \stackrel{\text{def}}{=} \sqrt{\frac{\langle \sigma' \rangle_\Omega : \langle \sigma' \rangle_\Omega}{\langle \varepsilon' \rangle_\Omega : \langle \varepsilon' \rangle_\Omega}}. \tag{4.4}$$

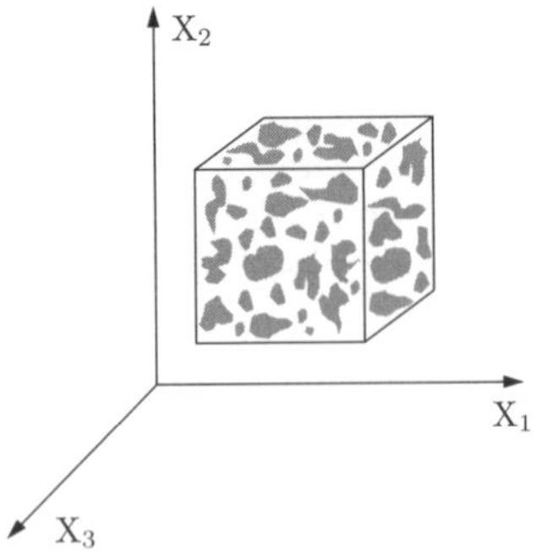

Fig. 4.1 A cubical sample of microheterogeneous material

In general, in order to determine structural scale material properties of microheterogeneous material, one computes 36 constitutive constants E^*_{ijkl} in the following relation between averages,

$$\begin{Bmatrix} \langle\sigma_{11}\rangle_\Omega \\ \langle\sigma_{22}\rangle_\Omega \\ \langle\sigma_{33}\rangle_\Omega \\ \langle\sigma_{12}\rangle_\Omega \\ \langle\sigma_{23}\rangle_\Omega \\ \langle\sigma_{13}\rangle_\Omega \end{Bmatrix} = \begin{bmatrix} E^*_{1111} & E^*_{1122} & E^*_{1133} & E^*_{1112} & E^*_{1123} & E^*_{1113} \\ E^*_{2211} & E^*_{2222} & E^*_{2233} & E^*_{2212} & E^*_{2223} & E^*_{2213} \\ E^*_{3311} & E^*_{3322} & E^*_{3333} & E^*_{3312} & E^*_{3323} & E^*_{3313} \\ E^*_{1211} & E^*_{1222} & E^*_{1233} & E^*_{1212} & E^*_{1223} & E^*_{1213} \\ E^*_{2311} & E^*_{2322} & E^*_{2333} & E^*_{2312} & E^*_{2323} & E^*_{2313} \\ E^*_{1311} & E^*_{1322} & E^*_{1333} & E^*_{1312} & E^*_{1323} & E^*_{1313} \end{bmatrix} \begin{Bmatrix} \langle\varepsilon_{11}\rangle_\Omega \\ \langle\varepsilon_{22}\rangle_\Omega \\ \langle\varepsilon_{33}\rangle_\Omega \\ 2\langle\varepsilon_{12}\rangle_\Omega \\ 2\langle\varepsilon_{23}\rangle_\Omega \\ 2\langle\varepsilon_{13}\rangle_\Omega \end{Bmatrix}. \tag{4.5}$$

As mentioned before, each independent loading leads to six equations and hence in total 36 equations are generated by the independent loadings, which are used to determine the tensor relation between average stress and strain, $\mathbb{E}^*$. $\mathbb{E}^*$ *is exactly what appears in engineering books as the "property" of a material.* The usual choices for the six independent load cases are

$$\mathscr{E} \text{ or } \mathscr{L} = \begin{bmatrix} \beta & 0 & 0 \\ 0 & 0 & 0 \\ 0 & 0 & 0 \end{bmatrix}, \begin{bmatrix} 0 & 0 & 0 \\ 0 & \beta & 0 \\ 0 & 0 & 0 \end{bmatrix}, \begin{bmatrix} 0 & 0 & 0 \\ 0 & 0 & 0 \\ 0 & 0 & \beta \end{bmatrix},$$
$$\begin{bmatrix} 0 & \beta & 0 \\ \beta & 0 & 0 \\ 0 & 0 & 0 \end{bmatrix}, \begin{bmatrix} 0 & 0 & 0 \\ 0 & 0 & \beta \\ 0 & \beta & 0 \end{bmatrix}, \begin{bmatrix} 0 & 0 & \beta \\ 0 & 0 & 0 \\ \beta & 0 & 0 \end{bmatrix}, \tag{4.6}$$

where β is a load parameter. Each independent loading state provides six equations, for a total of 36, which are used to determine the tensor relation between average stress and strain, $\mathbb{E}^*$. For completeness we record a few related fundamental results, which are useful in micro–macro mechanical analysis.

4.1.1 The Average Strain Theorem

If a heterogeneous body, see Fig. 4.2, has the following uniform loading on its surface: $\mathbf{u}|_{\partial\Omega} = \mathscr{E} \cdot \mathbf{x}$, then

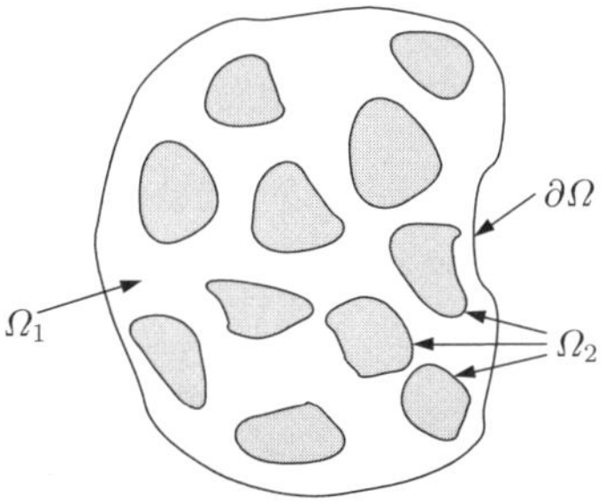

Fig. 4.2 Nomenclature for the averaging theorems

$$
\begin{aligned}
\langle \varepsilon \rangle_\Omega &= \frac{1}{2|\Omega|}\int_\Omega (\nabla \mathbf{u} + (\nabla \mathbf{u})^T)\,\mathrm{d}\Omega \\
&= \frac{1}{2|\Omega|}\left\{\int_{\Omega_1} (\nabla \mathbf{u} + (\nabla \mathbf{u})^T)\,\mathrm{d}\Omega + \int_{\Omega_2} (\nabla \mathbf{u} + (\nabla \mathbf{u})^T)\,\mathrm{d}\Omega\right\} \\
&= \frac{1}{2|\Omega|}\left\{\int_{\partial\Omega_1} (\mathbf{u}\otimes\mathbf{n} + \mathbf{n}\otimes\mathbf{u})\,\mathrm{d}A + \int_{\partial\Omega_2} (\mathbf{u}\otimes\mathbf{n} + \mathbf{n}\otimes\mathbf{u})\,\mathrm{d}A\right\} \\
&= \frac{1}{2|\Omega|}\left\{\int_{\partial\Omega} ((\mathscr{E}\cdot\mathbf{x})\otimes\mathbf{n} + \mathbf{n}\otimes(\mathscr{E}\cdot\mathbf{x}))\,\mathrm{d}A\right. \\
&\quad \left. + \int_{\partial\Omega_1\cap\partial\Omega_2} ([\![\mathbf{u}]\!]\otimes\mathbf{n} + \mathbf{n}\otimes[\![\mathbf{u}]\!])\,\mathrm{d}A\right\} \\
&= \frac{1}{2|\Omega|}\left\{\int_{\Omega} (\nabla(\mathscr{E}\cdot\mathbf{x}) + \nabla(\mathscr{E}\cdot\mathbf{x})^T)\,\mathrm{d}\Omega + \int_{\partial\Omega_1\cap\partial\Omega_2} ([\![\mathbf{u}]\!]\otimes\mathbf{n} + \mathbf{n}\otimes[\![\mathbf{u}]\!])\,\mathrm{d}A\right\} \\
&= \mathscr{E} + \frac{1}{2|\Omega|}\int_{\partial\Omega_1\cap\partial\Omega_2} ([\![\mathbf{u}]\!]\otimes\mathbf{n} + \mathbf{n}\otimes[\![\mathbf{u}]\!])\,\mathrm{d}A, \qquad (4.7)
\end{aligned}
$$

where $(\mathbf{u}\otimes\mathbf{n} \stackrel{\text{def}}{=} u_i n_j)$ is a tensor product of the vector $\mathbf{u}$ and vector $\mathbf{n}$. $[\![\mathbf{u}]\!]$ describes the displacement jumps at the interfaces between Ω_1 and Ω_2. *Therefore, only if the material is perfectly bonded, then*

$$
\langle \varepsilon \rangle_\Omega = \mathscr{E}. \qquad (4.8)
$$

Note that the presence of finite body forces does not effect this result. Also note that the third line in (4.7) is not an outcome of the divergence theorem, but of a generalization that can be found in a variety of places, for example Chandrasekharaiah and Debnath [24] or Malvern [133].

4.1.2 The Average Stress Theorem

Again we consider a body with $\mathbf{t}|_{\partial\Omega} = \mathscr{L}\cdot\mathbf{n}$. We make use of the identity $\nabla\cdot(\sigma\otimes\mathbf{x}) = (\nabla\cdot\sigma)\otimes\mathbf{x} + \sigma\cdot\nabla\mathbf{x} = -\mathbf{f}\otimes\mathbf{x} + \sigma$ and substitute this into the definition of the average stress

$$
\begin{aligned}
\langle \sigma \rangle_\Omega &= \frac{1}{|\Omega|}\int_\Omega \nabla\cdot(\sigma\otimes\mathbf{x})\,\mathrm{d}\Omega + \frac{1}{|\Omega|}\int_\Omega (\mathbf{f}\otimes\mathbf{x})\,\mathrm{d}\Omega \\
&= \frac{1}{|\Omega|}\int_{\partial\Omega} (\sigma\otimes\mathbf{x})\cdot\mathbf{n}\,\mathrm{d}A + \frac{1}{|\Omega|}\int_\Omega (\mathbf{f}\otimes\mathbf{x})\,\mathrm{d}\Omega \\
&= \frac{1}{|\Omega|}\int_{\partial\Omega} (\mathscr{L}\otimes\mathbf{x})\cdot\mathbf{n}\,\mathrm{d}A + \frac{1}{|\Omega|}\int_\Omega (\mathbf{f}\otimes\mathbf{x})\,\mathrm{d}\Omega \\
&= \mathscr{L} + \frac{1}{|\Omega|}\int_\Omega (\mathbf{f}\otimes\mathbf{x})\,\mathrm{d}\Omega. \qquad (4.9)
\end{aligned}
$$

If there are no body forces, $\mathbf{f} = \mathbf{0}$, then

$$\langle \sigma \rangle_\Omega = \mathscr{L}. \tag{4.10}$$

Note that debonding (interface separation) does not change this result.

4.1.3 Satisfaction of Hill's Energy Condition

Consider a body with a perfectly bonded microstructure and $\mathbf{f} = \mathbf{0}$, then $\int_{\partial\Omega} \mathbf{u} \cdot \mathbf{t} \, dA = \int_{\partial\Omega} \mathbf{u} \cdot \sigma \cdot \mathbf{n} \, dA = \int_\Omega \nabla \cdot (\mathbf{u} \cdot \sigma) \, d\Omega$. With $\nabla \cdot \sigma = \mathbf{0}$ follows $\int_\Omega \nabla \cdot (\mathbf{u} \cdot \sigma) \, d\Omega = \int_\Omega \nabla \mathbf{u} : \sigma \, d\Omega = \int_\Omega \varepsilon : \sigma \, d\Omega$. If $\mathbf{u}|_{\partial\Omega} = \mathscr{E} \cdot \mathbf{x}$ and $\mathbf{f} = \mathbf{0}$, then $\int_{\partial\Omega} \mathbf{u} \cdot \mathbf{t} \, dA = \int_{\partial\Omega} \mathscr{E} \cdot \mathbf{x} \cdot \sigma \cdot \mathbf{n} \, dA = \int_\Omega \nabla \cdot (\mathscr{E} \cdot \mathbf{x} \cdot \sigma) \, d\Omega = \int_\Omega \nabla(\mathscr{E} \cdot \mathbf{x}) : \sigma \, d\Omega = \mathscr{E} : \langle \sigma \rangle_\Omega |\Omega|$. Noting that $\langle \varepsilon \rangle_\Omega = \mathscr{E}$, we have

$$\langle \varepsilon \rangle_\Omega : \langle \sigma \rangle_\Omega = \langle \varepsilon : \sigma \rangle_\Omega. \tag{4.11}$$

If $\mathbf{t}|_{\partial\Omega} = \mathscr{L} \cdot \mathbf{n}$ and $\mathbf{f} = \mathbf{0}$, then $\int_{\partial\Omega} \mathbf{u} \cdot \mathbf{t} \, dA = \int_{\partial\Omega} \mathbf{u} \cdot \mathscr{L} \cdot \mathbf{n} \, dA = \int_\Omega \nabla \cdot (\mathbf{u} \cdot \mathscr{L}) \, d\Omega = \int_\Omega \nabla \mathbf{u} : \mathscr{L} \, d\Omega = \mathscr{L} : \int_\Omega \varepsilon \, d\Omega$. Therefore since $\langle \sigma \rangle_\Omega = \mathscr{L}$, as before we have $\langle \varepsilon \rangle_\Omega : \langle \sigma \rangle_\Omega = \langle \varepsilon : \sigma \rangle_\Omega$.

4.2 The Hill-Reuss-Voigt Bounds

Until recently, the direct computation of micromaterial responses was very difficult. Accordingly, classical approaches have sought to approximate or bound effective responses. Many classical approaches start by splitting the stress field within a sample into a volume average, and a purely fluctuating part $\varepsilon = \langle \varepsilon \rangle_\Omega + \tilde{\varepsilon}$ and we directly obtain

$$\begin{aligned} 0 \leq \int_\Omega \tilde{\varepsilon} : \mathbb{E} : \tilde{\varepsilon} \, d\Omega &= \int_\Omega (\varepsilon : \mathbb{E} : \varepsilon - 2\langle \varepsilon \rangle_\Omega : \mathbb{E} : \varepsilon + \langle \varepsilon \rangle_\Omega : \mathbb{E} : \langle \varepsilon \rangle_\Omega) \, d\Omega \\ &= (\langle \varepsilon \rangle_\Omega : \mathbb{E}^* : \langle \varepsilon \rangle_\Omega - 2\langle \varepsilon \rangle_\Omega : \langle \sigma \rangle_\Omega + \langle \varepsilon \rangle_\Omega : \langle \mathbb{E} \rangle_\Omega : \langle \varepsilon \rangle_\Omega)|\Omega| \\ &= \langle \varepsilon \rangle_\Omega : (\langle \mathbb{E} \rangle_\Omega - \mathbb{E}^*) : \langle \varepsilon \rangle_\Omega)|\Omega|. \end{aligned} \tag{4.12}$$

Similarly for the complementary case, with $\sigma = \langle \sigma \rangle_\Omega + \tilde{\sigma}$, and the following assumption

$$\underbrace{\langle \sigma : \mathbb{E}^{-1} : \sigma \rangle_\Omega}_{\text{micro energy}} = \underbrace{\langle \sigma \rangle_\Omega : \mathbb{E}^{*-1} : \langle \sigma \rangle_\Omega}_{\text{macro energy}} \qquad \text{where} \qquad \langle \varepsilon \rangle_\Omega = \mathbb{E}^{*-1} : \langle \sigma \rangle_\Omega \tag{4.13}$$

we have

$$
\begin{aligned}
0 &\leq \int_{\Omega} \tilde{\sigma} : \mathbb{E}^{-1} : \tilde{\sigma}\, d\Omega \\
&= \int_{\Omega} (\sigma : \mathbb{E}^{-1} : \sigma - 2\langle\sigma\rangle_{\Omega} : \mathbb{E}^{-1} : \sigma + \langle\sigma\rangle_{\Omega} : \mathbb{E}^{-1} : \langle\sigma\rangle_{\Omega})\, d\Omega \\
&= (\langle\sigma\rangle_{\Omega} : \mathbb{E}^{*-1} : \langle\sigma\rangle_{\Omega} - 2\langle\epsilon\rangle_{\Omega} : \langle\sigma\rangle_{\Omega} + \langle\sigma\rangle_{\Omega} : \langle\mathbb{E}^{-1}\rangle_{\Omega} : \langle\sigma\rangle_{\Omega})|\Omega| \\
&= \langle\sigma\rangle_{\Omega} : (\langle\mathbb{E}^{-1}\rangle_{\Omega} - \mathbb{E}^{*-1}) : \langle\sigma\rangle_{\Omega}|\Omega|. \qquad (4.14)
\end{aligned}
$$

Invoking Hill's condition, which is loading independent in this form, we have

$$
\underbrace{\langle\mathbb{E}^{-1}\rangle_{\Omega}^{-1}}_{\text{Reuss}} \leq \mathbb{E}^{*} \leq \underbrace{\langle\mathbb{E}\rangle_{\Omega}}_{\text{Voigt}}, \qquad (4.15)
$$

where we emphasize that this inequality means that the eigenvalues of the tensors $\mathbb{E}^{*} - \langle\mathbb{E}^{-1}\rangle_{\Omega}^{-1}$ and $\langle\mathbb{E}\rangle_{\Omega} - \mathbb{E}^{*}$ are non-negative. The practical outcome of the analysis is that bounds on effective properties are obtained. These bounds are commonly known as the Hill-Reuss-Voigt bounds, for historical reasons. Voigt [212], in 1889, assumed that the strain field within a sample of aggregate of polycrystalline material, was uniform (constant), under uniform strain exterior loading. If the constant strain Voigt field is assumed within the RVE, $\varepsilon = \varepsilon^{0}$, then $\langle\sigma\rangle_{\Omega} = \langle\mathbb{E} : \varepsilon\rangle_{\Omega} = \langle\mathbb{E}\rangle_{\Omega} : \varepsilon^{0}$, which implies $\mathbb{E}^{*} = \langle\mathbb{E}\rangle_{\Omega}$. The dual assumption was made by Reuss [173], in 1929, who approximated the stress fields within the aggregate of polycrystalline material as uniform (constant), $\sigma = \sigma^{0}$, leading to $\langle\varepsilon\rangle_{\Omega} = \langle\mathbb{E}^{-1} : \sigma\rangle_{\Omega} = \langle\mathbb{E}^{-1}\rangle_{\Omega} : \sigma^{0}$, and thus $\mathbb{E}^{*} = \langle\mathbb{E}^{-1}\rangle_{\Omega}^{-1}$. Equality is attained in the above bounds if the Reuss or Voigt assumptions hold, respectively.

Remark. Different boundary conditions (compared to the ones specified in (4.1) or (4.2)) are often used in computational homogenization analysis. For example, periodic boundary conditions are sometimes employed. Although periodicity conditions are really only appropriate for perfectly periodic media for many cases, it has been shown that, in some cases, their use can provide better effective responses than either linear displacement or uniform traction boundary conditions. For example, see Terada et al. [199] or Segurado and Llorca [176]. Periodic boundary conditions also satisfy Hill's condition a priori. Another related type of boundary conditions are so-called "uniform-mixed" types, whereby tractions are applied on some parts of the boundary and displacements on other parts, producing favorable results. For example, see Hazanov and Huet [77]. Another approach is "framing" whereby the traction or displacement boundary conditions are applied to a large sample of material, with the averaging being computed on an interior subsample, to avoid possible boundary-layer effects. Generally, the advantages of one boundary condition over another diminishes as the sample increases in size.

4.3 Observations

Consider a 1-dimensional bar of length l composed of random particles. There are a total of N bands, N_2 dark bands (E_2 particles) and N_1 white bands (E_1 particles), each of length $\Delta l = l/N$ (upper structure in Fig. 4.3). We apply uniform strains on the structure and obtain the following two point boundary value problem

$$\frac{d}{dx}(E(x)\frac{du(x)}{dx}) = 0, \qquad u(0) = 0,\; u(l) = \mathscr{E} \times l, \tag{4.16}$$

where $\mathscr{E}$ is a constant. Simple calculations reveal that

$$E^* = \frac{E_1 E_2}{E_2(1-v_2) + E_1 v_2} = \langle E^{-1} \rangle_\Omega^{-1}, \tag{4.17}$$

where $v_1 + v_2 = 1$, and where v_2 is the volume fraction which equals $v_2 = \frac{N_2 \Delta l}{l}$. This is exactly the Reuss bound, which is not surprising since the state of stress is constant throughout the rod. Alternatively consider the lower structure in Fig. 4.3. The state of strain is uniform and we have $E^* = (1-v_2)E_1 + v_2 E_2 = \langle E \rangle_\Omega$, which is the Voigt approximation. The interpretation is clear, the response is bounded from below (softer) by springs (moduli) in series (constant stress) and above (harder) by springs (moduli) in parallel (constant strain). The bounds provide a rough and quick way of determining approximate aggregate responses of microheterogeneous materials. The wideness of the bounds grows with volume fraction, for example illustrated for an aluminum/boron material combination (Fig. 4.6). Material combinations such as aluminum/boron are becoming more and more widely used in engineering because they are lightweight and stiff. The boron particles are used as a stiffener for the easy to form aluminum. Figure 4.6 illustrates the behavior of the bounds over the range of all admissible volume fractions.

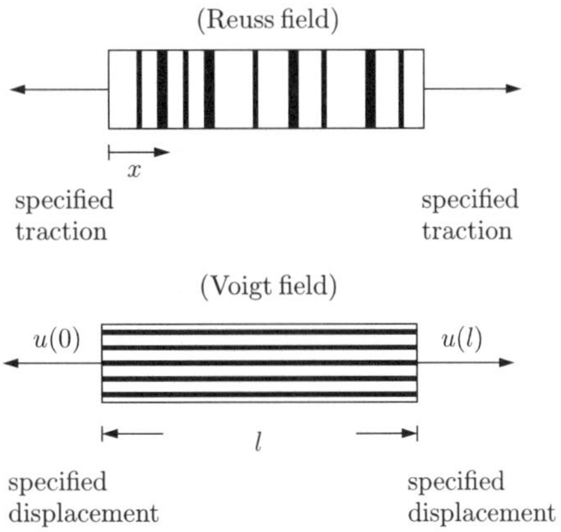

Fig. 4.3 Intuitive interpretations of the Reuss and Voigt fields with purely one dimensional effects assumed

4.4 Classical Micro–Macro Mechanical Approximations

For completeness we review a few more classical approximations in micro-macro mechanics.

4.4.1 The Asymptotic Hashin-Shtrikman Bounds

Improved bounds were developed in 1963 by Hashin and Shtrikman ([74, 75]) based on variational principles using the concept of polarization or "filtering" of micro-macro mechanical fields. Based on these formulations, they developed better asymptotic bounds on effective properties. *These bounds are sensitive to sample size and are strictly valid only asymptotically. Hence they are useful when the sample size is infinite in size relative to the microconstituent length scale.* The Hashin-Shtrikman bounds are the tightest possible bounds on isotropic effective responses, generated from isotropic microstructure, where the volumetric data and phase contrasts of the constituents are the only data known. The Hashin-Shtrikman principle represents a classical example of the filtering of scales for materials with microstructure. It essentially involves the *Principle of Minimum Potential Energy* (PMPE) written in terms of a filtered variable that admits to a straightforward approximation of the internal fields. With such an approximation one can bound the macroscopic response. One begins by writing $\sigma^0 = \mathrm{I\!E}^0 : \varepsilon^0$, where $\mathrm{I\!E}^0$ is spatially constant, and $\sigma = \mathrm{I\!E} : \varepsilon$, where $\mathrm{I\!E}$ is spatially nonconstant (heterogeneous). One defines polarization (filtered) variables, $\mathbf{p} = \sigma - \mathrm{I\!E}^0 : \varepsilon = (\mathrm{I\!E} - \mathrm{I\!E}^0) : \varepsilon \stackrel{\text{def}}{=} \Delta\mathrm{I\!E} : \varepsilon$, $\hat{\mathbf{u}} = \mathbf{u} - \mathbf{u}^0$ and $\hat{\varepsilon} = \varepsilon - \varepsilon^0$, where $\hat{\varepsilon}$ and $\mathbf{p}$ are unknowns in a "filtered" variational principle. The following potential is constructed

$$\Pi(\hat{\mathbf{u}}) = W_0 - \frac{1}{2}\int_\Omega (\mathbf{p} : (\Delta\mathrm{I\!E})^{-1} : \mathbf{p} - \mathbf{p} : \hat{\varepsilon} - 2\mathbf{p} : \varepsilon^0)\, d\Omega, \tag{4.18}$$

where $W_0 = \frac{1}{2}\int_\Omega \varepsilon^0 : \sigma^0 \, d\Omega$, $\nabla \cdot (\mathrm{I\!E}^0 : \hat{\varepsilon} + \mathbf{p}) = \mathbf{0}$ and $\mathbf{u}|_{\partial\Omega} = \mathbf{u}^0|_{\partial\Omega} \Rightarrow \hat{\mathbf{u}}|_{\partial\Omega} = \mathbf{0}$. One can make the following observations:

(1) The potential in (4.18) is stationary when $\mathbf{p} = \Delta\mathrm{I\!E} : \varepsilon$,
(2) The stationary value is a maximum if $\Delta\mathrm{I\!E} > \mathbf{0}$ and a minimum if $\Delta\mathrm{I\!E} < \mathbf{0}$,
(3) In the isotropic case this becomes: $\Pi(\hat{\mathbf{u}})$ is an absolute minimum if $\kappa > \kappa^0, \mu > \mu^0$, and $\Pi(\hat{\mathbf{u}})$ is an absolute maximum if $\kappa < \kappa^0, \mu < \mu^0$,
(4) When $\Pi(\hat{\mathbf{u}})$ is a minimum it is equal to the strain energy,
(5) When $\mathrm{I\!E}^0$ is vanishingly small compared to $\mathrm{I\!E}$, then the variational principle collapses to the PMCPE and
(6) When $\mathrm{I\!E}^0$ is infinitely large compared to $\mathrm{I\!E}$, then the variational principle collapses to the PMPE.

Similar properties hold for a pure traction loaded problem. The extreme values of the relations are then found by optimizing with respect to κ^0 and μ^0. Noting that when $\mathbf{u}|_{\partial\Omega} = \mathscr{E} \cdot \mathbf{x}$ the strain energy is $W = \frac{1}{2}(9\kappa^*(\frac{tr\mathscr{E}}{3})^2 + 2\mu^*\mathscr{E}' : \mathscr{E}')$, one can arrive at, for a two-phase microstructure:

$$\underbrace{\kappa_1 + \frac{v_2}{\frac{1}{\kappa_2-\kappa_1} + \frac{3(1-v_2)}{3\kappa_1+4\mu_1}}}_{\text{bulk modulus H−S lower bound} \stackrel{\text{def}}{=} \kappa^{*,-}} \leq \kappa^* \leq \underbrace{\kappa_2 + \frac{1-v_2}{\frac{1}{\kappa_1-\kappa_2} + \frac{3v_2}{3\kappa_2+4\mu_2}}}_{\text{bulk modulus H−S upper bound} \stackrel{\text{def}}{=} \kappa^{*,+}},$$

$$\underbrace{\mu_1 + \frac{v_2}{\frac{1}{\mu_2-\mu_1} + \frac{6(1-v_2)(\kappa_1+2\mu_1)}{5\mu_1(3\kappa_1+4\mu_1)}}}_{\text{shear modulus H−S lower bound} \stackrel{\text{def}}{=} \mu^{*,-}} \leq \mu^* \leq \underbrace{\mu_2 + \frac{(1-v_2)}{\frac{1}{\mu_1-\mu_2} + \frac{6v_2(\kappa_2+2\mu_2)}{5\mu_2(3\kappa_2+4\mu_2)}}}_{\text{shear modulus H−S upper bound} \stackrel{\text{def}}{=} \mu^{*,+}}, \quad (4.19)$$

where κ_1, μ_1 and κ_2, μ_2 are the bulk and shear moduli for the phases, while v_2 is the phase 2 volume fraction. The original proofs, which are algebraically complicated, can be found in Hashin and Shtrikman [74, 75]. *We emphasize that in the derivation of the bounds, the body is assumed to be infinite, the microstructure isotropic, and that the effective responses are isotropic.* Also, a further assumption is that $\kappa_2 \geq \kappa_1$ and $\mu_2 \geq \mu_1$. We remark that the bounds are the tightest possible, under the previous assumptions, when no geometric (micro-topological) information is included.

4.4.2 The Concentration Tensor: Microfield Behavior

Consider the following identities: (I) $\langle\varepsilon\rangle_\Omega = \frac{1}{|\Omega|}(\int_{\Omega_1} \varepsilon\, d\Omega + \int_{\Omega_2} \varepsilon\, d\Omega) = v_1\langle\varepsilon\rangle_{\Omega_1} + v_2\langle\varepsilon\rangle_{\Omega_2}$ and (II) $\langle\sigma\rangle_\Omega = \frac{1}{|\Omega|}(\int_{\Omega_1} \sigma\, d\Omega + \int_{\Omega_2} \sigma\, d\Omega) = v_1\langle\sigma\rangle_{\Omega_1} + v_2\langle\sigma\rangle_{\Omega_2}$. By direct manipulation we obtain

$$\begin{aligned}
\langle\sigma\rangle_\Omega &= v_1\langle\sigma\rangle_{\Omega_1} + v_2\langle\sigma\rangle_{\Omega_2} \\
&= v_1\mathbb{E}_1 : \langle\varepsilon\rangle_{\Omega_1} + v_2\mathbb{E}_2 : \langle\varepsilon\rangle_{\Omega_2} \\
&= \mathbb{E}_1 : (\langle\varepsilon\rangle_\Omega - v_2\langle\varepsilon\rangle_{\Omega_2}) + v_2\mathbb{E}_2 : \langle\varepsilon\rangle_{\Omega_2} \\
&= (\mathbb{E}_1 + v_2(\mathbb{E}_2 - \mathbb{E}_1) : \mathbf{C}) : \langle\varepsilon\rangle_\Omega
\end{aligned} \quad (4.20)$$

where

$$\underbrace{\left(\frac{1}{v_2}(\mathbb{E}_2 - \mathbb{E}_1)^{-1} : (\mathbb{E}^* - \mathbb{E}_1)\right)}_{\stackrel{\text{def}}{=} \mathbf{C}:\langle\varepsilon\rangle_\Omega} = \langle\varepsilon\rangle_{\Omega_2}. \quad (4.21)$$

Thereafter, we may write, for the variation in the stress $\mathbf{C} : \mathbb{E}^{*-1} : \langle\sigma\rangle_\Omega = \mathbb{E}_2^{-1} : \langle\sigma\rangle_{\Omega_2}$, which reduces to $\mathbb{E}_2 : \mathbf{C} : \mathbb{E}^{*-1} : \langle\sigma\rangle_\Omega \stackrel{\text{def}}{=} \overline{\mathbf{C}} : \langle\sigma\rangle_\Omega = \langle\sigma\rangle_{\Omega_2}$. $\overline{\mathbf{C}}$ is known as the stress concentration tensor. Therefore, once either $\overline{\mathbf{C}}$ or $\mathbb{E}^*$ are known, the other can be determined. In the case of isotropy we may write

$$\overline{C}_{\kappa} \stackrel{\text{def}}{=} \frac{1}{v_2} \frac{\kappa_2}{\kappa^*} \frac{\kappa^* - \kappa_1}{\kappa_2 - \kappa_1} \quad \text{and} \quad \overline{C}_{\mu} \stackrel{\text{def}}{=} \frac{1}{v_2} \frac{\mu_2}{\mu^*} \frac{\mu^* - \mu_1}{\mu_2 - \mu_1}. \tag{4.22}$$

Clearly, the microstress fields are minimally distorted when $\overline{C}_{\kappa} = \overline{C}_{\mu} = 1$.

Remark. There has been no approximation yet. The "burden" in the computations has shifted to the determination of **C**. Classical methods approximate **C**. For example, the simplest approximation is $\mathbf{C} = \mathbf{I}$, which is the Voigt approximation, $\mathbb{E}^* = (\mathbb{E}_1 + v_2(\mathbb{E}_2 - \mathbb{E}_1) : \mathbf{I})$. Also, for the Reuss approximation, $(\mathbb{E}^*)^{-1} = (\mathbb{E}_1)^{-1} + v_2((\mathbb{E}_2)^{-1} - (\mathbb{E}_1)^{-1}) : \mathbf{I})$, where $\overline{\mathbf{C}} = \mathbf{I}$.

4.4.3 The Eshelby Result

In 1957, Eshelby [36] developed an elegant formalism to determine the elasticity solution of a single inclusion embedded in an infinite matrix of material with uniform exterior loading (Fig. 4.4). Alone, this result is of little practical interest, however, this solution, which is relatively compact, has been the basis of approximation methods for effective properties for non-interacting and weakly interacting particulate solutions. The essentials of the Eshelby result are as follows. Consider a single linearly elastic particle, at infinitesimal strains, embedded in an infinite matrix: *If the shape of the particle is ellipsoidal, under uniform far field stress or strain, then the stress, and hence the strain, in the particle is constant.* As we have mentioned, this result is the foundation of most methods of approximation for effective properties and is used to approximate **C** for ellipsoidal geometries. Ellipsoidal shapes are qualitatively useful since the geometry can mimic a variety of microstructures (1) platelet behavior when the ellipsoid is oblate with a high aspect ratio, (2) needle-like microstructure when it is prolate with high aspect ratio, (3) crack-like behavior as the material in the particle is modelled as extremely soft with very high aspect ratios for the particle geometry, and (4) pores when the particle shapes are

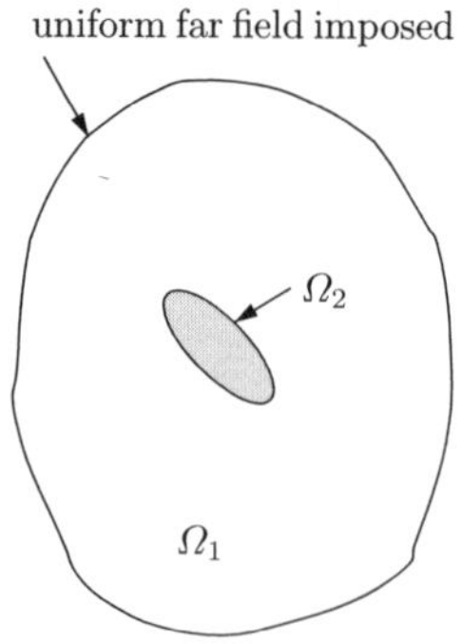

Fig. 4.4 The notation for the Eshelby formalism

spheres with very soft material inside. For a single particle in three dimensions we have, $\forall \mathbf{x} \in \Omega_2$,

$$\begin{aligned} \mathbb{E}_2 : \varepsilon &= \mathbb{E}_1 : (\mathscr{E} + \Delta\varepsilon - \mathbf{T}) = \mathbb{E}_1 : (\mathscr{E} + \mathbb{P} : \mathbf{T} - \mathbf{T}) \\ &= \mathbb{E}_1 : \mathscr{E} + \mathbb{E}_1 : (\mathbb{P} - \mathbf{I}) : \mathbf{T}, \end{aligned} \tag{4.23}$$

where $\mathscr{E}$ is the far field uniform strain, $\varepsilon = \mathscr{E} + \Delta\varepsilon$ is the strain in the particle, $\mathbf{T}$ is the transformation strain and $\mathbb{P}$ is the fourth-order Eshelby tensor which satisfies $\mathbb{P} : \mathbf{T} \stackrel{\text{def}}{=} \Delta\varepsilon$. The Eshelby result, stated in a compact way, is: $\mathbb{P}$ *and* $\mathbf{T}$ *are constants in an isolated ellipsoidal particle under uniform far field loading.*

4.4.4 Dilute Methods

The use of the Eshelby result is straightforward as a method of determining approximate effective properties for regularly shaped particles. When the particle is a sphere, $\mathbf{p}$ is isotropic and the relations are relatively easy to write down. Breaking the tensor relations into dilatational and deviatoric pieces yields, $\forall \mathbf{x} \in \Omega_2$ one has $\frac{tr\varepsilon}{3} = \alpha \frac{tr\mathbf{T}}{3} + \frac{tr\mathscr{E}}{3}$, $\alpha = \frac{1+\nu_1}{3(1-\nu_1)} = \frac{3\kappa_1}{3\kappa_1+4\mu_1}$, $\varepsilon' = \beta \mathbf{T}' + \mathscr{E}'$ and $\beta = \frac{2}{15}\frac{4-5\nu_1}{1-\nu_1}$. The constitutive relations are, $\forall \mathbf{x} \in \Omega_2$, $\frac{tr\sigma}{3} = 3\kappa_2 \frac{tr\varepsilon}{3} = 3\kappa_1 (\frac{tr\varepsilon}{3} - \frac{tr\mathbf{T}}{3})$ and $\sigma' = 2\mu_2 \varepsilon' = 2\mu_1(\varepsilon' - \mathbf{T}')$, where $\nu = \frac{3\kappa - 2\mu}{6\kappa+2\mu}$. Combining the previous expressions yields, $\forall \mathbf{x} \in \Omega_2$, $tr\varepsilon = \gamma tr\mathscr{E}$, $\varepsilon' = \rho \mathscr{E}'$, $\gamma \stackrel{\text{def}}{=} \frac{\kappa_1}{\alpha\kappa_2 + \kappa_1(1-\alpha)}$ and $\rho \stackrel{\text{def}}{=} \frac{\mu_1}{\beta\mu_2 + \mu_1(1-\beta)}$. The approximations allow the determination of the strains in the particles as a function of the loading and geometry. Notice that the deviatoric and dilatational strain components in the particle are the far field corresponding strain components multiplied by a factor composed by the corresponding components of the matrix material divided by a convex combination of the matrix and particle materials. The transformation relations are tabulated for various simple particulate shapes in Mura [150]. *It is important to realize that the stress fields in the matrix* are not constant, *and are extremely complicated. Fortunately, in developing approximate expressions for the effective response it unnecessary to know the fields in the matrix.* Consequently, the concentration tensor, under the non-interacting particle assumption, is isotropic and can be written as

$$\mathbf{C} \stackrel{\text{def}}{=} \begin{bmatrix} \frac{\gamma}{3} + \frac{4}{3}\frac{\rho}{2} & \frac{\gamma}{3} - \frac{2}{3}\frac{\rho}{2} & \frac{\gamma}{3} - \frac{2}{3}\frac{\rho}{2} & 0 & 0 & 0 \\ \frac{\gamma}{3} - \frac{2}{3}\frac{\rho}{2} & \frac{\gamma}{3} + \frac{4}{3}\frac{\rho}{2} & \frac{\gamma}{3} - \frac{2}{3}\frac{\rho}{2} & 0 & 0 & 0 \\ \frac{\gamma}{3} - \frac{2}{3}\frac{\rho}{2} & \frac{\gamma}{3} - \frac{2}{3}\frac{\rho}{2} & \frac{\gamma}{3} + \frac{4}{3}\frac{\rho}{2} & 0 & 0 & 0 \\ 0 & 0 & 0 & \frac{\rho}{2} & 0 & 0 \\ 0 & 0 & 0 & 0 & \frac{\rho}{2} & 0 \\ 0 & 0 & 0 & 0 & 0 & \frac{\rho}{2} \end{bmatrix}. \tag{4.24}$$

Therefore, in the case of noninteracting spheres, $\mathbf{C}$ is isotropic with deviatoric and dilatational components given by γ and ρ, which are furnished by the Eshelby solution. This is the framework for the Dilute method, which postulates the complete non interaction between microstructural components. Therefore we have $\mathbb{E}^* \approx \mathbb{E}_1 + v_2(\mathbb{E}_2 - \mathbb{E}_1) : \mathbf{C}$ which implies

$$\kappa^* \approx \kappa_1 + v_2(\kappa_2 - \kappa_1)\gamma \qquad \text{and} \qquad \mu^* \approx \mu_1 + v_2(\mu_2 - \mu_1)\rho. \tag{4.25}$$

Dilute approximations are usually only accurate at extremely low volume fractions.

4.4.5 The Mori-Tanaka Method

Non-interaction of particulates is an unrealistic assumption for materials with randomly dispersed particulate microstructure at even a few percent volume fraction. Weak interaction between particles, via a sensitivity to increased volume fraction, involves slight modification of the dilute method. This involves assuming that there exists another tensor, such that $\langle \varepsilon \rangle_{\Omega_2} = \mathbf{A} : \langle \varepsilon \rangle_{\Omega}$, thus implying $\mathbb{E}^* : \langle \varepsilon \rangle_{\Omega} = (\mathbb{E}_1 + v_2(\mathbb{E}_2 - \mathbb{E}_1) : \mathbf{A}) : \langle \varepsilon \rangle_{\Omega}$. The tensor $\mathbf{A} = \mathbf{A}(v_2)$ must be approximated. There are two requirements:

(1) $\lim_{v_2 \to 0} \mathbf{A} = \mathbf{C}$ and
(2) $\lim_{v_2 \to 1} \mathbf{A} = \mathbf{I}$.

The most widely used approach is that of Mori-Tanaka [148]. The Mori-Tanaka method proceeds by constructing a tensor $\mathbf{G}$ such that

$$\mathbf{G} : \langle \varepsilon \rangle_{\Omega_1} = \langle \varepsilon \rangle_{\Omega_2}, \tag{4.26}$$

thus implying that $\langle \varepsilon \rangle_{\Omega} = v_1 \langle \varepsilon \rangle_{\Omega_1} + v_2 \langle \varepsilon \rangle_{\Omega_2} = v_1 \mathbf{G}^{-1} : \langle \varepsilon \rangle_{\Omega_2} + v_2 \langle \varepsilon \rangle_{\Omega_2} = (v_1 \mathbf{G}^{-1} + v_2 \mathbf{I}) : \langle \varepsilon \rangle_{\Omega_2} = \mathbf{A}^{-1} : \langle \varepsilon \rangle_{\Omega_2}$. The Mori-Tanaka method approximates $\mathbf{G}$ by $\mathbf{C}$. The physical implications are that the particle "sees" a matrix material with an average stress state provided by the average in the matrix. One can consider the Mori-Tanaka as *one possible* curve fit (Fig. 4.5) that satisfies the two mentioned conditions, i.e. $\mathbf{A} = \mathbf{I}$ when $v_2 = 1$ and $\mathbf{A} = \mathbf{C}$ when $v_2 = 0$. The result is

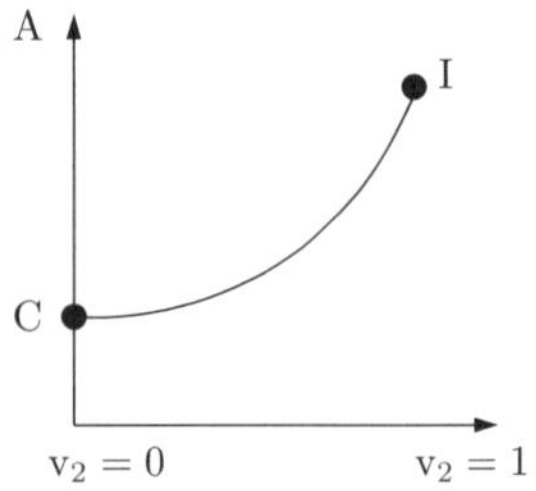

Fig. 4.5 The Mori-Tanaka "extrapolation"

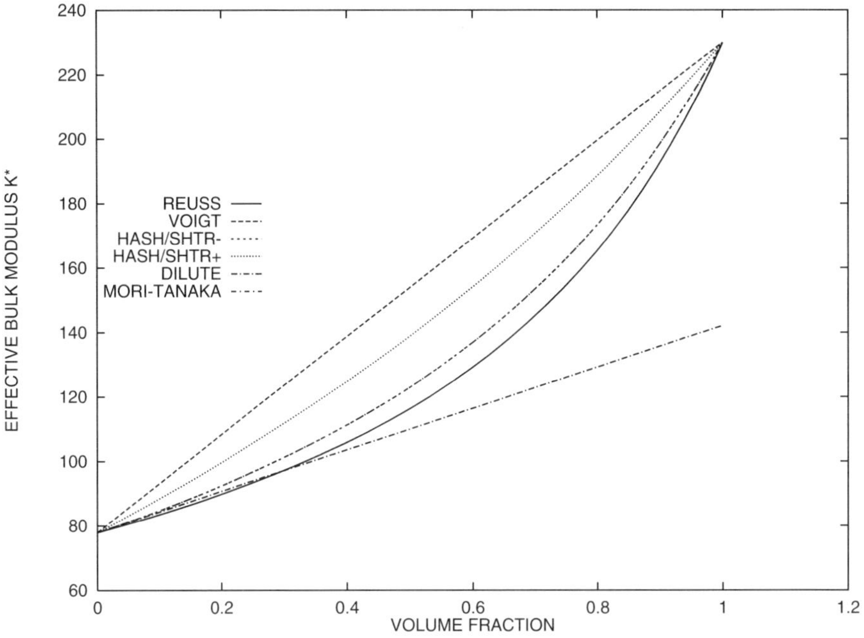

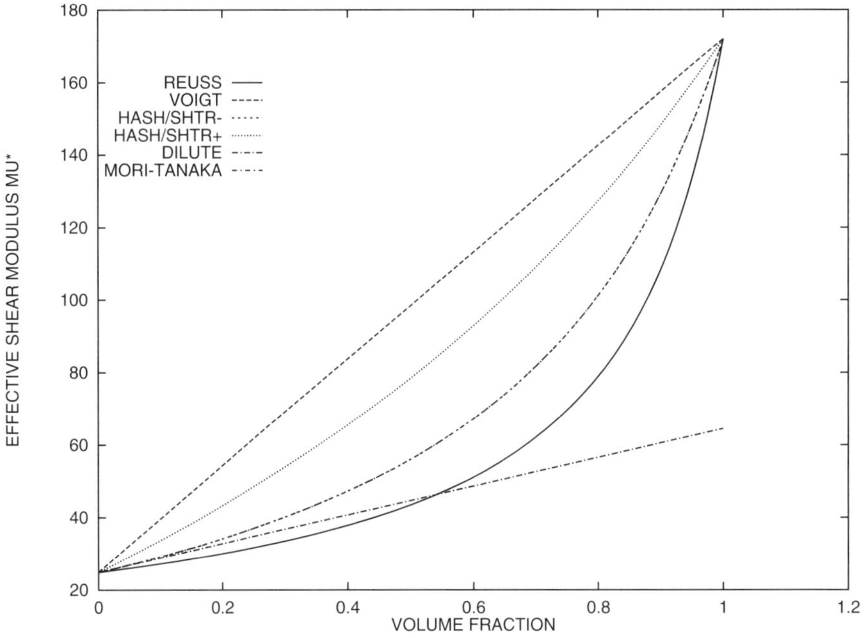

Fig. 4.6 The predicted effective bulk (κ^*) and shear moduli (μ^*) for an Aluminum matrix ($\kappa = 77.9\,GPa$, $\mu = 24.9\,GPa$)/Boron particle ($\kappa = 230\,GPa$, $\mu = 172\,GPa$)

$$\kappa^* \stackrel{\text{def}}{=} \kappa_1 + v_2(\kappa_2 - \kappa_1)\eta \qquad \text{and} \qquad \mu^* \stackrel{\text{def}}{=} \mu_1 + v_2(\mu_2 - \mu_1)\zeta, \tag{4.27}$$

where $\eta \stackrel{\text{def}}{=} \frac{\Theta}{\Theta+(1-v_2)(\kappa_2-\kappa_1)}$, $\zeta \stackrel{\text{def}}{=} \frac{\Phi}{\Phi+(1-v_2)(\mu_2-\mu_1)}$, $\Theta = \kappa_1 + \frac{4}{3}\mu_1$, $\Phi = \mu_1 + \frac{\mu_1(9\kappa_1+8\mu_1)}{6(\kappa_1+2\mu_1)}$, where $\forall \mathbf{x} \in \Omega_2$, $tr\varepsilon = \eta tr\mathscr{E}$ and $\varepsilon' = \zeta\mathscr{E}'$. The interaction, via a volume fraction dependency, is explicitly clear in the expressions for η and ζ. It is noted that for spherical particles the Mori-Tanaka matches the well known Hashin-Shtrikman lower bound [75] on effective properties.

Remark. Another approach to incorporate particulate interaction is the Self Consistent Method. The idea is simply that the particles "see" the effective medium instead of the matrix in the computation of the transformation strain. In other words, the "matrix material" in the Eshelby tensor is simply the effective medium. Therefore, one simply replaces the matrix material in the concentration formulae (only in **C**) with the effective property to obtain

$$\kappa^* = \kappa_1(1 - v_2\frac{\gamma^*-1}{\alpha^*}) \qquad \text{and} \qquad \mu^* = \mu_1(1 - v_2\frac{\rho^*-1}{\beta^*}), \tag{4.28}$$

where $\alpha^* = \frac{1+v^*}{3(1-v^*)} = \frac{3\kappa^*}{3\kappa^*+4\mu^*}$ and where $\beta^* = \frac{2}{15}\frac{4-5v^*}{1-v^*}$. The approach is identical to the dilute approximation, with the result being somewhat more complicated expressions for the approximate effective property, however, which can be computed in closed form in many cases. We refer the reader to the books of Mura [150], Aboudi [1] and Nemat-Nasser and Hori [151]. The self-consistent method produces negative effective bulk and shear responses for voids above volume fractions of 50%. For rigid inclusions it produces infinite effective bulk responses for any volume fraction and infinite effective shear responses above 40%. For proofs and discussions on variations of this method, see Aboudi [1]. With these facts in mind, one can safely employ such methods only at very low volume fraction levels.

4.4.6 Further Methods

There exist a multitude of other approaches which seek to estimate or bound the aggregate responses microheterogeneous materials. A complete survey is outside the scope of the present work. We refer the reader to the works of Hashin [76], Mura [150], Aboudi [1], Nemat-Nasser and Hori [151] and recently Torquato [204] for such reviews. Also, for indepth analyses, with extensions into nonlinear behavior, blending analytical, semi-analytical and numerical techniques, we refer the reader to the extensive works of Llorca and co-workers: Segurado and Llorca [176], González and Llorca [65], Segurado et al. [177], González and Llorca [66], Llorca [125], González and Llorca [67], Poza and Llorca [170], Llorca and González [127], Llorca [128], Llorca et al. [129].

4.5 Micro-Geometrical (Manufacturing) Idealizations

Obviously it is virtually impossible to manufacture particles with perfectly smooth geometries. However, if one were to attempt to simulate a sample's response with highly irregular shapes, a problem of such complexity would occur that virtually no analytical, semi-analytical or standard numerical technique would suffice. Fortunately, in the computation of effective properties, the shapes can be safely approximated by much smoother geometrical idealizations, which is the subject of this section.

4.5.1 Upper and Lower Variational Bounds

Variational inequalities are now derived that govern the growth or decrease in system energy due to the local stiffness reduction. With these inequalities, bounds can be developed for the effective response for irregularly shaped particles by creating outer and inner smooth envelopes, as shown in the Fig. 4.7. In order to do this, the principle of virtual work and the complementary principle are employed.

Consider two symmetric positive definite material property (elasticity tensor) distributions, $\mathbb{E}_{(I)}$ and $\mathbb{E}_{(II)}$ used in same boundary value problem. The corresponding stress and strain states using these materials are denoted $(\varepsilon^{(I)}, \sigma^{(I)})$ and $(\varepsilon^{(II)}, \sigma^{(II)})$ respectively. Consider the respective strain energies of both material systems:

$$W_I \stackrel{\text{def}}{=} \underbrace{\int_\Omega \varepsilon^{(I)} : \mathbb{E}_{(I)} : \varepsilon^{(I)} \, d\Omega}_{\text{I's energy}} \quad \text{and} \quad W_{II} \stackrel{\text{def}}{=} \underbrace{\int_\Omega (\varepsilon^{(I)} + \delta\varepsilon) : \mathbb{E}_{(II)} : (\varepsilon^{(I)} + \delta\varepsilon) \, d\Omega}_{\text{II's energy}} . \tag{4.29}$$

Denoting $\delta\varepsilon = \varepsilon^{(II)} - \varepsilon^{(I)}$, one obtains, assuming, $(\mathbb{E}_{(I)} - \mathbb{E}_{(II)}) \geq \mathbf{0}$

$$\underbrace{\int_\Omega \varepsilon^{(I)} : \mathbb{E}_{(I)} : \varepsilon^{(I)} \, d\Omega}_{\text{I's energy}} - \underbrace{\int_\Omega (\varepsilon^{(I)} + \delta\varepsilon) : \mathbb{E}_{(II)} : (\varepsilon^{(I)} + \delta\varepsilon) \, d\Omega}_{\text{II's energy}} \geq 0. \tag{4.30}$$

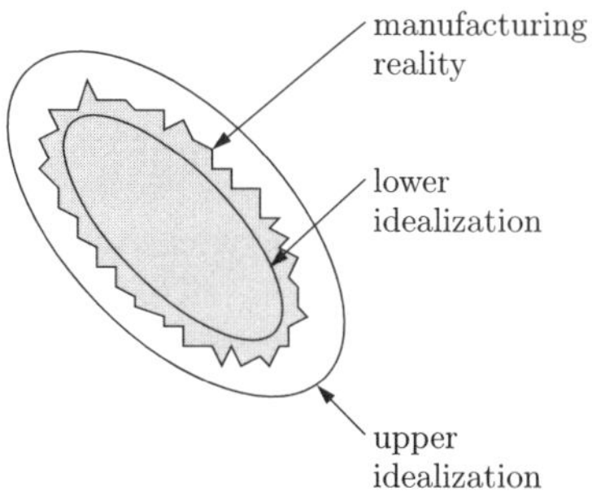

Fig. 4.7 Upper and lower idealizations of shapes that can be manufactured

Similarly, defining $\delta\sigma = \sigma^{(II)} - \sigma^{(I)}$, if $\Gamma_t = \partial\Omega$

$$W_I \stackrel{\text{def}}{=} \underbrace{\int_\Omega \sigma^{(I)} : \mathbb{E}_{(I)}{}^{-1} : \sigma^{(I)}\,d\Omega}_{\text{I's energy}} \text{ and } W_{II} \stackrel{\text{def}}{=} \underbrace{\int_\Omega (\sigma^{(I)} + \delta\sigma) : \mathbb{E}_{(II)}{}^{-1} : (\sigma^{(I)} + \delta\sigma)\,d\Omega}_{\text{II's energy}}, \tag{4.31}$$

and thus, assuming $(\mathbb{E}_{(I)}{}^{-1} - \mathbb{E}_{(II)}{}^{-1}) < \mathbf{0}$,

$$\underbrace{\int_\Omega \sigma^{(I)} : \mathbb{E}_{(I)}{}^{-1} : \sigma^{(I)}\,d\Omega}_{\text{I's energy}} - \underbrace{\int_\Omega (\sigma^{(I)} + \delta\sigma) : \mathbb{E}_{(II)}{}^{-1} : (\sigma^{(I)} + \delta\sigma)\,d\Omega}_{\text{II's energy}} \leq 0. \tag{4.32}$$

The general proofs are constructive, and we believe add extra insight to the analysis that follows later.

4.5.2 *Proof of Energetic Ordering*

Again, consider two symmetric positive definite material property (elasticity tensor) distributions, $\mathbb{E}_{(I)}$ and $\mathbb{E}_{(II)}$ used in same boundary value problem, with corresponding stress and strain states using these materials denoted by $(\varepsilon^{(I)}, \sigma^{(I)})$ and $(\varepsilon^{(II)}, \sigma^{(II)})$ respectively, with the following respective strain energies:

$$W_I \stackrel{\text{def}}{=} \underbrace{\int_\Omega \varepsilon^{(I)} : \mathbb{E}_{(I)} : \varepsilon^{(I)}\,d\Omega}_{\text{I's energy}} \quad \text{and} \quad W_{II} \stackrel{\text{def}}{=} \underbrace{\int_\Omega (\varepsilon^{(I)} + \delta\varepsilon) : \mathbb{E}_{(II)} : (\varepsilon^{(I)} + \delta\varepsilon)\,d\Omega}_{\text{II's energy}}. \tag{4.33}$$

We have, denoting $\delta\varepsilon = \varepsilon^{(II)} - \varepsilon^{(I)}$,

$$\begin{aligned} W_I - W_{II} &= \int_\Omega \varepsilon^{(I)} : \mathbb{E}_{(I)} : \varepsilon^{(I)}\,d\Omega - \int_\Omega \varepsilon^{(I)} : \mathbb{E}_{(II)} : (\varepsilon^{(I)} + \delta\varepsilon)\,d\Omega \\ &\quad \underbrace{- \int_\Omega \delta\varepsilon : \mathbb{E}_{(II)} : (\varepsilon^{(I)} + \delta\varepsilon)\,d\Omega}_{=\mathbf{0} \text{ if } \Gamma_u = \partial\Omega} \\ &= \int_\Omega \varepsilon^{(I)} : (\mathbb{E}_{(I)} - \mathbb{E}_{(II)}) : \varepsilon^{(I)}\,d\Omega - \int_\Omega \varepsilon^{(I)} : \mathbb{E}_{(II)} : \delta\varepsilon\,d\Omega \\ &= \int_\Omega \varepsilon^{(I)} : (\mathbb{E}_{(I)} - \mathbb{E}_{(II)}) : \varepsilon^{(I)}\,d\Omega - \underbrace{\int_\Omega (\varepsilon^{(I)} + \delta\varepsilon) : \mathbb{E}_{(II)} : \delta\varepsilon\,d\Omega}_{=\mathbf{0}\ if\ \Gamma_u = \partial\Omega} \\ &\quad + \int_\Omega \delta\varepsilon : \mathbb{E}_{(II)} : \delta\varepsilon\,d\Omega \end{aligned}$$

$$= \underbrace{\int_{\Omega} \varepsilon^{(I)} : (\mathbb{E}_{(I)} - \mathbb{E}_{(II)}) : \varepsilon^{(I)} \, d\Omega}_{\text{pos. def. for } (\mathbb{E}_{(I)} - \mathbb{E}_{(II)}) > 0 \, \forall \, \mathbf{x}} + \underbrace{\int_{\Omega} \delta\varepsilon : \mathbb{E}_{(II)} : \delta\varepsilon \, d\Omega}_{\geq 0} . \tag{4.34}$$

The vanishing terms in the preceding analysis terms are due to a direct application of the principle of virtual work. The end result is that $(\mathbb{E}_{(I)} - \mathbb{E}_{(II)}) \geq \mathbf{0}$ implies

$$\underbrace{\int_{\Omega} \varepsilon^{(I)} : \mathbb{E}_{(I)} : \varepsilon^{(I)} \, d\Omega}_{\text{I's energy}} - \underbrace{\int_{\Omega} (\varepsilon^{(I)} + \delta\varepsilon) : \mathbb{E}_{(II)} : (\varepsilon^{(I)} + \delta\varepsilon) \, d\Omega}_{\text{II's energy}} \geq 0. \tag{4.35}$$

Similarly, defining $\delta\sigma = \sigma^{(II)} - \sigma^{(I)}$, if $\Gamma_t = \partial\Omega$

$$W_I \stackrel{\text{def}}{=} \underbrace{\int_{\Omega} \sigma^{(I)} : \mathbb{E}_{(I)}^{-1} : \sigma^{(I)} \, d\Omega}_{\text{I's energy}}, \tag{4.36}$$

and

$$W_{II} \stackrel{\text{def}}{=} \underbrace{\int_{\Omega} (\sigma^{(I)} + \delta\sigma) : \mathbb{E}_{(II)}^{-1} : (\sigma^{(I)} + \delta\sigma) \, d\Omega}_{\text{II's energy}}, \tag{4.37}$$

and we have

$$\begin{aligned}
W_I - W_{II} &= \int_{\Omega} \sigma^{(I)} : \mathbb{E}_{(I)}^{-1} : \sigma^{(I)} \, d\Omega - \int_{\Omega} \sigma^{(I)} : \mathbb{E}_{(II)}^{-1} : (\sigma^{(I)} + \delta\sigma) \, d\Omega \\
&\quad + \underbrace{\int_{\Omega} \delta\sigma : \mathbb{E}_{(II)}^{-1} : (\sigma^{(I)} - \delta\sigma) \, d\Omega}_{=\mathbf{0} \ if \ \Gamma_t = \partial\Omega} \\
&= \int_{\Omega} \sigma^{(I)} : (\mathbb{E}_{(I)}^{-1} - \mathbb{E}_{(II)}^{-1}) : \sigma^{(I)} \, d\Omega - \int_{\Omega} \sigma^{(I)} : \mathbb{E}_{(II)}^{-1} : \delta\sigma \, d\Omega \\
&= \int_{\Omega} \sigma^{(I)} : (\mathbb{E}_{(I)}^{-1} - \mathbb{E}_{(II)}^{-1}) : \sigma^{(I)} \, d\Omega + \underbrace{\int_{\Omega} (\sigma^{(I)} + \delta\sigma) : \mathbb{E}_{(II)}^{-1} : \delta\sigma \, d\Omega}_{=\mathbf{0} \text{ if } \Gamma_t = \partial\Omega} \\
&\quad - \int_{\Omega} \delta\sigma : \mathbb{E}_{(II)}^{-1} : \delta\sigma \, d\Omega \\
&= \underbrace{\int_{\Omega} \sigma^{(I)} : (\mathbb{E}_{(I)}^{-1} - \mathbb{E}_{(II)}^{-1}) : \sigma^{(I)} \, d\Omega}_{\text{neg. def. for } (\mathbb{E}_{(I)}^{-1} - \mathbb{E}_{(II)}^{-1}) < 0 \, \forall \, \mathbf{x}} - \underbrace{\int_{\Omega} \delta\sigma : \mathbb{E}_{(II)}^{-1} : \delta\sigma \, d\Omega}_{\leq 0} .
\end{aligned} \tag{4.38}$$

Therefore $(\mathbb{E}_{(I)}^{-1} - \mathbb{E}_{(II)}^{-1}) < \mathbf{0}$ implies

$$\underbrace{\int_{\Omega} \sigma^{(I)} : \mathbb{E}_{(I)}^{-1} : \sigma^{(I)} \, d\Omega}_{\text{I's energy}} - \underbrace{\int_{\Omega} (\sigma^{(I)} + \delta\sigma) : \mathbb{E}_{(II)}^{-1} : (\sigma^{(I)} + \delta\sigma) \, d\Omega}_{\text{II's energy}} \leq 0. \tag{4.39}$$

Similar results, however for more restrictive (uniform) loading cases, and not in the context of a microfailure analysis, can be found in Huet, Navi, and Roelfstra [94].

4.5.3 Uses to Approximate the Effective Property

Under the special case of uniform test boundary loadings, one has

- if $\mathbf{u}|_{\partial\Omega} = \mathscr{E} \cdot \mathbf{x}$ and $\mathbb{E}_{(I)} - \mathbb{E}_{(II)} \geq \mathbf{0} \Rightarrow 2W_I - 2W_{II} = \mathscr{E} : (\mathbb{E}_{(I)}{}^{*} - \mathbb{E}_{(II)}{}^{*}) : \mathscr{E}|\Omega| \geq 0$,
- if $\mathbf{t}|_{\partial\Omega} = \mathscr{L} \cdot \mathbf{n}$ and $\mathbb{E}_{(I)}{}^{-1} - \mathbb{E}_{(II)}{}^{-1} \leq \mathbf{0} \Rightarrow 2W_I - 2W_{II} = \mathscr{L} : (\mathbb{E}_{(I)}{}^{-1*} - \mathbb{E}_{(II)}{}^{-1*}) : \mathscr{L}|\Omega| \leq 0$.

Therefore, if $\mathbb{E}_{(II)}$ is assigned the real microstructure composed of irregular shapes, $\mathbb{E}_{(II)} = \mathbb{E}$, and $\mathbb{E}_{(I)}$ represents a microstructure with a smoother particulate geometry that *envelopes* the real particles (Fig. 4.7), denoted $\mathbb{E}^{+}$, with corresponding effective property $\mathbb{E}^{*+}$, then $\forall \mathbf{x} \in \Omega$ and $\mathbb{E}^{+} - \mathbb{E} \geq \mathbf{0}$, thus $\mathbb{E}^{*} \leq \mathbb{E}^{*+}$. Alternatively, if $\mathbb{E}_{(I)}$ is assigned the real microstructure composed of irregular shapes, $\mathbb{E}_{(I)} = \mathbb{E}$, and $\mathbb{E}_{(II)}$ represents a microstructure with a smoother particulate geometry that is *enveloped* by the real particles (Fig. 4.7), denoted $\mathbb{E}^{-}$, with corresponding effective property $\mathbb{E}^{*-}$, then $\forall \mathbf{x} \in \Omega$ and $\mathbb{E} - \mathbb{E}^{-} \geq \mathbf{0}$, thus $\mathbb{E}^{*-} \leq \mathbb{E}^{*}$. Therefore,

$$\underbrace{\mathbb{E}^{*-}}_{\text{using inner envelopes}} \leq \mathbb{E}^{*} \leq \underbrace{\mathbb{E}^{*+}}_{\text{using outer envelopes}}. \tag{4.40}$$

These relations, for uniform boundary conditions, appear to have been first developed in Huet et al. [94], based on the results of Hill [83]. The result has been coined by Huet et al. [94] as the "Hill Modification Theorem". Of critical importance is the fact that one can safely use smooth idealizations to approximate rougher more geometrically complicated shapes that occur during manufacturing (or naturally).

Chapter 5
A Basic Finite Element Implementation

Classical numerical techniques construct approximations from globally kinematically admissible functions. Two main obstacles arise (1) it may be very difficult to find a simple kinematically admissible function over the entire domain and (2) if such functions are found they lead to large, strongly coupled, and complicated systems of equations. These problems have been overcome by the fact that local approximations, i.e. over a very small portion of the domain are possible that deliver adequate solutions, and simultaneously lead to systems of equations which have an advantageous structure amenable to large scale computation by high speed computers. This piece-wise or "element-wise" approximation technique had been recognized at least 60 years ago by Courant [29]. There have been a variety of such approximation methods to solve equations of elasticity. The most popular is the Finite Element Method (FEM). The central feature of the method is to partition the body in a physically sound and systematic manner into an assembly of discrete subdomains or "elements". The process is designed to keep the resulting algebraic systems as computationally manageable, and memory efficient as possible. The implementation, theory and application of FEM is a subject of immense literature. We will not attempt to review this huge subject. For general references on the subject see the well-known books of Bathe [14], Becker et al. [15], Hughes [101], Szabo and Babúska [196] and Wriggers [214].

5.1 Finite Element Method Implementation

Consider the following general weak form, stated already in (3.4):

Find $\mathbf{u} \in \mathbf{H}^1(\Omega)$ such that $\forall \mathbf{v} \in \mathbf{H}^1(\Omega)$

$$\int_{\Omega} \nabla \mathbf{v} : \mathbb{E} : \nabla \mathbf{u} \, d\Omega = \int_{\Omega} \mathbf{f} \cdot \mathbf{v} \, d\Omega + \int_{\Gamma_t} \mathbf{t} \cdot \mathbf{v} \, dA + P^{\star} \int_{\Gamma_u} (\mathbf{d} - \mathbf{u}) \cdot \mathbf{v} \, dA, \quad (5.1)$$

where the last term is to be thought of as a penalty term to enforce the applied displacement (kinematic) boundary conditions $\mathbf{u} = \mathbf{w}$ on Γ_u. The (penalty) parameter, $P^\star$, is a large number. A penalty formulation has a variety of interpretations. For example, since we relax the restriction that the test function vanish in the displacement part of the boundary, we have

$$\int_{\Gamma_u} \mathbf{t} \cdot \mathbf{v} dA \approx P^\star \int_{\Gamma_u} (\mathbf{d} - \mathbf{u}) \cdot \mathbf{v} dA, \tag{5.2}$$

thus, the term $P^\star(\mathbf{d} - \mathbf{u})$ takes on the physical interpretation as a very stiff ("spring") resisting force which is proportional to the amount of violation from the true boundary displacement.

Remark. The ability to change the boundary data quickly will be very important later in the finite element computations, and is the hallmark of the penalty method in the context of a virtual work formulation. This is done by relaxing kinematic assumptions on the members of the space of admissible functions and adding a term to "account for the violation" on the exterior boundary. This is widely used in practice and, therefore, in order to keep the formulation as general as possible, we include penalty terms, *although this implementation is not mandatory*. Obviously, one could simply condense out the known (imposed) values of boundary displacements. Nevertheless we consider the (exterior) penalty method formulation for generality, although one does not necessarily need to use it.

5.2 FEM Approximation

It is convenient to write the bilinear form in the following (matrix) manner

$$\int_{\Omega} ([\mathbf{D}]\{\mathbf{v}\})^T [\mathbb{E}]([\mathbf{D}]\{\mathbf{u}\})\, d\Omega = \int_{\Omega} \{\mathbf{v}\}^T \{\mathbf{f}\}\, d\Omega + \int_{\Gamma_t} \{\mathbf{v}\}^T \{\mathbf{t}\}\, dA + P^\star \int_{\Gamma_u} \{\mathbf{v}\}^T \{\mathbf{d} - \mathbf{u}\}\, dA, \tag{5.3}$$

where $[\mathbf{D}]$, the deformation tensor, is

$$[\mathbf{D}] \stackrel{\text{def}}{=} \begin{bmatrix} \frac{\partial}{\partial x_1} & 0 & 0 \\ 0 & \frac{\partial}{\partial x_2} & 0 \\ 0 & 0 & \frac{\partial}{\partial x_3} \\ \frac{\partial}{\partial x_2} & \frac{\partial}{\partial x_1} & 0 \\ 0 & \frac{\partial}{\partial x_3} & \frac{\partial}{\partial x_2} \\ \frac{\partial}{\partial x_3} & 0 & \frac{\partial}{\partial x_1} \end{bmatrix}, \{\mathbf{u}\} \stackrel{\text{def}}{=} \begin{Bmatrix} u_1 \\ u_2 \\ u_3 \end{Bmatrix}, \{\mathbf{f}\} \stackrel{\text{def}}{=} \begin{Bmatrix} f_1 \\ f_2 \\ f_3 \end{Bmatrix}, \{\mathbf{t}\} \stackrel{\text{def}}{=} \begin{Bmatrix} t_1 \\ t_2 \\ t_3 \end{Bmatrix}. \tag{5.4}$$

It is clear that in an implementation of the finite element method, the sparsity of **D** (and other terms) should be taken into account. It is also convenient to write

$$u_1(x_1,x_2,x_3) = \sum_{i=1}^{N} a_i \phi_i(x_1,x_2,x_3),$$

$$u_2(x_1,x_2,x_3) = \sum_{i=1}^{N} a_{i+N} \phi_i(x_1,x_2,x_3),$$

$$u_3(x_1,x_2,x_3) = \sum_{i=1}^{N} a_{i+2N} \phi_i(x_1,x_2,x_3), \tag{5.5}$$

or, compactly, $\{\mathbf{u}\} = [\phi]\{\mathbf{a}\}$, where, for example, for trilinear shape functions

$$[\phi] \stackrel{\text{def}}{=} \begin{bmatrix} \phi_1\,\phi_2\,\phi_3\,\phi_4\,\phi_5\,\phi_6 \ldots \phi_N & 0\,0\,0\,0\,0\,0\,0 \ldots & 0\,0\,0\,0\,0\,0\,0 \ldots \\ 0\,0\,0\,0\,0\,0\,0 \ldots & \phi_1\,\phi_2\,\phi_3\,\phi_4\,\phi_5\,\phi_6 \ldots \phi_N & 0\,0\,0\,0\,0\,0\,0 \ldots \\ 0\,0\,0\,0\,0\,0\,0 \ldots & 0\,0\,0\,0\,0\,0\,0 \ldots & \phi_1\,\phi_2\,\phi_3\,\phi_4\,\phi_5\,\phi_6 \ldots \phi_N \end{bmatrix}. \tag{5.6}$$

It is advantageous to write

$$\{\mathbf{a}\} \stackrel{\text{def}}{=} \begin{Bmatrix} a_1 \\ a_2 \\ a_3 \\ \cdot \\ \cdot \\ \cdot \\ a_{3N} \end{Bmatrix}, \quad \underbrace{\{\phi_i\} \stackrel{\text{def}}{=} \begin{Bmatrix} \phi_i \\ 0 \\ 0 \end{Bmatrix}}_{\text{for } 1 \le i \le N}, \quad \underbrace{\{\phi_i\} \stackrel{\text{def}}{=} \begin{Bmatrix} 0 \\ \phi_i \\ 0 \end{Bmatrix}}_{\text{for } N+1 \le i \le 2N}, \quad \underbrace{\{\phi_i\} \stackrel{\text{def}}{=} \begin{Bmatrix} 0 \\ 0 \\ \phi_i \end{Bmatrix}}_{\text{for } 2N+1 \le i \le 3N}, \tag{5.7}$$

and $\{\mathbf{u}\} = \sum_{i=1}^{3N} a_i\{\phi_i\}$. If we choose $\mathbf{v}$ with the same basis, but a different linear combination $\{\mathbf{v}\} = [\phi]\{\mathbf{b}\}$, then we may write

$$\underbrace{\int_{\Omega} ([\mathbf{D}][\phi]\{\mathbf{b}\})^T [\mathbb{E}]([\mathbf{D}][\phi]\{\mathbf{a}\})\, d\Omega}_{\{\mathbf{b}\}^T[\mathbf{K}]\ \{\mathbf{a}\}\ \text{stiffness}} = \underbrace{\int_{\Omega} ([\phi]\{\mathbf{b}\})^T \{\mathbf{f}\}\, d\Omega}_{\text{body load}} + \underbrace{\int_{\Gamma_t} ([\phi]\{\mathbf{b}\})^T \{\mathbf{t}\}\, \mathrm{d}A}_{\text{traction load}}$$

$$\underbrace{+P^{\star} \int_{\Gamma_u} ([\phi]\{\mathbf{b}\})^T \{\mathbf{d} - ([\phi]\{\mathbf{a}\})\}\, \mathrm{d}A}_{\text{boundary penalty term}}. \tag{5.8}$$

Since $\{\mathbf{b}\}$ is arbitrary, i.e. the weak statement implies $\forall \mathbf{v} \Rightarrow \forall\{\mathbf{b}\}$, therefore

$$
\begin{aligned}
&\{\mathbf{b}\}^T\{[\mathbf{K}]\{\mathbf{a}\}-\{\mathbf{R}\}\}=0 \Rightarrow [\mathbf{K}]\{\mathbf{a}\}=\{\mathbf{R}\},\\
&[\mathbf{K}] \stackrel{\text{def}}{=} \int_\Omega([\mathbf{D}][\phi])^T[\mathbb{E}]([\mathbf{D}][\phi])\,d\Omega + P^\star \int_{\Gamma_u}[\phi]^T[\phi]\,\mathrm{d}A,\\
&\{\mathbf{R}\} \stackrel{\text{def}}{=} \int_\Omega[\phi]^T\{\mathbf{f}\}\,d\Omega + \int_{\Gamma_t}[\phi]^T\{\mathbf{t}\}\,\mathrm{d}A + P^\star \int_{\Gamma_u}[\phi]^T\{\mathbf{d}\}\,\mathrm{d}A.
\end{aligned}
\tag{5.9}
$$

This is the system of equations that is to be solved.

5.3 Global/local Transformations

One strength of the finite element method is that most of the computations can be done in an element by element manner. We define the entries of $[\mathbf{K}]$,

$$
K_{ij} = \int_\Omega([\mathbf{D}][\phi_i])^T[\mathbb{E}]([\mathbf{D}][\phi_j])\,d\Omega + P^\star \int_{\Gamma_u}[\phi_i]^T[\phi_j]\,\mathrm{d}A \tag{5.10}
$$

and

$$
R_i = \int_\Omega[\phi_i]^T\{\mathbf{f}\}\,d\Omega + \int_{\Gamma_t}[\phi_i]^T\{\mathbf{t}\}\,\mathrm{d}A + P^\star \int_{\Gamma_u}[\phi_i]^T\{\mathbf{d}\}\,\mathrm{d}A. \tag{5.11}
$$

Breaking the calculations into elements, $K_{ij} = \sum_e K^e_{ij}$, where

$$
K^e_{ij} = \int_{\Omega_e}([\mathbf{D}][\phi_i])^T[\mathbb{E}]([\mathbf{D}][\phi_j])\,d\Omega + P^\star \int_{\Gamma_u}[\phi_i]^T[\phi_j]\,\mathrm{d}A. \tag{5.12}
$$

In order to make the calculations systematic we wish to use the generic or master element defined in a local coordinate system $(\zeta_1,\zeta_2,\zeta_3)$. Accordingly, we need the following mapping functions (Fig. 5.1), from the master coordinates to the real space coordinates, $M:(x_1,x_2,x_3)\mapsto(\zeta_1,\zeta_2,\zeta_3)$(for example trilinear bricks):

$$
\begin{aligned}
x_1 &= \textstyle\sum_{i=1}^8 X_{1i}\hat{\phi}_i \stackrel{\text{def}}{=} M_{x_1}(\zeta_1,\zeta_2,\zeta_3),\\
x_2 &= \textstyle\sum_{i=1}^8 X_{2i}\hat{\phi}_i \stackrel{\text{def}}{=} M_{x_2}(\zeta_1,\zeta_2,\zeta_3),\\
x_3 &= \textstyle\sum_{i=1}^8 X_{3i}\hat{\phi}_i \stackrel{\text{def}}{=} M_{x_3}(\zeta_1,\zeta_2,\zeta_3),
\end{aligned}
\tag{5.13}
$$

where (X_{1i},X_{2i},X_{3i}) are the spatial coordinates of the ith node in the mesh, and where $\hat{\phi}(\zeta_1,\zeta_2,\zeta_3) \stackrel{\text{def}}{=} \phi(x_1(\zeta_1,\zeta_2,\zeta_3),x_2(\zeta_1,\zeta_2,\zeta_3),x_3(\zeta_1,\zeta_2,\zeta_3))$. These types of

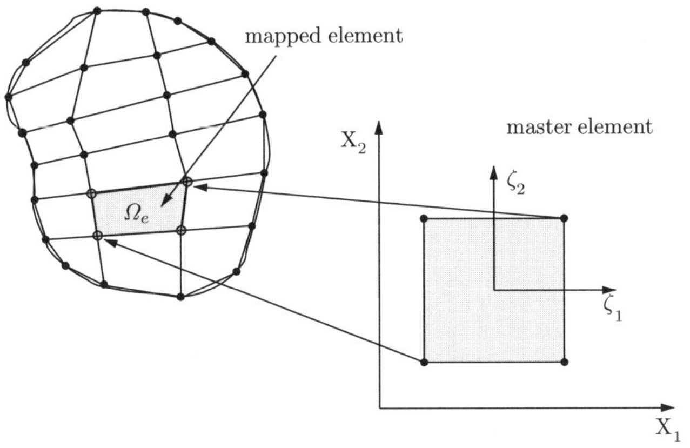

Fig. 5.1 A two-dimensional finite element mapping

mappings are usually termed parametric maps. If (a) the polynomial order of the shape functions is as high as the element, it is an isoparametric map, (b) if lower, then subparametric map and (c) if higher, then superparametric.

5.4 Differential Properties of Shape Functions

The master element shape functions form a nodal bases of trilinear approximation given by:

$$
\begin{aligned}
\hat{\phi}_1 &= \frac{1}{8}(1-\zeta_1)(1-\zeta_2)(1-\zeta_3), & \hat{\phi}_2 &= \frac{1}{8}(1+\zeta_1)(1-\zeta_2)(1-\zeta_3),\\
\hat{\phi}_3 &= \frac{1}{8}(1+\zeta_1)(1+\zeta_2)(1-\zeta_3), & \hat{\phi}_4 &= \frac{1}{8}(1-\zeta_1)(1+\zeta_2)(1-\zeta_3),\\
\hat{\phi}_5 &= \frac{1}{8}(1-\zeta_1)(1-\zeta_2)(1+\zeta_3), & \hat{\phi}_6 &= \frac{1}{8}(1+\zeta_1)(1-\zeta_2)(1+\zeta_3),\\
\hat{\phi}_7 &= \frac{1}{8}(1+\zeta_1)(1+\zeta_2)(1+\zeta_3), & \hat{\phi}_8 &= \frac{1}{8}(1-\zeta_1)(1+\zeta_2)(1+\zeta_3).
\end{aligned}
\tag{5.14}
$$

- For trilinear elements: Here we have a nodal basis consisting of 8 nodes, and since it is vector valued, 24 total degrees of freedom (3 degrees of freedom for each node, Fig. 5.2).
- For triquadratic elements: Here we have a nodal basis consisting of 27 nodes, and since it is vector valued, 81 total degrees of freedom (3 degrees of freedom for

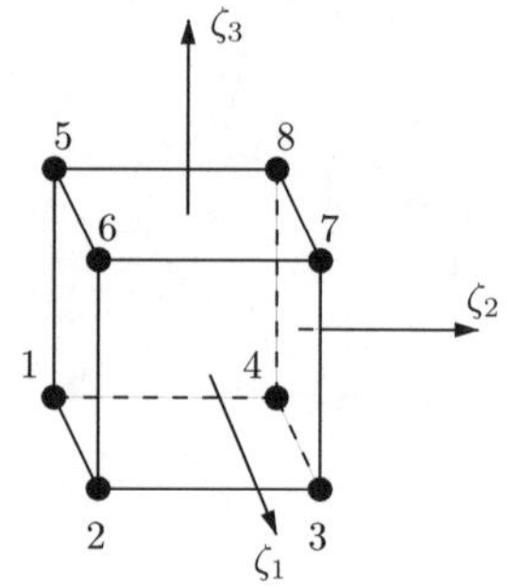

Fig. 5.2 A trilinear hexahedron or "brick"

each node). The nodal shape functions can be derived quite easily, by realizing that it is a nodal basis, i.e. they are unity at the corresponding node, and zero at all other nodes., etc...

We note that the ϕ_i's are never really computed, we actually start with the $\hat{\phi}$'s. Therefore in the stiffness matrix and righthand side element calculations, all terms must be defined in terms of the local coordinates. With this in mind we lay down some fundamental relations, which are directly related to the concepts of deformation presented in our discussion in continuum mechanics. It is not surprising that a deformation gradient reappears in the following form:

$$|\mathbf{F}| \stackrel{\text{def}}{=} \left|\frac{\partial \mathbf{x}(x_1,x_2,x_3)}{\partial \zeta(\zeta_1,\zeta_2,\zeta_3)}\right| \quad \text{where} \quad \mathbf{F} \stackrel{\text{def}}{=} \begin{bmatrix} \frac{\partial x_1}{\partial \zeta_1} & \frac{\partial x_1}{\partial \zeta_2} & \frac{\partial x_1}{\partial \zeta_3} \\ \frac{\partial x_2}{\partial \zeta_1} & \frac{\partial x_2}{\partial \zeta_2} & \frac{\partial x_2}{\partial \zeta_3} \\ \frac{\partial x_3}{\partial \zeta_1} & \frac{\partial x_3}{\partial \zeta_2} & \frac{\partial x_3}{\partial \zeta_3} \end{bmatrix}. \tag{5.15}$$

The corresponding determinant is

$$|\mathbf{F}| = \frac{\partial x_1}{\partial \zeta_1}\left(\frac{\partial x_2}{\partial \zeta_2}\frac{\partial x_3}{\partial \zeta_3} - \frac{\partial x_3}{\partial \zeta_2}\frac{\partial x_2}{\partial \zeta_3}\right) - \frac{\partial x_1}{\partial \zeta_2}\left(\frac{\partial x_2}{\partial \zeta_1}\frac{\partial x_3}{\partial \zeta_3} - \frac{\partial x_3}{\partial \zeta_1}\frac{\partial x_2}{\partial \zeta_3}\right) + \frac{\partial x_1}{\partial \zeta_3}\left(\frac{\partial x_2}{\partial \zeta_1}\frac{\partial x_3}{\partial \zeta_2} - \frac{\partial x_3}{\partial \zeta_1}\frac{\partial x_2}{\partial \zeta_2}\right). \tag{5.16}$$

The differential relations $\zeta \to \mathbf{x}$, are

$$\begin{aligned} \frac{\partial}{\partial \zeta_1} &= \frac{\partial}{\partial x_1}\frac{\partial x_1}{\partial \zeta_1} + \frac{\partial}{\partial x_2}\frac{\partial x_2}{\partial \zeta_1} + \frac{\partial}{\partial x_3}\frac{\partial x_3}{\partial \zeta_1}, \\ \frac{\partial}{\partial \zeta_2} &= \frac{\partial}{\partial x_1}\frac{\partial x_1}{\partial \zeta_2} + \frac{\partial}{\partial x_2}\frac{\partial x_2}{\partial \zeta_2} + \frac{\partial}{\partial x_3}\frac{\partial x_3}{\partial \zeta_2}, \\ \frac{\partial}{\partial \zeta_3} &= \frac{\partial}{\partial x_1}\frac{\partial x_1}{\partial \zeta_3} + \frac{\partial}{\partial x_2}\frac{\partial x_2}{\partial \zeta_3} + \frac{\partial}{\partial x_3}\frac{\partial x_3}{\partial \zeta_3}. \end{aligned} \tag{5.17}$$

The inverse differential relations $\mathbf{x} \rightarrow \zeta$, are

$$
\begin{aligned}
\frac{\partial}{\partial x_1} &= \frac{\partial}{\partial \zeta_1}\frac{\partial \zeta_1}{\partial x_1} + \frac{\partial}{\partial \zeta_2}\frac{\partial \zeta_2}{\partial x_1} + \frac{\partial}{\partial \zeta_3}\frac{\partial \zeta_3}{\partial x_1},\\
\frac{\partial}{\partial x_2} &= \frac{\partial}{\partial \zeta_1}\frac{\partial \zeta_1}{\partial x_2} + \frac{\partial}{\partial \zeta_2}\frac{\partial \zeta_2}{\partial x_2} + \frac{\partial}{\partial \zeta_3}\frac{\partial \zeta_3}{\partial x_2},\\
\frac{\partial}{\partial x_3} &= \frac{\partial}{\partial \zeta_1}\frac{\partial \zeta_1}{\partial x_3} + \frac{\partial}{\partial \zeta_2}\frac{\partial \zeta_2}{\partial x_3} + \frac{\partial}{\partial \zeta_3}\frac{\partial \zeta_3}{\partial x_3},
\end{aligned}
\tag{5.18}
$$

and

$$
\begin{Bmatrix} dx_1 \\ dx_2 \\ dx_3 \end{Bmatrix} = \underbrace{\begin{bmatrix} \frac{\partial x_1}{\partial \zeta_1} & \frac{\partial x_1}{\partial \zeta_2} & \frac{\partial x_1}{\partial \zeta_3} \\ \frac{\partial x_2}{\partial \zeta_1} & \frac{\partial x_2}{\partial \zeta_2} & \frac{\partial x_2}{\partial \zeta_3} \\ \frac{\partial x_3}{\partial \zeta_1} & \frac{\partial x_3}{\partial \zeta_2} & \frac{\partial x_3}{\partial \zeta_3} \end{bmatrix}}_{\mathbf{F}} \begin{Bmatrix} d\zeta_1 \\ d\zeta_2 \\ d\zeta_3 \end{Bmatrix}
\tag{5.19}
$$

and the inverse form

$$
\begin{Bmatrix} d\zeta_1 \\ d\zeta_2 \\ d\zeta_3 \end{Bmatrix} = \underbrace{\begin{bmatrix} \frac{\partial \zeta_1}{\partial x_1} & \frac{\partial \zeta_1}{\partial x_2} & \frac{\partial \zeta_1}{\partial x_3} \\ \frac{\partial \zeta_2}{\partial x_1} & \frac{\partial \zeta_2}{\partial x_2} & \frac{\partial \zeta_2}{\partial x_3} \\ \frac{\partial \zeta_3}{\partial x_1} & \frac{\partial \zeta_3}{\partial x_2} & \frac{\partial \zeta_3}{\partial x_3} \end{bmatrix}}_{\mathbf{F}^{-1}} \begin{Bmatrix} dx_1 \\ dx_2 \\ dx_3 \end{Bmatrix}.
\tag{5.20}
$$

Noting the following relationship

$$
\mathbf{F}^{-1} = \frac{adj\mathbf{F}}{|\mathbf{F}|} \qquad \text{where} \qquad adj\mathbf{F} \stackrel{\text{def}}{=} \begin{bmatrix} A_{11} & A_{12} & A_{13} \\ A_{21} & A_{22} & A_{23} \\ A_{31} & A_{32} & A_{33} \end{bmatrix}^T,
\tag{5.21}
$$

where

$$A_{11}=\left[\frac{\partial x_2}{\partial \zeta_2}\frac{\partial x_3}{\partial \zeta_3}-\frac{\partial x_3}{\partial \zeta_2}\frac{\partial x_2}{\partial \zeta_3}\right]=|\mathbf{F}|\frac{\partial \zeta_1}{\partial x_1},\quad A_{12}=-\left[\frac{\partial x_2}{\partial \zeta_1}\frac{\partial x_3}{\partial \zeta_3}-\frac{\partial x_3}{\partial \zeta_1}\frac{\partial x_2}{\partial \zeta_3}\right]=|\mathbf{F}|\frac{\partial \zeta_2}{\partial x_1},$$

$$A_{13}=\left[\frac{\partial x_2}{\partial \zeta_1}\frac{\partial x_3}{\partial \zeta_2}-\frac{\partial x_3}{\partial \zeta_1}\frac{\partial x_2}{\partial \zeta_2}\right]=|\mathbf{F}|\frac{\partial \zeta_3}{\partial x_1},\quad A_{21}=-\left[\frac{\partial x_1}{\partial \zeta_2}\frac{\partial x_3}{\partial \zeta_3}-\frac{\partial x_3}{\partial \zeta_2}\frac{\partial x_1}{\partial \zeta_3}\right]=|\mathbf{F}|\frac{\partial \zeta_1}{\partial x_2},$$

$$A_{22}=\left[\frac{\partial x_1}{\partial \zeta_1}\frac{\partial x_3}{\partial \zeta_3}-\frac{\partial x_3}{\partial \zeta_1}\frac{\partial x_1}{\partial \zeta_3}\right]=|\mathbf{F}|\frac{\partial \zeta_2}{\partial x_2},\quad A_{23}=-\left[\frac{\partial x_1}{\partial \zeta_1}\frac{\partial x_3}{\partial \zeta_2}-\frac{\partial x_3}{\partial \zeta_1}\frac{\partial x_1}{\partial \zeta_2}\right]=|\mathbf{F}|\frac{\partial \zeta_3}{\partial x_2},$$

$$A_{31}=\left[\frac{\partial x_1}{\partial \zeta_2}\frac{\partial x_2}{\partial \zeta_3}-\frac{\partial x_2}{\partial \zeta_2}\frac{\partial x_1}{\partial \zeta_3}\right]=|\mathbf{F}|\frac{\partial \zeta_1}{\partial x_3},\quad A_{32}=-\left[\frac{\partial x_1}{\partial \zeta_1}\frac{\partial x_2}{\partial \zeta_3}-\frac{\partial x_2}{\partial \zeta_1}\frac{\partial x_1}{\partial \zeta_3}\right]=|\mathbf{F}|\frac{\partial \zeta_2}{\partial x_3},$$

$$A_{33}=\left[\frac{\partial x_1}{\partial \zeta_1}\frac{\partial x_2}{\partial \zeta_2}-\frac{\partial x_2}{\partial \zeta_1}\frac{\partial x_1}{\partial \zeta_2}\right]=|\mathbf{F}|\frac{\partial \zeta_3}{\partial x_3}. \tag{5.22}$$

With these relations, one can then solve for the components of $\mathbf{F}$ and $\mathbf{F}^{-1}$.

5.5 Differentiation in the Referential Coordinates

We now need to express $[\mathbf{D}]$ in terms $\zeta_1, \zeta_2, \zeta_3$, via

$$[\mathbf{D}(\phi(x_1,x_2,x_3))]=[\hat{\mathbf{D}}(\hat{\phi}(M_{x_1}(\zeta_1,\zeta_2,\zeta_3),M_{x_2}(\zeta_1,\zeta_2,\zeta_3),M_{x_3}(\zeta_1,\zeta_2,\zeta_3))]. \tag{5.23}$$

Therefore we write for the first column of $[\hat{\mathbf{D}}]$[1]

$$\begin{bmatrix} \frac{\partial}{\partial \zeta_1}\frac{\partial \zeta_1}{\partial x_1}+\frac{\partial}{\partial \zeta_2}\frac{\partial \zeta_2}{\partial x_1}+\frac{\partial}{\partial \zeta_3}\frac{\partial \zeta_3}{\partial x_1} \\ 0 \\ 0 \\ \frac{\partial}{\partial \zeta_1}\frac{\partial \zeta_1}{\partial x_2}+\frac{\partial}{\partial \zeta_2}\frac{\partial \zeta_2}{\partial x_2}+\frac{\partial}{\partial \zeta_3}\frac{\partial \zeta_3}{\partial x_2} \\ 0 \\ \frac{\partial}{\partial \zeta_1}\frac{\partial \zeta_1}{\partial x_3}+\frac{\partial}{\partial \zeta_2}\frac{\partial \zeta_2}{\partial x_3}+\frac{\partial}{\partial \zeta_3}\frac{\partial \zeta_3}{\partial x_3} \end{bmatrix}, \tag{5.24}$$

[1] This is for illustration purposes only. For computational efficiency, one should not program such operations in this way. Clearly, the needless multiplication of zeros is to be avoided.

for the second column

$$\begin{bmatrix} 0 \\ \frac{\partial}{\partial \zeta_1}\frac{\partial \zeta_1}{\partial x_2} + \frac{\partial}{\partial \zeta_2}\frac{\partial \zeta_2}{\partial x_2} + \frac{\partial}{\partial \zeta_3}\frac{\partial \zeta_3}{\partial x_2} \\ 0 \\ \frac{\partial}{\partial \zeta_1}\frac{\partial \zeta_1}{\partial x_1} + \frac{\partial}{\partial \zeta_2}\frac{\partial \zeta_2}{\partial x_1} + \frac{\partial}{\partial \zeta_3}\frac{\partial \zeta_3}{\partial x_1} \\ \frac{\partial}{\partial \zeta_1}\frac{\partial \zeta_1}{\partial x_3} + \frac{\partial}{\partial \zeta_2}\frac{\partial \zeta_2}{\partial x_3} + \frac{\partial}{\partial \zeta_3}\frac{\partial \zeta_3}{\partial x_3} \\ 0 \end{bmatrix}, \tag{5.25}$$

and for the last column

$$\begin{bmatrix} 0 \\ 0 \\ \frac{\partial}{\partial \zeta_1}\frac{\partial \zeta_1}{\partial x_3} + \frac{\partial}{\partial \zeta_2}\frac{\partial \zeta_2}{\partial x_3} + \frac{\partial}{\partial \zeta_3}\frac{\partial \zeta_3}{\partial x_3} \\ 0 \\ \frac{\partial}{\partial \zeta_1}\frac{\partial \zeta_1}{\partial x_2} + \frac{\partial}{\partial \zeta_2}\frac{\partial \zeta_2}{\partial x_2} + \frac{\partial}{\partial \zeta_3}\frac{\partial \zeta_3}{\partial x_2} \\ \frac{\partial}{\partial \zeta_1}\frac{\partial \zeta_1}{\partial x_1} + \frac{\partial}{\partial \zeta_2}\frac{\partial \zeta_2}{\partial x_1} + \frac{\partial}{\partial \zeta_3}\frac{\partial \zeta_3}{\partial x_1} \end{bmatrix}. \tag{5.26}$$

For an element, our shape function matrix ($\stackrel{\text{def}}{=} [\hat{\phi}]$) has the following form for linear shape functions, for the first eight columns

$$\begin{bmatrix} \hat{\phi}_1 & \hat{\phi}_2 & \hat{\phi}_3 & \hat{\phi}_4 & \hat{\phi}_5 & \hat{\phi}_6 & \hat{\phi}_7 & \hat{\phi}_8 \\ 0 & 0 & 0 & 0 & 0 & 0 & 0 & 0 \\ 0 & 0 & 0 & 0 & 0 & 0 & 0 & 0 \end{bmatrix}, \tag{5.27}$$

for the second eight columns

$$\begin{bmatrix} 0 & 0 & 0 & 0 & 0 & 0 & 0 & 0 \\ \hat{\phi}_1 & \hat{\phi}_2 & \hat{\phi}_3 & \hat{\phi}_4 & \hat{\phi}_5 & \hat{\phi}_6 & \hat{\phi}_7 & \hat{\phi}_8 \\ 0 & 0 & 0 & 0 & 0 & 0 & 0 & 0 \end{bmatrix}, \tag{5.28}$$

for the last eight columns

$$\begin{bmatrix} 0 & 0 & 0 & 0 & 0 & 0 & 0 & 0 \\ 0 & 0 & 0 & 0 & 0 & 0 & 0 & 0 \\ \hat{\phi}_1 & \hat{\phi}_2 & \hat{\phi}_3 & \hat{\phi}_4 & \hat{\phi}_5 & \hat{\phi}_6 & \hat{\phi}_7 & \hat{\phi}_8 \end{bmatrix}, \tag{5.29}$$

which in total is a 3×24 matrix. Therefore the product $[\hat{\mathbf{D}}][\hat{\phi}]$ is a 6×24 matrix of the form, for the first eight columns

$$\begin{bmatrix} \frac{\partial\hat{\phi}_1}{\partial\zeta_1}\frac{\partial\zeta_1}{\partial x_1}+\frac{\partial\hat{\phi}_1}{\partial\zeta_2}\frac{\partial\zeta_2}{\partial x_1}+\frac{\partial\hat{\phi}_1}{\partial\zeta_3}\frac{\partial\zeta_3}{\partial x_1},\ldots 8 \\ 0\quad 0\quad 0\; 0\quad 0\quad 0\quad 0\quad 0 \\ 0\quad 0\quad 0\; 0\quad 0\quad 0\quad 0\quad 0 \\ \frac{\partial\hat{\phi}_1}{\partial\zeta_1}\frac{\partial\zeta_1}{\partial x_2}+\frac{\partial\hat{\phi}_1}{\partial\zeta_2}\frac{\partial\zeta_2}{\partial x_2}+\frac{\partial\hat{\phi}_1}{\partial\zeta_3}\frac{\partial\zeta_3}{\partial x_2},\ldots 8 \\ 0\quad 0\quad 0\; 0\quad 0\quad 0\quad 0\quad 0 \\ \frac{\partial\hat{\phi}_1}{\partial\zeta_1}\frac{\partial\zeta_1}{\partial x_3}+\frac{\partial\hat{\phi}_1}{\partial\zeta_2}\frac{\partial\zeta_2}{\partial x_3}+\frac{\partial\hat{\phi}_1}{\partial\zeta_3}\frac{\partial\zeta_3}{\partial x_3},\ldots 8 \end{bmatrix}, \tag{5.30}$$

and for the second eight columns

$$\begin{bmatrix} 0\quad 0\quad 0\; 0\quad 0\quad 0\quad 0\quad 0 \\ \frac{\partial\hat{\phi}_1}{\partial\zeta_1}\frac{\partial\zeta_1}{\partial x_2}+\frac{\partial\hat{\phi}_1}{\partial\zeta_2}\frac{\partial\zeta_2}{\partial x_2}+\frac{\partial\hat{\phi}_1}{\partial\zeta_3}\frac{\partial\zeta_3}{\partial x_2},\ldots 8 \\ 0\quad 0\quad 0\; 0\quad 0\quad 0\quad 0\quad 0 \\ \frac{\partial\hat{\phi}_1}{\partial\zeta_1}\frac{\partial\zeta_1}{\partial x_1}+\frac{\partial\hat{\phi}_1}{\partial\zeta_2}\frac{\partial\zeta_2}{\partial x_1}+\frac{\partial\hat{\phi}_1}{\partial\zeta_3}\frac{\partial\zeta_3}{\partial x_1},\ldots 8 \\ \frac{\partial\hat{\phi}_1}{\partial\zeta_1}\frac{\partial\zeta_1}{\partial x_3}+\frac{\partial\hat{\phi}_1}{\partial\zeta_2}\frac{\partial\zeta_2}{\partial x_3}+\frac{\partial\hat{\phi}_1}{\partial\zeta_3}\frac{\partial\zeta_3}{\partial x_3},\ldots 8 \\ 0\quad 0\quad 0\; 0\quad 0\quad 0\quad 0\quad 0 \end{bmatrix}, \tag{5.31}$$

and for the last eight columns

$$\begin{bmatrix} 0\quad 0\quad 0\; 0\quad 0\quad 0\quad 0\quad 0 \\ 0\quad 0\quad 0\; 0\quad 0\quad 0\quad 0\quad 0 \\ \frac{\partial\hat{\phi}_1}{\partial\zeta_1}\frac{\partial\zeta_1}{\partial x_3}+\frac{\partial\hat{\phi}_1}{\partial\zeta_2}\frac{\partial\zeta_2}{\partial x_3}+\frac{\partial\hat{\phi}_1}{\partial\zeta_3}\frac{\partial\zeta_3}{\partial x_3},\ldots 8 \\ 0\quad 0\quad 0\; 0\quad 0\quad 0\quad 0\quad 0 \\ \frac{\partial\hat{\phi}_1}{\partial\zeta_1}\frac{\partial\zeta_1}{\partial x_2}+\frac{\partial\hat{\phi}_1}{\partial\zeta_2}\frac{\partial\zeta_2}{\partial x_2}+\frac{\partial\hat{\phi}_1}{\partial\zeta_3}\frac{\partial\zeta_3}{\partial x_2},\ldots 8 \\ \frac{\partial\hat{\phi}_1}{\partial\zeta_1}\frac{\partial\zeta_1}{\partial x_1}+\frac{\partial\hat{\phi}_1}{\partial\zeta_2}\frac{\partial\zeta_2}{\partial x_1}+\frac{\partial\hat{\phi}_1}{\partial\zeta_3}\frac{\partial\zeta_3}{\partial x_1},\ldots 8 \end{bmatrix}. \tag{5.32}$$

Finally with quadrature for each element

$$K_{ij}^e = \underbrace{\sum_{q=1}^{g}\sum_{r=1}^{g}\sum_{s=1}^{g} w_q w_r w_s ([\hat{\mathbf{D}}]\{\hat{\phi}_i\})^T [\mathbb{E}]([\hat{\mathbf{D}}]\{\hat{\phi}_j\})|\mathbf{F}|}_{\text{standard}}$$
$$+ \underbrace{\sum_{q=1}^{g}\sum_{r=1}^{g} w_q w_r P^{\star} [\hat{\phi}_i]^T \{\hat{\phi}_j\}|\mathbf{F}_s|}_{\text{penalty for } \Gamma_u \cap \Omega_e \neq 0}, \tag{5.33}$$

and

$$R_i^e = \underbrace{\sum_{q=1}^{g}\sum_{r=1}^{g}\sum_{s=1}^{g} w_q w_r w_s \{\hat{\phi}_i\}^T \{\mathbf{f}\}|\mathbf{F}|}_{\text{standard}}$$
$$+ \underbrace{\sum_{q=1}^{g}\sum_{r=1}^{g} w_q w_r [\hat{\phi}_i]^T \{\mathbf{t}\}|\mathbf{F}_s|}_{\text{for } \Gamma_t \cap \Omega_e \neq 0} + \underbrace{\sum_{q=1}^{g}\sum_{r=1}^{g} w_q w_r P^{\star} [\hat{\phi}_i]^T \{\mathbf{d}\}|\mathbf{F}_s|}_{\text{penalty for } \Gamma_u \cap \Omega_e \neq 0}, \tag{5.34}$$

where w_q, etc, are Gauss weights and where $|\mathbf{F}_s|$ represents the (surface) Jacobians of element faces on the exterior surface of the body where, depending on the surface on which it is to be evaluated upon, one of the ζ components will be +1 or -1. These surface Jacobians can be evaluated in a variety of ways, for example using the Nanson formula derived earlier.

Remark. It is permitted to have material discontinuities within the finite elements. On the implementation level, the system of equations to be solved are $[\mathbf{K}]\{\mathbf{a}\} = \{\mathbf{R}\}$, where the stiffness matrix is represented by $\mathbf{K}(I,J) = \mathbf{k}(ELEM\#, i, j)$, where I, J are the global entries, and i, j are the local entries. The amount of memory required with this relatively simple storage system is, for trilinear hexahedra, $\mathbf{k}(ELEM\#, 24, 24) = 576\times$ the number of finite elements, where the $\mathbf{k}$ are the individual element stiffness matrices. If matrix symmetry is taken into account, the memory requirements are $\mathbf{k}(ELEM\#, 24, 24) = 300\times$ the number of finite elements. This simple approach is so-called element by element storage. The element by element storage is critical in this regard to reduce the memory requirements.[2] Here a global/local index relation must be made to connect the local entry to the global entry when solution time begins. This is a relatively simple and efficient storage system to encode. The element by element strategy has other advantages with regard to element by element system CG-solvers. This is discussed later.

[2] If a direct storage of the finite element storage of the stiffness matrix were attempted, $\mathbf{K}(DOF, DOF) = DOF \times DOF$ memory would be required, which, for large problems is an impossibility for most workstations.

5.6 Post Processing

Post processing for the stress, strain and energy from the existing displacement solution, i.e., the values of the nodal displacements and the shape functions, is straightforward. Essentially the process is the same as the formation of the virtual energy in the system $[\mathbf{D}]\{\mathbf{u}^h\} = \{\varepsilon^h\}$. Therefore, for each element

$$\begin{Bmatrix} \varepsilon_{11}^h \\ \varepsilon_{22}^h \\ \varepsilon_{33}^h \\ 2\varepsilon_{12}^h \\ 2\varepsilon_{23}^h \\ 2\varepsilon_{13}^h \end{Bmatrix} = \begin{bmatrix} \frac{\partial}{\partial x_1} & 0 & 0 \\ 0 & \frac{\partial}{\partial x_2} & 0 \\ 0 & 0 & \frac{\partial}{\partial x_3} \\ \frac{\partial}{\partial x_2} & \frac{\partial}{\partial x_1} & 0 \\ 0 & \frac{\partial}{\partial x_3} & \frac{\partial}{\partial x_2} \\ \frac{\partial}{\partial x_3} & 0 & \frac{\partial}{\partial x_1} \end{bmatrix} \underbrace{\begin{Bmatrix} \sum_{i=1}^8 u_{1i}^h \phi_i \\ \sum_{i=1}^8 u_{2i}^h \phi_i \\ \sum_{i=1}^8 u_{3i}^h \phi_i \end{Bmatrix}}_{\text{known values}}, \tag{5.35}$$

where u_i^h and the nodal values (the a_i's). *The global coordinates must be transformed to the master system, in both the deformation tensor, and the displacement representation.* At each Gauss point, we add up all eight contributions for each of the six components, then multiply by the corresponding nodal displacements that have previously been calculated. The following expressions must be evaluated at the Gauss points, multiplied by the appropriate weights and added together:

$$\begin{aligned}
&\frac{\partial u_1^h}{\partial x_1} = \sum_{i=1}^{8} u_{1i}^h \frac{\partial \phi_i}{\partial x_1}, \quad \frac{\partial u_2^h}{\partial x_1} = \sum_{i=1}^{8} u_{1i}^h \frac{\partial \phi_i}{\partial x_1}, \quad \frac{\partial u_3^h}{\partial x_1} = \sum_{i=1}^{8} u_{1i}^h \frac{\partial \phi_i}{\partial x_1}, \\
&\frac{\partial u_1^h}{\partial x_2} = \sum_{i=1}^{8} u_{1i}^h \frac{\partial \phi_i}{\partial x_2}, \quad \frac{\partial u_2^h}{\partial x_2} = \sum_{i=1}^{8} u_{1i}^h \frac{\partial \phi_i}{\partial x_2}, \quad \frac{\partial u_3^h}{\partial x_2} = \sum_{i=1}^{8} u_{1i}^h \frac{\partial \phi_i}{\partial x_2}, \\
&\frac{\partial u_1^h}{\partial x_3} = \sum_{i=1}^{8} u_{1i}^h \frac{\partial \phi_i}{\partial x_3}, \quad \frac{\partial u_2^h}{\partial x_3} = \sum_{i=1}^{8} u_{1i}^h \frac{\partial \phi_i}{\partial x_3}, \quad \frac{\partial u_3^h}{\partial x_3} = \sum_{i=1}^{8} u_{1i}^h \frac{\partial \phi_i}{\partial x_3},
\end{aligned} \tag{5.36}$$

where u_{1i}^h denotes the x_1 component of the displacement of the ith node. Combining the numerical derivatives to form the strains we obtain $\varepsilon_{11}^h = \frac{\partial u_1^h}{\partial x_1}$, $\varepsilon_{22}^h = \frac{\partial u_2^h}{\partial x_2}$, $\varepsilon_{33}^h = \frac{\partial u_3^h}{\partial x_3}$ and $2\varepsilon_{12}^h = \gamma_{12} = \frac{\partial u_1^h}{\partial x_2} + \frac{\partial u_2^h}{\partial x_1}$, $2\varepsilon_{23}^h = \gamma_{23} = \frac{\partial u_2^h}{\partial x_3} + \frac{\partial u_3^h}{\partial x_2}$, and $2\varepsilon_{13}^h = \gamma_{13} = \frac{\partial u_1^h}{\partial x_3} + \frac{\partial u_3^h}{\partial x_1}$.

5.7 Accuracy of the Finite Element Method

As we have seen, the essential idea in the finite element method is to select a finite dimensional subspatial approximation of the true solution and form the following weak boundary problem:

$$\text{Find } \mathbf{u}^h \in \mathbf{H}_u^h(\Omega) \subset \mathbf{H}^1(\Omega), \mathbf{u}^h|_{\Gamma_u} = \mathbf{d}, \text{ such that}$$

$$\underbrace{\int_{\Omega} \nabla \mathbf{v}^h : \mathbb{E} : \nabla \mathbf{u}^h \, d\Omega}_{\mathscr{B}(\mathbf{u}^h,\mathbf{v}^h)} = \underbrace{\int_{\Omega} \mathbf{f} \cdot \mathbf{v}^h \, d\Omega + \int_{\Gamma_t} \mathbf{t} \cdot \mathbf{v}^h \, dA}_{\mathscr{F}(\mathbf{v}^h)}, \tag{5.37}$$

$$\forall \mathbf{v}^h \in \mathbf{H}_v^h(\Omega) \subset \mathbf{H}^1(\Omega), \mathbf{v}^h|_{\Gamma_u} = \mathbf{0}.$$

The critical point is that $\mathbf{H}_u^h(\Omega), \mathbf{H}_v^h(\Omega) \subset \mathbf{H}^1(\Omega)$. This "inner" approximation allows the development of straightforward subspatial error estimates. In most cases $\mathbf{H}_u^h(\Omega)$ and $\mathbf{H}_v^h(\Omega)$ coincide. We have for any kinematically admissible function $\mathbf{w}$ a definition of the so-called energy norm

$$||\mathbf{u}-\mathbf{w}||_{E(\Omega)}^2 \stackrel{\text{def}}{=} \int_{\Omega} (\nabla \mathbf{u} - \nabla \mathbf{w}) : \mathbb{E} : (\nabla \mathbf{u} - \nabla \mathbf{w}) \, d\Omega = \mathscr{B}(\mathbf{u}-\mathbf{w}, \mathbf{u}-\mathbf{w}). \tag{5.38}$$

Note that in the event that nonconstant displacements are specified on the boundary, then $\mathbf{u} - \mathbf{w} = \textit{constant}$ is unobtainable unless $\mathbf{u} - \mathbf{w} = \mathbf{0}$, and the semi-norm in (5.38) is a norm in the strict mathematical sense. Under standard assumptions the fundamental a-priori error estimate for the finite element method is

$$||\mathbf{u}-\mathbf{u}^h||_{E(\Omega)} \leq \mathscr{C}(\mathbf{u}, p) h^{\min(r-1,p) \stackrel{\text{def}}{=} \gamma}, \tag{5.39}$$

where p is the (complete) polynomial order of the finite element method used, r is the regularity of the exact solution, $\mathscr{C}$ is a global constant dependent on the exact solution and the polynomial approximation. $\mathscr{C}$ is independent of h, the maximum element diameter. For details see, for example, Hughes [101]. Related forms of this estimate holds locally, but with constants that are element dependent. To prove this, select any $\mathbf{z}^h \in \mathbf{H}^s(\Omega)$, $\mathbf{H}^r(\Omega) \subset \mathbf{H}^s(\Omega)$, therefore

$$||\mathbf{u}-\mathbf{z}^h||_{H^s(\Omega)} \leq Ch^{\theta} ||\mathbf{u}||_{H^r(\Omega)} \quad \text{where} \quad \theta = \min(p+1-s, r-s), \tag{5.40}$$

is a generalization of the result with γ. We have $\mathscr{B}(\mathbf{u},\mathbf{v}) = \mathscr{F}(\mathbf{v})$, $\forall \mathbf{v} \in \mathbf{H}^1(\Omega)$ and $\mathscr{B}(\mathbf{u}^h,\mathbf{v}^h) = \mathscr{F}(\mathbf{v}^h)$, $\forall \mathbf{v}^h \in \mathbf{H}_v^h(\Omega) \subset \mathbf{H}^1(\Omega)$, implying a Galerkin (Fig. 5.3) orthogonality property of "inner approximations":

$$\mathscr{B}(\mathbf{u}-\mathbf{u}^h,\mathbf{v}^h)=\mathscr{B}(\mathbf{e}^h,\mathbf{v}^h)=0, \qquad \forall \mathbf{v}^h \in \mathbf{H}^h_v(\Omega) \subset \mathbf{H}^1(\Omega), \quad (5.41)$$

where the error is defined by $\mathbf{e}^h \stackrel{\text{def}}{=} \mathbf{u}-\mathbf{u}^h$. Therefore any member of the subspace can be represented by $\mathbf{e}^h - \mathbf{v}^h = \mathbf{u}-\mathbf{u}^h-\mathbf{v}^h = \mathbf{u}-\mathbf{z}^h$. Using this representation we have

$$\mathscr{B}(\mathbf{e}^h-\mathbf{v}^h,\mathbf{e}^h-\mathbf{v}^h)=\mathscr{B}(\mathbf{e}^h,\mathbf{e}^h)-2\mathscr{B}(\mathbf{e}^h,\mathbf{v}^h)+\mathscr{B}(\mathbf{v}^h,\mathbf{v}^h), \quad (5.42)$$

which implies

$$\mathscr{B}(\mathbf{u}-\mathbf{u}^h,\mathbf{u}-\mathbf{u}^h)\leq \mathscr{B}(\mathbf{u}-\mathbf{z}^h,\mathbf{u}-\mathbf{z}^h). \quad (5.43)$$

We recall that $\mathbb{E}$ is bounded in the following sense: $a_+\varepsilon:\varepsilon \geq \varepsilon : \mathbb{E}(\mathbf{x}):\varepsilon \geq a_-\varepsilon:\varepsilon$, $\forall \varepsilon \in \mathbb{R}^{3\times 3}$, $\varepsilon=\varepsilon^T$, $\infty > a_-, a_+ > 0$. With this we finally have

$$||\mathbf{u}-\mathbf{u}^h||^2_{E(\Omega)} \leq \frac{1}{a_-}\mathscr{B}(\mathbf{e}^h,\mathbf{e}^h) \leq \frac{1}{a_-}\mathscr{B}(\mathbf{u}-\mathbf{z}^h,\mathbf{u}-\mathbf{z}^h)$$

$$\leq \frac{a_+}{a_-}||\mathbf{u}-\mathbf{z}^h||^2_{E(\Omega)} \leq C^2 h^{2\theta}||\mathbf{u}||^2_{E(\Omega)}. \quad (5.44)$$

This result holds locally as well, however with the constant and exponent being different for each element.

5.8 Local Adaptive Mesh Refinement

One possible way to save computational expense when using the FEM is local adaptive mesh refinement. Although such techniques are outside the scope of this monograph, we briefly review two of the most popular error estimation procedures, which are used to guide mesh refinement: *Recovery Methods and Residual Methods*.

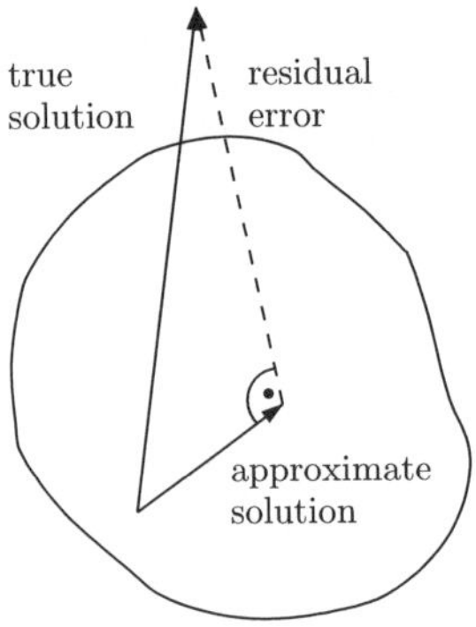

Fig. 5.3 Orthogonality of the approximation error

5.8.1 A-Posteriori Recovery Methods

Recovery methods are based on the assumption that there is a function $\gamma(\mathbf{u}^h)$ that is closer to $\nabla\mathbf{u}$ than $\nabla\mathbf{u}^h$, which can be used to estimate the error. The most popular of these methods is the Zienkiewicz and Zhu [222] error estimation technique, which is effective for a wide class of problems. It is based on the notion that gradients of the solution obtained on a given mesh can be smoothed and compared with the original solution to assess the error. The sampling points at which the gradient's error are to be evaluated are so-called superconvergent points where the convergence is optimal. However, these points must be searched for, and may not even exist, since superconvergence occurs only in very special situations. By superconvergence, we mean that the exponent is higher that the standard theoretical estimate (w):

$$||\mathbf{u} - \mathbf{u}^h||_{H^s(\Omega)} \leq \mathscr{C}(\mathbf{u}, p)h^{\min(p+1-s, r-s)} \stackrel{\text{def}}{=} w \tag{5.45}$$

The function γ is obtained by calculating a least squares fit to the gradient of, potentially several hundred sample points in three dimensions (superconvergent points of elements surrounding a finite element node). The new gradient then serves to estimate the error locally over a "patch" of elements, i.e. a group of element sharing a common node,

$$||\gamma(\mathbf{u}^h) - \nabla\mathbf{u}^h||_{\text{patch}} \propto \text{error}. \tag{5.46}$$

This is by far the most popular method in the engineering community to estimate the error, and also has the benefit of postprocessing the approximate solution as a by-product.

5.8.2 A-Posteriori Residual Methods

Residual methods require no a-posteriori system of equations to be solved. Such methods make use of

- the FEM solution itself,
- the data on the boundary,
- the error equation, and
- the Galerkin orthogonality property,

to bound the error. The basic residual equation can be written and bounded as follows:

$$\mathscr{B}(\mathbf{e},\mathbf{v}) = \mathscr{F}(\mathbf{v}) - \mathscr{B}(\mathbf{u}^h,\mathbf{v}) = \sum_{e=1}^{N} \left(\int_{\Omega_e} \underbrace{(\mathbf{f} + \nabla \cdot (\mathbb{E} : \nabla \mathbf{u}^h))}_{\mathbf{r}} \cdot \mathbf{v} d\Omega \right.$$
$$+ \int_{\partial\Omega_e \cap \Gamma_t} (\mathbf{t} - (\mathbb{E} : \nabla \mathbf{u}^h) \cdot \mathbf{n}) \cdot \mathbf{v} dA$$
$$\left. - \int_{\partial\Omega_e} (\mathbb{E} : \nabla \mathbf{u}^h) \cdot \mathbf{n} \cdot \mathbf{v} dA \right) \tag{5.47}$$

The approach is to then form the following bound

$$||\mathbf{u} - \mathbf{u}^h||^2_{E(\Omega)} \leq C \underbrace{\sum_{e=1}^{N} \left(h_e^2 ||\mathbf{r}||^2_{L^2(\Omega_e)} + \frac{1}{2} h_e ||\mathbf{r}||^2_{L^2(\partial\Omega_e)} \right)}_{\zeta_e}. \tag{5.48}$$

The local quantities ζ_e, are used to decide whether an element is to be refined. Such estimates, used to guide local adaptive finite element mesh refinement techniques, were first developed in Babúska and Rheinboldt [10] for one-dimensional problems and in Babúska and Miller [11] and Kelly et al. [106] for two-dimensional problems. For reviews see Ainsworth and Oden [2].

5.9 Solution of Algebraic Equations

There are two main approaches to solving systems of equations resulting from numerical discretization of solid mechanics problems, direct and iterative. There are a large number of variants of each. Direct solvers are usually employed when the number of unknowns is not very large, and there are multiple load vectors. Basically, one can operate on the multiple righthand sides simultaneously via Gaussian elimination. Gaussian elimination costs approximately $2(N + (N-1)^2 + (N-2)^2 + N-3)^2 + ...1^2) = 2\sum_{k=1}^{N} k^2 \approx 2\int_0^N k^2 dk = \frac{2}{3}N^3$. For a back substitution the cost is $2(0+1+2+3...+N-1) = 2\sum_{k=1}^{N}(k-1) = N(N-1)$. Therefore, the total cost of solving such a system is the cost to reduce the system to upper triangular form plus the cost of back substitution, i.e. $\frac{2}{3}N^3 + N(N-1)$. However since the operation counts to factor and solve an $N \times N$ system are $\mathscr{O}(N^3)$, iterative solvers are preferred when the systems are very large.[3]

Many modern solvers, for large symmetric, positive-definite, systems, like the ones of interest here, employ Conjugate Gradient (CG) type iterative techniques, which can deliver solutions in $\mathscr{O}(N)^2$ operations. It is inescapable, for almost all variants of Gaussian elimination, unless they involve sparsity tracking to eliminate unneeded operations on zero entries, that the operation costs are $\mathscr{O}(N^3)$. However,

[3] An operation such as the addition of two numbers, or multiplication of two numbers, is defined as one operation count.

banded solvers, which exploit the bandedness of the stiffness matrix, can reduce the number of operation counts to Nb^2, where b is the bandwidth. Many schemes exist for the optimal ordering of nodes in order to make the bandwidth as small as possible. Skyline solvers locate the uppermost nonzero elements starting from the diagonal and concentrate only on elements below the skyline. Frontal methods, which are analogous to a moving front in the finite element mesh, perform Gaussian elimination element by element, before the element is incorporated into the global stiffness matrix. In this procedure, memory is reduced, and the elimination process can be done for all elements simultaneously, at least in theory. Upon assembly of the stiffness matrix and righthand side, back substitution can be started immediately. If the operations are performed in an optimal order, it can be shown that the number of operations behaves proportionally to N^2. Such a process is, of course, nontrivial. We note that with direct methods, zeros within the band, below the skyline, and in the front are generally filled and must be tracked in the operations. In very large problems the storage requirements and the number of operation counts can become so large that solution by direct methods is not feasible. The data structures, and I/O are also non-trivial concerns. However, we note that a matrix/vector multiplication involves $2N^2$ operation counts, and that a method based on repeated vector multiplication, if convergent in less than N iterations, could be very attractive. This usually the premise in using iterative methods, such as the Conjugate Gradient Method (CG).

A very important feature of iterative methods is that the memory requirements remain constant during the solution process. It is important to note that modern computer architectures are based on

(1) *registers*, which can perform very fast operations on computer "words", but have virtually no memory capabilities
(2) *cache*, which slightly larger memory capabilities, with a slight reduction in speed, but are thermally very "hot", and are thus limited for physical, as well as manufacturing, reasons
(3) *main memory*, which is slower since I/O is required, but still within a workstation and
(4) *disk and tape or magnetic drums*, which are out of core, and thus require a huge I/O component, and are very slow.

Therefore, one point that we emphasize is that advantage should be taken of the element by element structure inherent in the finite element method for data storage and matrix vector multiplication in the CG method. The element by element data structure is also critical for the ability to fit matrix/vector multiplications into the computer cache, which essentially is a low memory/high floating point operation per second portion of the computer hardware. Of course this can also be achieved when compact storage is used for the matrix **K**, however which much more complexity in data structure.

Remark. One singularly distinguishing feature of iterative solvers is the fact that since they are based on successive updates of a starting guess solution vector, they can be given a tremendous head start by a good solution guess, for example, provided by an analytical or semi-analytical solution.

5.9.1 Krylov Searches and Minimum Principles

We now discuss minimum principles, which play a key role in the construction of a certain class of iterative solvers. *By itself, the PMPE is a powerful theoretical result. However it can be used to develop methods to solve systems of equations arising from a finite element discretization of a infinitesimal strain linearly elastic structure.* This result is the essence of the so-called Krylov family of searches. Suppose we wish to solve the discrete system $[\mathbf{K}]\{\mathbf{a}\} = \{\mathbf{R}\}$. $[\mathbf{K}]$ is a symmetric positive definite $N \times N$ matrix, $\{\mathbf{a}\}$ is the $N \times 1$ solution vector, and $\{\mathbf{R}\}$ is the $N \times 1$ righthand side. We define a potential

$$\Pi \stackrel{\text{def}}{=} \frac{1}{2}\{\mathbf{a}\}^T[\mathbf{K}]\{\mathbf{a}\} - \{\mathbf{a}\}^T\{\mathbf{R}\}. \tag{5.49}$$

Correspondingly, from basic calculus we have

$$\nabla\Pi \stackrel{\text{def}}{=} \{\frac{\partial \Pi}{\partial a_1}, \frac{\partial \Pi}{\partial a_2}, \dots \frac{\partial \Pi}{\partial a_N}\}^T = \mathbf{0} \Rightarrow [\mathbf{K}]\{\mathbf{a}\} - \{\mathbf{R}\} = \mathbf{0}. \tag{5.50}$$

Therefore, the minimizer of the potential Π is also the solution to the discrete system. A family of iterative solving techniques for symmetric systems based upon minimizing Π by successively updating a starting vector are the Krylov class. The minimization takes place over vector spaces called the Krylov spaces. These methods are based on the hope that a solution to a tolerable accuracy can be achieved in much less than $\mathcal{O}(N^3)$ operations, as required with most Gaussian-type techniques. The simplest of this family is the method of steepest descent, which is a precursor to the widely used Conjugate Gradient Method.

5.9.2 The Method of Steepest Descent

The Method of Steepest Descent is based upon the following simple idea: if the gradient of the potential is not zero at a possible solution vector, then the greatest increase of the scalar function is in the direction of the gradient, therefore we move in the opposite direction $-\nabla\Pi$. The ingredients in the methods are the residual vector,

$$\{\mathbf{r}\}^i \stackrel{\text{def}}{=} -\nabla\Pi = \{\mathbf{R}\} - [\mathbf{K}]\{\mathbf{a}\}^i, \tag{5.51}$$

and the successive iterates,

$$\{\mathbf{a}\}^{i+1} \stackrel{\text{def}}{=} \{\mathbf{a}\}^i + \lambda^i\{\mathbf{r}\}^i. \tag{5.52}$$

We seek a λ^i relation such that Π is a global minimum. Directly we have $\Pi = \frac{1}{2}\{\mathbf{a}\}^{T,i}[\mathbf{K}]\{\mathbf{a}\}^i + \lambda^i\{\mathbf{a}\}^{T,i}[\mathbf{K}]\{\mathbf{r}\}^i + \frac{1}{2}\lambda^{i2}\{\mathbf{r}\}^{T,i}[\mathbf{K}]\{\mathbf{r}\}^i - \{\mathbf{a}\}^{T,i}\{\mathbf{R}\} - \lambda^i\{\mathbf{r}\}^{T,i}\{\mathbf{R}\}$, where it was assumed that $[\mathbf{K}]$ is symmetric. Forcing $\frac{\partial \Pi}{\partial \lambda^i} = \mathbf{0}$, and solving for λ^i yields

$$\lambda^i = \frac{\{\mathbf{r}\}^{T,i}(\{\mathbf{R}\} - [\mathbf{K}]\{\mathbf{a}\}^i)}{\{\mathbf{r}\}^{T,i}[\mathbf{K}]\{\mathbf{r}\}^i} = \frac{\{\mathbf{r}\}^{T,i}\{\mathbf{r}\}^i}{\{\mathbf{r}\}^{T,i}[\mathbf{K}]\{\mathbf{r}\}^i}. \tag{5.53}$$

Therefore, the method of Steepest Descent consists of the following:

STEP 1 : SELECT A STARTING GUESS $\{\mathbf{a}\}^1$

STEP 2 : COMPUTE :

$$\{\mathbf{r}\}^i = \{\mathbf{R}\} - [\mathbf{K}]\{\mathbf{a}\}^i \qquad \lambda^i = \frac{\{\mathbf{r}\}^{T,i}\{\mathbf{r}\}^i}{\{\mathbf{r}\}^{T,i}[\mathbf{K}]\{\mathbf{r}\}^i} \qquad \{\mathbf{a}\}^{i+1} = \{\mathbf{a}\}^i + \lambda^i\{\mathbf{r}\}^i$$

STEP 3 : COMPUTE :

$$\lambda^i||\{\mathbf{r}\}^i||_K^2 = ||\{\mathbf{a}\}^{i+1} - \{\mathbf{a}\}^i||_K^2 \stackrel{\text{def}}{=} (\{\mathbf{a}\}^{T,i+1} - \{\mathbf{a}\}^{T,i+1})[\mathbf{K}](\{\mathbf{a}\}^{i+1} - \{\mathbf{a}\}^{i+1})$$

$$\text{IF } \frac{||\{\mathbf{a}\}^{i+1} - \{\mathbf{a}\}^{i}||_K}{||\{\mathbf{a}\}^{i}||_K} < \tau = \text{TOL} \Rightarrow \text{STOP}$$

$$\text{IF } \frac{||\{\mathbf{a}\}^{i+1} - \{\mathbf{a}\}^{i}||_K}{||\{\mathbf{a}\}^{i}||_K} \geq \tau = \text{TOL} \Rightarrow \text{GO TO STEP 2.} \tag{5.54}$$

where $||\{\cdot\}||_K \stackrel{\text{def}}{=} \sqrt{\{\cdot\}^T[\mathbf{K}]\{\cdot\}}$. The rate of convergence of the method is related to the condition number of the stiffness matrix[4]

$$||\{\mathbf{a}\}^{ex} - \{\mathbf{a}\}^i||_K = (1 - \frac{1}{C([\mathbf{K}])})^{i/2}||\{\mathbf{a}\}^{ex} - \{\mathbf{a}\}^1||_K, \tag{5.55}$$

where $C([\mathbf{K}]) \stackrel{\text{def}}{=} \frac{\max\ [\mathbf{K}]\ \text{eigenvalue}}{\min\ [\mathbf{K}]\ \text{eigenvalue}}$. The rate of convergence of the method is typically quite slow, however, a variant, the Conjugate Gradient Method, is guaranteed to converge in, at most, N iterations, provided the algebra is performed exactly.

[4] The term $\mathbf{a}^{ex}$ denotes the exact solution to the algebraic equation.

5.9.3 The Conjugate Gradient Method

In the Conjugate Gradient Method at each iteration the computational cost is $\mathcal{O}(N)$. We refer the reader to Axelsson [9] for details. We define the residual,

$$\{\mathbf{r}\}^{i} \stackrel{\text{def}}{=} -\nabla\Pi = \{\mathbf{R}\} - [\mathbf{K}]\{\mathbf{a}\}^{i}, \tag{5.56}$$

and the successive iterates,

$$\{\mathbf{a}\}^{i+1} \stackrel{\text{def}}{=} \{\mathbf{a}\}^{i} + \lambda^{i}\{\mathbf{z}\}^{i}. \tag{5.57}$$

We seek a λ^i such that Π is a global minimum. The ith iteration, $\{\mathbf{a}\}^{i+1}$, is computed by $\{\mathbf{a}\}^{i+1} = \{\mathbf{a}^{i}\}^{i} + \lambda^{i}\{\mathbf{z}\}^{i}$, $i = 1,2,3...$, where the update vector is defined as $\{\mathbf{z}\}^{i} \stackrel{\text{def}}{=} \{\mathbf{r}\}^{i} + \theta^{i}\{\mathbf{z}\}^{i-1}$, for $i = 1,2,3....$ The coefficient θ^i is chosen so that $\{\mathbf{z}\}$ is $[\mathbf{K}] - conjugate$ to $\{\mathbf{z}\}^{i-1}$, i.e. $\{\mathbf{z}\}^{T,i}[\mathbf{K}]\{\mathbf{z}\}^{i-1} = 0 \Rightarrow \theta^{i} = -\frac{\{\mathbf{r}\}^{T,i}[\mathbf{K}]\{\mathbf{z}\}^{i-1}}{\{\mathbf{z}\}^{T,i-1}[\mathbf{K}]\{\mathbf{z}\}^{i-1}}$. Therefore, the value of λ^i which minimizes

$$\Pi = \frac{1}{2}(\{\mathbf{a}\}^{i} + \lambda^{i}\{\mathbf{z}\}^{i})^{T}[\mathbf{K}](\{\mathbf{a}\}^{i} + \lambda^{i}\{\mathbf{z}\}^{i}) - (\{\mathbf{a}\}^{i} + \lambda^{i}\{\mathbf{z}\}^{i})^{T}\{\mathbf{R}\}, \tag{5.58}$$

is $(i = 1,2,3...)$,

$$\lambda^{i} = \frac{\{\mathbf{z}\}^{T,i}(\{\mathbf{R}\} - [\mathbf{K}]\{\mathbf{a}\}^{i})}{\{\mathbf{z}\}^{T,i}[\mathbf{K}]\{\mathbf{z}\}^{i}} = \frac{\{\mathbf{z}\}^{T,i}\{\mathbf{r}\}^{i}}{\{\mathbf{z}\}^{T,i}[\mathbf{K}]\{\mathbf{z}\}^{i}}. \tag{5.59}$$

The solution steps are:

STEP 1 : FOR i = 1 :*SELECT* $\{\mathbf{a}\}^1 \Rightarrow \{\mathbf{r}\}^1 = \{\mathbf{R}\} - [\mathbf{K}]\{\mathbf{a}\}^1 = \{\mathbf{z}\}^1$

STEP 2 : COMPUTE (WITH $\{\mathbf{z}\}^1 = \{\mathbf{r}\}^1$)

$$\lambda^{1} = \frac{\{\mathbf{z}\}^{T,1}(\{\mathbf{R}\} - [\mathbf{K}]\{\mathbf{a}\}^{1})}{\{\mathbf{z}\}^{T,1}[\mathbf{K}]\{\mathbf{z}\}^{1}} = \frac{\{\mathbf{z}\}^{T,1}\{\mathbf{r}\}^{1}}{\{\mathbf{z}\}^{T,1}[\mathbf{K}]\{\mathbf{z}\}^{1}} \qquad i = 1,2...$$

STEP 3 : COMPUTE $\{\mathbf{a}\}^2 = \{\mathbf{a}\}^1 + \lambda^1\{\mathbf{z}\}^1$

STEP 4 : (FOR i > 1) COMPUTE $\{\mathbf{r}\}^{i} = \{\mathbf{R}\} - [\mathbf{K}]\{\mathbf{a}\}^{i}$

$$\theta^{i} = -\frac{\{\mathbf{r}\}^{T,i}[\mathbf{K}]\{\mathbf{z}\}^{i-1}}{\{\mathbf{z}\}^{T,i-1}[\mathbf{K}]\{\mathbf{z}\}^{i-1}} \qquad \{\mathbf{z}\}^{i} \stackrel{\text{def}}{=} \{\mathbf{r}\}^{i} + \theta^{i}\{\mathbf{z}\}^{i-1}$$

$$\lambda^i = \frac{\{\mathbf{z}\}^{T,i}(\{\mathbf{R}\} - [\mathbf{K}]\{\mathbf{a}\}^i)}{\{\mathbf{z}\}^{T,i}[\mathbf{K}]\{\mathbf{z}\}^i} = \frac{\{\mathbf{z}\}^{T,i}\{\mathbf{r}\}^i}{\{\mathbf{z}\}^i[\mathbf{K}]\{\mathbf{z}\}^i}$$

$$\text{COMPUTE } \{\mathbf{a}\}^{\mathrm{i}+1} = \{\mathbf{a}\}^{\mathrm{i}} + \lambda^{\mathrm{i}}\{\mathbf{z}\}^{\mathrm{i}}$$

$$\text{STEP 5 : COMPUTE } \mathbf{e} \stackrel{\text{def}}{=} \frac{||\{\mathbf{a}\}^{\mathrm{i}+1} - \{\mathbf{a}\}^{\mathrm{i}}||_{\mathrm{K}}}{||\{\mathbf{a}\}^{\mathrm{i}}||_{\mathrm{K}}} = \frac{|\lambda^{\mathrm{i}}|||\{\mathbf{z}\}^{\mathrm{i}}||_{\mathrm{K}}}{||\{\mathbf{a}\}^{\mathrm{i}}||_{\mathrm{K}}} \leq \tau \ (\tau = \text{TOL})$$

$$\text{IF } \mathbf{e} < \tau \Rightarrow \text{STOP}$$

$$\text{IF } \mathbf{e} \geq \tau \Rightarrow \text{GO TO STEP 4.} \tag{5.60}$$

5.9.4 Accelerating Computations

The rate of convergence of the CG method is related to the condition number ($\{\mathbf{a}\}^{ex}$ = exact algebraic solution)

$$||\{\mathbf{a}\}^{ex} - \{\mathbf{a}\}^i||_K \leq \left(\frac{\sqrt{C([\mathbf{K}])} - 1}{\sqrt{C([\mathbf{K}])} + 1}\right)^i ||\{\mathbf{a}\}^{ex} - \{\}^1||_K. \tag{5.61}$$

Proofs of the various characteristics of the method can be found in Axelsson [9]. As is standard, in an attempt to reduce the condition number and hence increase the rate of convergence, typically, preconditioning of $[\mathbf{K}]$ is done by forming the following transformation of variables, $\underline{\{\mathbf{a}\}} = [\mathbf{T}]\{\mathscr{A}\}$, which produces a preconditioned system, with stiffness matrix $\underline{[\mathbf{K}]} = [\mathbf{T}]^T[\mathbf{K}][\mathbf{T}]$. Ideally we would like $[\mathbf{T}] = [\mathbf{L}]^{-T}$ where $[\mathbf{L}][\mathbf{L}]^T = [\mathbf{K}]$, and where $[\mathbf{L}]$ is a lower triangular matrix, thus forcing $[\mathbf{T}]^T[\mathbf{K}][\mathbf{T}] = [\mathbf{L}]^{-1}[\mathbf{L}][\mathbf{L}]^T[\mathbf{L}]^{-T} = \mathbf{I}$. However, the reduction of the stiffness matrix into a lower triangular matrix and its transpose is comparable in the number of operations to solving the system by Gaussian elimination. Therefore, only an approximation to $[\mathbf{L}]^{-1}$ is computed. Thus, inexpensive preconditioners are usually used. For example, diagonal preconditioning, which is the least expensive, involves defining $[\mathbf{T}]$ as a diagonal matrix with entries $p_{ii} = \frac{1}{\sqrt{K_{ii}}}, i, j = 1, ..., ndof$, where the K_{ii} are the diagonal entries of $[\mathbf{K}]$. In this case the resulting terms in the preconditioned stiffness matrix are unity on the diagonal. The off diagonal terms, K_{ij} are divided by $\frac{1}{\sqrt{K_{ii}K_{jj}}}$. There are a variety of other preconditioning techniques, of widely ranging expense to compute. For more details see Axelsson [9]. It is strongly suggested to precondition the system. For example with the simple diagonal preconditioner, we obtain the following stiffness matrix

$$
\begin{bmatrix}
1 & \frac{K_{12}}{\sqrt{K_{11}}\sqrt{K_{22}}} & \frac{K_{13}}{\sqrt{K_{11}}\sqrt{K_{33}}} & \frac{K_{14}}{\sqrt{K_{11}}\sqrt{K_{44}}} & \ldots\ldots \\
\frac{K_{21}}{\sqrt{K_{11}}\sqrt{K_{22}}} & 1 & \frac{K_{23}}{\sqrt{K_{22}}\sqrt{K_{33}}} & \frac{K_{24}}{\sqrt{K_{22}}\sqrt{K_{44}}} & \ldots\ldots \\
\frac{K_{31}}{\sqrt{K_{33}}\sqrt{K_{11}}} & \frac{K_{32}}{\sqrt{K_{33}}\sqrt{K_{22}}} & 1 & . & \ldots\ldots \\
\frac{K_{41}}{\sqrt{K_{44}}\sqrt{K_{11}}} & \frac{K_{42}}{\sqrt{K_{44}}\sqrt{K_{22}}} & . & 1 & \ldots\ldots \\
. & . & . & . & .\,1\ldots \\
. & . & . & . & \ldots\ldots \\
. & . & . & . & \ldots\ldots
\end{bmatrix}. \tag{5.62}
$$

5.10 Remark on Penalty Methods

As we have mentioned earlier, the penalty method can be used to enforce displacement boundary conditions. For example, consider an augmented potential $\mathscr{J}(\mathbf{u},P^{\star}) \stackrel{\text{def}}{=} \mathscr{J}(\mathbf{u}) + P^{\star}\int_{\Gamma_u}(\mathbf{d}-\mathbf{u})\cdot(\mathbf{d}-\mathbf{u})\,dA$, $\mathbf{u}\in\mathbf{H}^1(\Omega)$, whose variation (Euler-Lagrange equation) is

$$
\begin{gathered}
\text{Find } \mathbf{u}\in\mathbf{H}^1(\Omega) \text{ such that } \forall\mathbf{v}\in\mathbf{H}^1(\Omega) \\
\int_{\Omega}\nabla\mathbf{v}:\sigma\,d\Omega = \int_{\Omega}\mathbf{f}\cdot\mathbf{v}\,d\Omega + \int_{\Gamma_t}\mathbf{t}\cdot\mathbf{v}\,dA + P^{\star}\int_{\Gamma_u}(\mathbf{d}-\mathbf{u})\cdot\mathbf{v}\,dA.
\end{gathered} \tag{5.63}
$$

The penalty term can be thought of as a quadratic augmentation of the potential energy. When no potential exists, the exterior point method can be considered simply as an (exterior-method) enforcement of a constraint. In many of the numerical studies to follow, the penalty parameters have been varied repeatedly in simulations and the solutions have been checked to be independent of such parameters, beyond threshold values. These results have been checked against directly enforcing the boundary conditions by eliminating rows and columns in the stiffness matrix of nodes where displacement boundary conditions are applied, with the results being essentially indistinguishable. Iterative solvers that employ preconditioning, are also insensitive to the penalty parameters. This is primarily due to the fact that they produce isolated increases in the spectrum of eigenvalues of the stiffness matrix, and thus affect the solver minimally. Generally, it is a distributed increase in the eigenvalue spectrum (not the case with the penalty method) which can pose a problem. At any rate, such penalty augmentations have minimal effect on iterative solver performance, due to the fact that preconditioning is used. In short, the penalty parameters are not an issue of importance in the simulations presented later.

Chapter 6
Computational/Statistical Testing Methods

The steady rise in computer power has led reseachers in the field to consider direct simulation, whereby the effective responses can be obtained by volumetrically averaging (post processing) numerical solutions of boundary value problems representing the response of samples of heterogeneous material. In this chapter, we investigate topics related to the numerical simulation of the testing of mechanical responses of samples of microheterogeneous solid materials formed by aggregates of particulates suspended in a binding matrix. The microstructures considered are generated by randomly distributing particles throughout an otherwise homogeneous matrix material. The resulting microstructures are irregular and nonperiodic. A primary issue in the simulation of such materials is the fact that only finite sized samples can be tested, leading to no single response, but a distribution of responses. Accordingly, a technique employing potential energy principles is developed to interpret the results of testing groups of samples. Three dimensional numerical examples employing the finite element method are given to illustrate the overall analysis and computational testing process. The basis for this chapter follows from work found in Zohdi and Wriggers [229].

6.1 A Boundary Value Formulation

We consider a sample of heterogeneous material (Fig. 6.1), with domain Ω, under a given set of specified boundary loadings. The weak form boundary value problem is

$$\text{Find } \mathbf{u} \in \mathbf{H}^1(\Omega), \mathbf{u}|_{\Gamma_u} = \mathbf{d}, \text{ such that}$$
$$\int_{\Omega} \nabla \mathbf{v} : \mathbb{E} : \nabla \mathbf{u}\, d\Omega = \int_{\Omega} \mathbf{f} \cdot \mathbf{v}\, d\Omega + \int_{\Gamma_t} \mathbf{t} \cdot \mathbf{v}\, dA \qquad \forall \mathbf{v} \in \mathbf{H}^1(\Omega), \mathbf{v}|_{\Gamma_u} = \mathbf{0}. \tag{6.1}$$

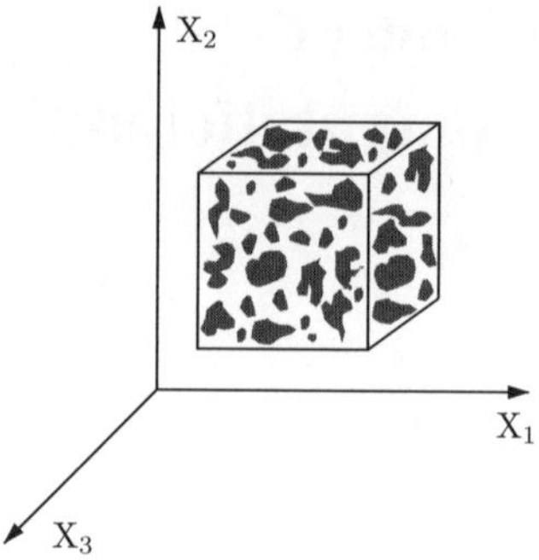

Fig. 6.1 A cubical sample of microheterogeneous material

When we perform material tests satisfying Hill's condition, we have, in the case of displacement controlled tests (loading case (1)) $\Gamma_u = \partial\Omega$ and $\mathbf{u}|_{\partial\Omega} = \mathscr{E}\cdot\mathbf{x}$ or for traction controlled tests (case (2)) $\Gamma_t = \partial\Omega$ and $\mathbf{t}|_{\partial\Omega} = \mathscr{L}\cdot\mathbf{n}$. In either case we consider $\mathbf{f} = \mathbf{0}$ and that the material is perfectly bonded. We note that (1) Hill's condition is satisfied with $\mathbf{f} = \mathbf{0}$ (and no debonding) and in case (2) it is satisfied even with debonding, however only if $\mathbf{f} = \mathbf{0}$. The boundary value problem in Box 6.1 must be solved for each new sample, each possessing a different random microstructure ($\mathbb{E}(\mathbf{x})$). The solution is then post processed for the effective quantities. It is convenient to consider the RVE domain Ω as a cube, and we shall do so for the remainder of the work.

6.2 Numerical Discretization

In order to computationally simulate effective responses, our choice for spatial discretization is the finite element method. There are essentially two choices to mesh the microstructure with the finite element method, a microstructure-nonconforming or a microstructure-conforming approach. We refer to a nonconforming approach as one which does not require the finite element boundaries to coincide with material interfaces when meshing the internal geometry (Fig. 6.2). This leads to material discontinuities within the finite elements. A conforming approach would impose that the element boundaries coincide with material interfaces and therefore the elements have no material discontinuities within them. There are advantages and disadvantages to both approaches. Nonconforming meshing has the advantage of rapid generation of structured internal meshes and consequently no finite element distortion arising from the microstructure. This is critical to computational performance if iterative solvers are to be used. The conforming meshing usually will require less finite elements than the nonconforming approach for the same pointwise accuracy. However, the disadvantages are the (extremely difficult) mesh generation for irregular microstructures in three dimensions. Even if such microstructures can be meshed in a conforming manner, the finite element distortion leads to stiffness matrix ill conditioning and possible element instability (element nonconvexity). For numerical

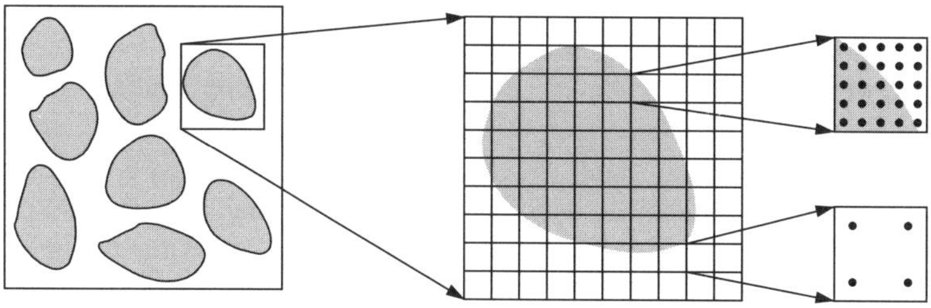

Fig. 6.2 Microstructure-nonconforming meshing with material discontinuities within an element

studies comparing the meshing approaches, see Zohdi et al. [224]. Our emphasis is on studying somewhat irregular microstructures, specifically randomly dispersed particulates, and rapidly evaluating them during the testing process. Therefore, we have adopted the nonconforming approach, which we discuss in more detail next.

6.2.1 Topological Resolution

Inherent in the nonconforming approach is the integration of discontinuous integrands (Fig. 6.3). The topology is not embedded into the finite element a priori, as it would be in a conforming approach, via isoparametric maps onto material interfaces. To some extent, if the elements are much smaller than the particle length scales, the topology will be approximately captured. However, one can improve this representation (Fig. 6.2). Since the finite element method is an integral-based method, the quadrature rules can be increased in an element by element fashion to better capture the geometry in elements with material discontinuities. A primary

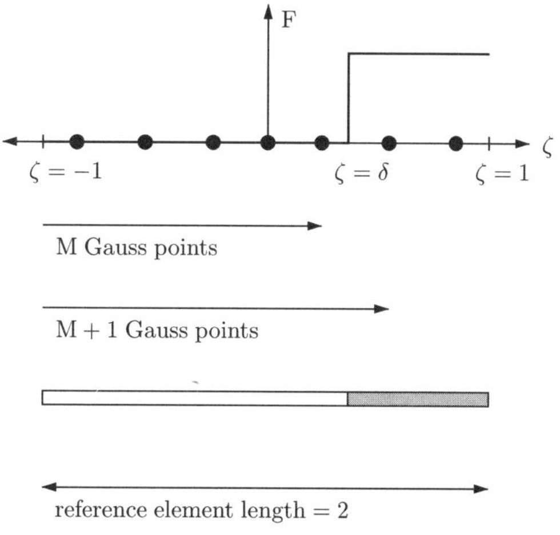

Fig. 6.3 A material discontinuity in a reference finite element

question is: If there is an integrand discontinuity in a finite element, how high should the quadrature rules be to perform accurate integration?

Throughout the simulations, we shall employ the finite element method, along with a technique of Gauss point oversampling to accurately resolve the topological features of the microstructure. In order to introduce oversampling, first consider a one dimensional reference finite element domain, and an associated integrand, $F(\zeta)$ with a jump discontinuity at $\zeta = \delta$ (Fig. 6.3), which admits to a decomposition of the function into continuous and discontinuous parts,

$$F(\zeta) = C(\zeta) + [\![F(\delta)]\!] H(\zeta - \delta), \tag{6.2}$$

where H is the Heaviside function, ζ is the local coordinate, and $[\![(\cdot)]\!] \stackrel{\text{def}}{=} (\cdot)|_+ - (\cdot)|_-$ is the jump operator. We assume that the elements are small compared to the length scales of the particulate matter and as a consequence that there is at most one discontinuity within the element. However, this assumption makes no difference at the implementation level. Integrating over a reference element, we have

$$\int_{-1}^{1} F(\zeta)d\zeta = \int_{-1}^{1} (C(\zeta) + [\![F(\delta)]\!] H(\zeta - \delta))\, d\zeta. \tag{6.3}$$

We perform a straightforward Gauss-Legendre quadrature, with a total of g quadrature points, where m points lie before the discontinuity at δ. We assume that the continuous function, possibly a polynomial, can be integrated exactly, or nearly exactly, with standard quadrature $\int_{-1}^{1} C(\zeta)d\zeta \approx \sum_{i=1}^{g} C(\zeta_i)w_i$, where w_i are the Gauss weights and ζ_i are the Gauss point locations. We note the property $\sum_{i=1}^{g} w_i = 2$. The remainder is the discontinuous part $[\![F(\delta)]\!] \int_{-1}^{1} H(\zeta - \delta)d\zeta = [\![F(\delta)]\!](1 - \delta)$. We may write $[\![F(\delta)]\!] \sum_{i=1}^{g} H(\zeta_i - \delta)w_i = [\![F(\delta)]\!] \sum_{i=1}^{m} 0 \times w_i + [\![F(\delta)]\!] \sum_{i=m+1}^{g} 1 \times w_i$. As a consequence, the maximum amount of variation in the computed integration is

$$VARIATION = |[\![F(\delta)]\!]| \left(\sum_{i=m}^{g} w_i - \sum_{i=m+1}^{g} w_i \right) \leq |[\![F(\delta)]\!]| \max_i w_i. \tag{6.4}$$

In order to bound the dependence of the largest quadrature weight, w_i in the interval $(-1, 1)$, with the Gauss-Legendre rule, we least squares curve-fit (Gauss rules of $1 \leq g \leq 10$) with the following, $\max_{i \leq g} w_i \approx 1.93 g^{-0.795}$ with $R^2 = 0.99$, where $R^2 = 1.0$ indicates a perfect regression value of the curve fit. Simple three dimensional estimates can be made by applying this procedure in all three directions on a reference element. For example, consider a 3-D step function discontinuity over a reference finite element $((-1, 1) \times (-1, 1) \times (-1, 1))$. Denoting the volumetrically normalized (by the reference element) bound by $B \stackrel{\text{def}}{=} \frac{(1.93 g^{-0.795})^3}{8}$, since $8 = 2 \times 2 \times 2$ = volume of the reference element, one has for (I) a $2 \times 2 \times 2$ Gauss rule $B = 0.1720$, (II) a $3 \times 3 \times 3$ Gauss rule $B = 0.0654$, (III) a $4 \times 4 \times 4$ Gauss rule $B = 0.0329$ and (IV) a $5 \times 5 \times 5$ Gauss rule $B = 0.0193$. Consequently, the amount

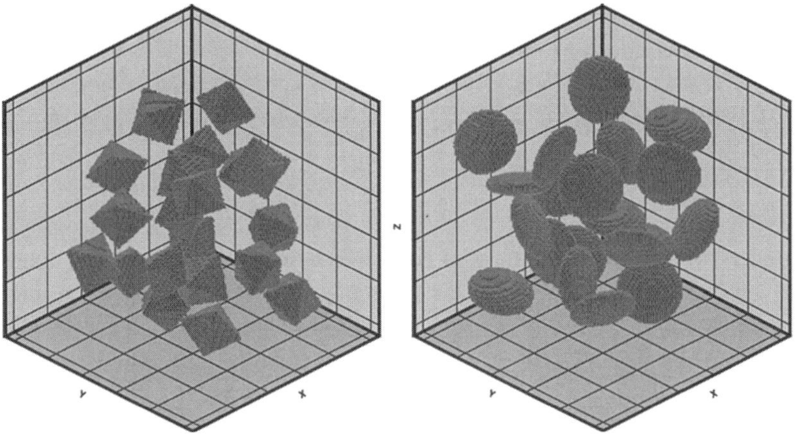

Fig. 6.4 A random microstructure consisting of 20 non-intersecting particles. **Left:** a diamond-type microstructure ($s_1 = s_2 = s_3 = 1$). **Right:** an oblate disk-type microstructure (aspect ratio of 3:1). Both microstructures contain particles which occupy approximately 7% of the volume

of variation in the integral is at most approximately 2%. Therefore, for efficient implementation, a 2/5 rule should be used, whereby a $2 \times 2 \times 2$ Gauss rule if there is no material discontinuity in the element, and a $5 \times 5 \times 5$ rule if there is a material discontinuity. We emphasize that this procedure is used simply to accurately integrate elemental quantities with discontinuities.

For a variety of numerical tests in Zohdi and Wriggers [233], the typical mesh density to deliver mesh insensitive results, for the quantities of interest in the upcoming simulations, was $9 \times 9 \times 9$ trilinear finite element hexahedra (approximately 2200–3000 degrees of freedom (DOF)) *per particle*. For example, disk-type and a diamond-type microstructures, as resolved by the meshing algorithm with a $24 \times 24 \times 24$ trilinear hexahedra mesh density, with a total of 46875 degrees of freedom (approximately $9 \times 9 \times 9$ hexahedra or 2344 degrees of freedom per element), are shown in Fig. 6.4.

6.3 Elementally Averaged Quantities

Using meshes of the type mentioned, elements with material discontinuities will arise, and thus r in (5.39) becomes $r = 1.5$ (see Szabo and Babúska [196]). Therefore, the error in the energy norm is bounded from above by $Ch^{\frac{1}{2}}$. However, if we ask only to compare the average of the solutions over each element $(\langle(\cdot)\rangle_\Omega \stackrel{\text{def}}{=} \frac{1}{|\Omega|}\int_\Omega (\cdot)\, d\Omega)$ we have

$$\int_\Omega \left(\langle\frac{du}{dx}\rangle_\Omega - \langle\frac{du^h}{dx}\rangle_\Omega\right)\left(\langle\sigma\rangle_\Omega - \langle\sigma^h\rangle_\Omega\right)dx$$
$$\leq \underbrace{\left(\int_\Omega (\langle\frac{du}{dx}\rangle_\Omega - \langle\frac{du^h}{dx}\rangle_\Omega)^2\,dx\right)^{\frac{1}{2}}}_{\leq C_1 h^a}\underbrace{\left(\int_\Omega \left(\langle\sigma\rangle_\Omega - \langle\sigma^h\rangle_\Omega\right)^2 dx\right)^{\frac{1}{2}}}_{\leq C_2 h^b} \leq C_3 h^{a+b}, \tag{6.5}$$

where $a, b \geq 1$. In short, the rates of convergence between the averaged quantities are much higher than between the unaveraged quantities. Consider a one-dimensional example. Suppose the true solution contained in an element domain containing material discontinuities is $u = Ax \Rightarrow \frac{du}{dx} = A$, if $x \leq x^\star$ and $u = Bx + (A-B)x^\star \Rightarrow \frac{du}{dx} = B$ if $x \geq x^\star$, as shown in Fig. 6.5. Now consider a field produced by a linear finite element approximation The field will be a purely linear function, which we denote $u^h = Dx$. The difference between the true and the approximate solution is

$$\sqrt{\int_\Omega \left(\underbrace{\frac{1}{h}\int_0^h \frac{du}{dx}\,dx}_{\langle\frac{du}{dx}\rangle_\Omega dx} - \underbrace{\frac{1}{h}\int_0^h \frac{du^h}{dx}\,dx}_{\langle\frac{du^h}{dx}\rangle_\Omega dx}\right)^2 dx} = h|(A\frac{x^*}{h} + B\frac{h-x^*}{h} - D)| \leq Ch. \tag{6.6}$$

For example, take the special case when $x^* = \frac{h}{2}$

$$\sqrt{\int_\Omega \left((\frac{1}{h}\int_0^h \frac{du}{dx}dx - \frac{1}{h}\int_0^h \frac{du^h}{dx}dx)^2\right)dx} = h|\frac{A+B}{2} - D|. \tag{6.7}$$

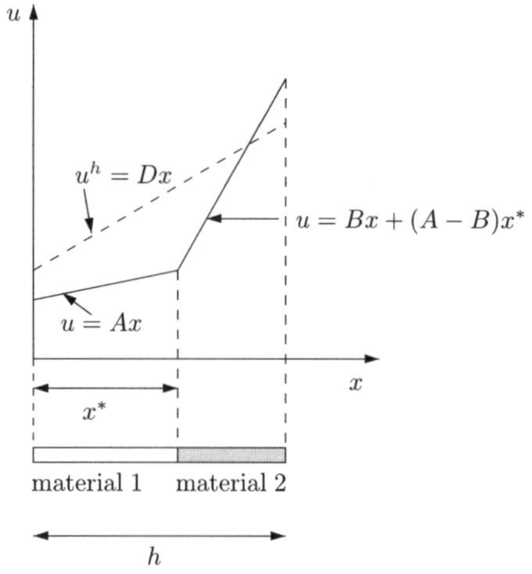

Fig. 6.5 An intuitive characterization of heterogeneous element averaged quantities

Therefore, we observe that the rate of convergence is at least linear. However, this can be increased further, for example via more sophisticated post processing "extraction" techniques (see Szabo and Babúska [196]). Essentially the error in the element averaged gradient of the primal fields scale at least linearly with element size.

6.4 Iterative Krylov Solvers/ Microstructural "Correctors"

For micro–macro mechanics problems, one important feature of iterative solvers, such as the CG method, is the fact that since they are based on successive updates of a starting guess solution vector. Consequently, they can be given a head start by a good solution guess provided by an inexpensive analytical approximation solution. In a micro-macro mechanical context, the Conjugate Gradient (CG) method can be interpreted as making fine scale (high frequency) corrections with each iteration. Importantly, one should use inexpensive approximate solutions as initial guesses for the iterative solver. This further enhances the effectiveness of the CG searches for this class of problems. A well known fact, with regard to iterative solvers, is that they are quite adept at capturing local high-frequency responses. Localized effects are resolved quickly during CG iterations. However, Krylov methods typically may require many iterations to capture "long-wave" modes. For a discussion see Briggs [18]. In other words, a starting vector that captures a-priori the "long wave" components of the solution is advantageous.

For example, if we assume that either of the two classical boundary conditions that satisfy Hill's conditions hold: (1) $\mathbf{u}|_{\partial\Omega} = \mathscr{E} \cdot \mathbf{x} \Rightarrow \langle \varepsilon \rangle_\Omega = \mathscr{E}$ and (2) $\mathbf{t}|_{\partial\Omega} = \mathscr{L} \cdot \mathbf{n} \Rightarrow \langle \sigma \rangle_\Omega = \mathscr{L}$. The classical assumptions of Voigt [212] (uniform strain) and Reuss [173] (uniform stress) produce the following:

- $\mathbf{u}|_{\partial\Omega} = \mathscr{E} \cdot \mathbf{x}$, which implies

  (1) VOIGT: $\sigma = \mathbb{E} : \mathscr{E}, \varepsilon = \mathscr{E}$
  (2) REUSS: $\sigma = \langle \mathbb{E}^{-1} \rangle_\Omega : \mathscr{E}, \varepsilon = \mathbb{E}^{-1} : \langle \mathbb{E}^{-1} \rangle_\Omega^{-1} : \mathscr{E}$

- $\mathbf{t}|_{\partial\Omega} = \mathscr{L} \cdot \mathbf{n}$, which implies

  (1) VOIGT: $\varepsilon = \langle \mathbb{E} \rangle^{-1} : \mathscr{L}, \sigma = \mathbb{E} : \langle \mathbb{E}^{-1} \rangle_\Omega : \mathscr{L}$
  (2) REUSS: $\sigma = \mathscr{L}, \varepsilon = \mathbb{E}^{-1} : \mathscr{L}$.

The Reuss solution, while statically admissible, is highly oscillatory in the displacement field variable and kinematically inadmissible. The Voigt assumption, which is statically inadmissible and kinematically admissible, is not oscillatory in the displacement field variable. Therefore, from an iterative solver point of view, the Reuss field assumption is not of much value as a starting vector for a displacement based method (FEM). Furthermore, such a guess cannot be projected onto the initial nodal values because it is double valued at material discontinuities (not kinematically admissible). However, it is important to note that if a complementary based numerical method were to be used, where the primary variable is the stress, the Reuss guess would be advantageous due to the smooth nature of

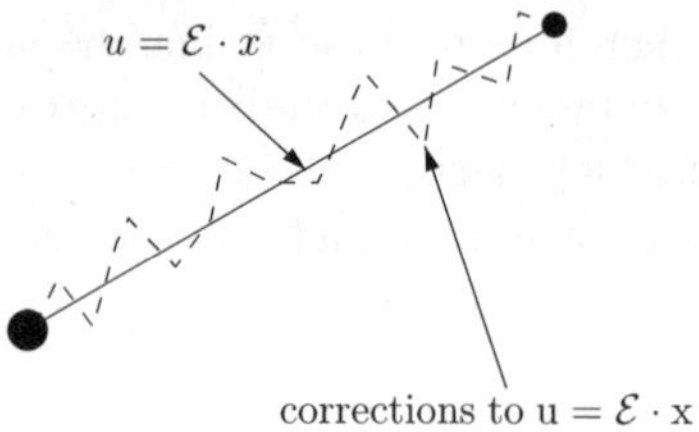

Fig. 6.6 Krylov "corrections" to a smooth "Voigt" starting vector

the assumed stress field. The Voigt (constant strain) assumption produces a low frequency initial guess. Therefore, from an iterative solver point of view, it is of value when using displacement based weak form such as the standard FEM. Furthermore, such a guess can easily be projected onto initial nodal values, because it is single valued. The central point is that each Conjugate Gradient iteration can be viewed as a microstructural "corrector" to an initially statically inadmissible initial guess (Fig. 6.6), and that $\mathbf{u}|_{\Omega} = \mathscr{E} \cdot \mathbf{x}$ should be used for the displacement tests.

6.5 Overall Testing Process: Numerical Examples

As considered before, a typical example of a composite material combination is that of an aluminum matrix (77.9, 24.9 GPa) embedded with (stiffening) boron particles (230, 172 GPa). We chose Aluminum/Boron as a material combination which exhibits significant enough mismatch in the mechanical properties to be representative of a wide range of cases. All tests were run on a single workstation. Such standard hardware is available in most academic and industrial work places, therefore such simulations are easily reproducible elsewhere for other parameter selections.

6.5.1 Successive Sample Enlargement

In a first set of tests, the number of particles contained in a sample were increased holding the volume fraction constant. During the tests, we repeatedly refined the mesh to obtain mesh-invariant macroscopic responses. A sample/particle size ratio was used as a microstructural control parameter. This was done by defining a subvolume size $V \stackrel{\text{def}}{=} \frac{L \times L \times L}{N}$, where N is the number of particles in the entire sample and where L is the length of the (cubical $L \times L \times L$) sample. A generalized diameter (and radius) was defined, $d = 2r$, which was the diameter of the smallest sphere that can enclose a single particle, of possibly non-spherical shape (if desired). The ratio between the generalized radius and the subvolume was defined by $\xi \stackrel{\text{def}}{=} \frac{r}{V^{\frac{1}{3}}}$.

Table 6.1 Results of successive sample "enlargements". Five tests, each with a different random distribution, were performed at each sample/particulate size ratio level to obtain somewhat representative data

Part	$\frac{d}{L}$	DOF	W (GPa)	κ^* (GPa)	μ^* (GPa)
2	0.595	5184	0.001444	98.2	46.7
4	0.472	10125	0.001408	97.3	44.3
8	0.375	20577	0.001386	96.5	43.2
16	0.298	41720	0.001375	96.2	42.5
32	0.236	81000	0.001365	95.9	41.6
64	0.188	151959	0.001358	95.7	41.4

For a variety of numerical tests, discussed momentarily, the typical mesh density to deliver invariant volumetrically averaged responses was $9 \times 9 \times 9$ trilinear finite element hexahedra (approximately 2200–3000 degrees of freedom) *per particle*. We used $\xi = 0.375$, which resulted in a (fixed) volume fraction of approximately 22%. The following particle per sample sequence was used to study the dependence of the effective responses on the sample size: 2 (5184 DOF), 4 (10125 DOF), 8 (20577 DOF), 16 (41720 DOF), 32 (81000 DOF) and 64 (151959 DOF) particles. In order to obtain more reliable response data for each particle number set, the tests were performed five times (each time with a different particulate distribution) and the responses averaged. Throughout the tests, we considered a single combined boundary loading satisfying Hill's condition, $\mathbf{u}|_{\partial\Omega} = \mathscr{E} \cdot \mathbf{x}$, $\mathscr{E}_{ij} = 0.001, i, j = 1,2,3$. We tracked the strain energy, as well as κ^* and μ^*, as defined in (4.4). Table 6.1 and Fig. 6.7 depict the dependency of the responses with growth in particle number.

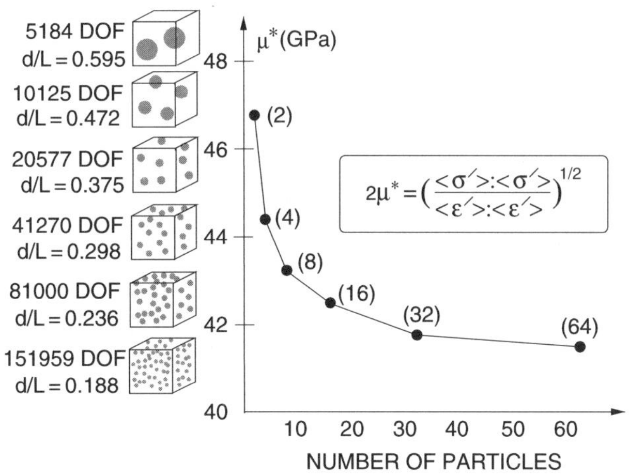

Fig. 6.7 The values of the effective shear responses for samples containing increasingly larger numbers of particles. One hundred tests were performed per particle/sample combination and the results averaged

Justified by the somewhat ad-hoc fact that for three successive enlargements of the number of particles, i.e. 16, 32 and 64 particle samples, the responses differed from one another, on average, by less than 1%, we selected the 20-particle microstructures for further tests. We remark that we applied a "2/5" rule, i.e. a $2 \times 2 \times 2$ Gauss rule if there is no discontinuity in the element, and a $5 \times 5 \times 5$ rule if there is a discontinuity, which is consistent with the earlier derivation in the work. The microstructure, as seen by this mesh density, is shown in Fig. 6.8.

6.5.2 Multiple Sample Tests

For further tests, we simulated 100 different samples, each time with a different random distribution of 20 nonintersecting particles occupying 22% ($\xi = 0.375$). Consistent with the previous test's mesh densities per particle, we used a $24 \times 24 \times 24$ mesh ($9 \times 9 \times 9$ trilinear hexahedra or 2344 DOF per particle, 46875 DOF per test sample). The plots of the behavior of the various quantities of interest are shown in Figs. 6.11, 6.12, 6.13, 6.14, and 6.15. The averages, standard deviations and maximum/minimum of these quantities are tabulated in Table 6.2. For the 100 sample tests, with 20 particles per sample, the results for the effective responses were

$$\begin{aligned} 91.37 = \langle \kappa^{-1} \rangle_{\Omega}^{-1} \leq \tilde{\kappa}^* = 96.17 \leq \langle \kappa \rangle_{\Omega} = 111.79, \\ 30.76 = \langle \mu^{-1} \rangle_{\Omega}^{-1} \leq \tilde{\mu}^* = 42.35 \leq \langle \mu \rangle_{\Omega} = 57.68, \end{aligned} \quad (6.8)$$

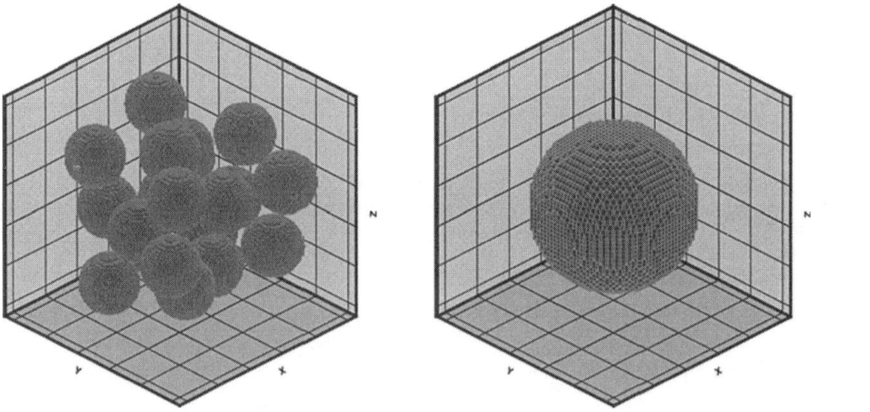

Fig. 6.8 Left: a random microstructure consisting of 20 non-intersecting boron spheres, occupying approximately 22% of the volume in an aluminum matrix, as seen by the algorithm with a $24 \times 24 \times 24$ trilinear hexahedra mesh density for a total of 46875 degrees of freedom (approximately $9 \times 9 \times 9$ hexahedra or 2344 degrees of freedom per element). A "2/5" rule, i.e. a $2 \times 2 \times 2$ Gauss rule if there is no discontinuity in the element, and a $5 \times 5 \times 5$ rule if there is a discontinuity, was used. **Right:** a zoom on one particle

Table 6.2 Results of 100 material tests for randomly distributed particulate microstructures (20 spheres). We note that for a zero starting guess for the iterative solver, the average number of CG-iterations was 55, as opposed to 49, therefore using the initial Voigt ($\mathbf{u} = \mathscr{E} \cdot \mathbf{x}$) guess saved approximately 12% of the computational solution effort

Quan.	Average	Stan. Dev.	Max–Min
W(GPa)	0.001373	7.4×10^{-6}	3.6×10^{-5}
κ^*(GPa)	96.171	0.2025	0.950
μ^* (GPa)	42.350	0.4798	2.250
$CG-IT$	49.079	1.5790	9

where $\tilde{\kappa}^*$ and $\tilde{\mu}^*$ are the averaged effective responses from the 100 tests, and where the lower and upper bounds are, respectively, the classical Reuss [173] and Voigt [212] bounds. We also compared the computed results to the well-known Hashin-Shtrikman bounds ([74, 75]) which are, strictly speaking, *only applicable* to asymptotic cases of an infinite (sample length)/(particulate length) ratio and purely isotropic macroscopic responses. The "bounds" were as follows:

$$94.32 = \kappa^{(-)} \leq \tilde{\kappa}^* = 96.17 \leq \kappa^{(+)} = 102.38,$$

$$35.43 = \mu^{(-)} \leq \tilde{\mu}^* = 42.35 \leq \mu^{(+)} = 45.64,$$

$$\kappa^{(-)} \stackrel{\text{def}}{=} \underbrace{\kappa_1 + \frac{v_2}{\frac{1}{\kappa_2-\kappa_1} + \frac{3(1-v_2)}{3\kappa_1+4\mu_1}}}_{\text{bulk modulus H/S lower bound}}, \quad \kappa^{(+)} \stackrel{\text{def}}{=} \underbrace{\kappa_2 + \frac{1-v_2}{\frac{1}{\kappa_1-\kappa_2} + \frac{3v_2}{3\kappa_2+4\mu_2}}}_{\text{bulk modulus H/S upper bound}},$$

$$\mu^{(-)} \stackrel{\text{def}}{=} \underbrace{\mu_1 + \frac{v_2}{\frac{1}{\mu_2-\mu_1} + \frac{6(1-v_2)(\kappa_1+2\mu_1)}{5\mu_1(3\kappa_1+4\mu_1)}}}_{\text{shear modulus H/S lower bound}},$$

$$\mu^{(+)} \stackrel{\text{def}}{=} \underbrace{\mu_2 + \frac{(1-v_2)}{\frac{1}{\mu_1-\mu_2} + \frac{6v_2(\kappa_2+2\mu_2)}{5\mu_2(3\kappa_2+4\mu_2)}}}_{\text{shear modulus H/S upper bound}}, \tag{6.9}$$

where κ_1, μ_1 and κ_2, μ_2 are the bulk and shear moduli for the matrix and particle phases. Despite the fact that the bounds are technically inapplicable for finite sized samples, the computed results did fall within them. The time to preprocess, solve and postprocess each 20 particle finite element test took no more than one minute on a single RISC 6000 workstation. Therefore, as before, 100 of such tests lasted approximately 1.5 hours. A primary question is now: "What other information can we extract from the average of the responses of many samples?" This is discussed next.

6.6 A Minimum Principle Interpretation

Consider the following process for a large sample of material with $\mathbf{u}|_{\partial\Omega} = \mathscr{E} \cdot \mathbf{x}$:

1. **Step 1:** Take the sample, and cut it into N pieces, $\Omega = \cup_{K=1}^{N} \Omega_K$. The pieces do not have to be the same size or shape, although for illustration purposes it is convenient to take a uniform (regular) partitioning (Fig. 6.9)
2. **Step 2:** Test each piece (solve the subdomain BVP) with the loading: $\mathbf{u}|_{\partial\Omega_K} = \mathscr{E} \cdot \mathbf{x}$. The function $\tilde{\mathbf{u}}_K$ is the solution to the BVP posed over subsample Ω_K
3. **Step 3:** One is guaranteed the following properties:

$$\langle\tilde{\sigma}\rangle_{\Omega_K} \stackrel{\text{def}}{=} \tilde{\mathbb{E}}_K^* : \langle\tilde{\varepsilon}\rangle_{\Omega_K}, \qquad \tilde{\mathbb{E}}^* \stackrel{\text{def}}{=} \Sigma_{K=1}^{N} \tilde{\mathbb{E}}_K^* \frac{|\Omega_K|}{|\Omega|},$$

$$||\mathbf{u} - \tilde{\mathbf{u}}||^2_{E(\Omega)} = \mathscr{E} : (\tilde{\mathbb{E}}^* - \mathbb{E}^*) : \mathscr{E}|\Omega| \leq \mathscr{E} : (\tilde{\mathbb{E}}^* - \langle\mathbb{E}^{-1}\rangle_{\Omega}^{-1}) : \mathscr{E}|\Omega|,$$

$$\langle\mathbb{E}^{-1}\rangle_{\Omega}^{-1} \leq \mathbb{E}^* \leq \tilde{\mathbb{E}}^* \leq \langle\mathbb{E}\rangle_{\Omega},$$

$$\tilde{\mathbf{u}} \stackrel{\text{def}}{=} \tilde{\mathbf{u}}_1|_{\overline{\Omega}_1} + \tilde{\mathbf{u}}_2|_{\overline{\Omega}_2} ... \tilde{\mathbf{u}}_N|_{\overline{\Omega}_N}. \tag{6.10}$$

The same process can be done for traction test loading cases: $\mathbf{t}|_{\partial\Omega_K} = \mathscr{L} \cdot \mathbf{n}$. The effective material ordering, line three in Box 6.10, has been derived by Huet [92]. The second line of Box 6.10, and generalizations to nonuniform loading, were developed

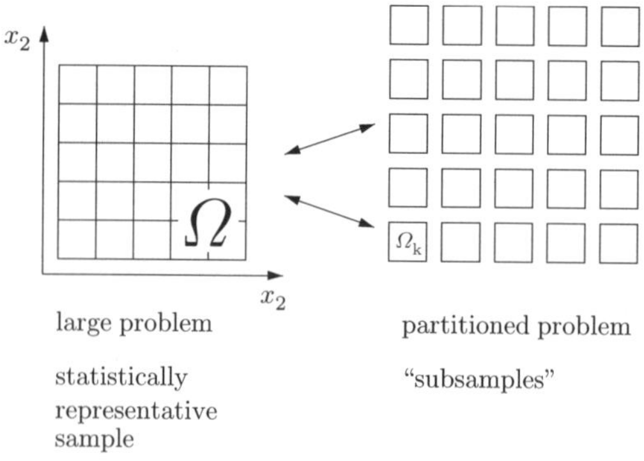

Fig. 6.9 The idea of partitioning a sample into smaller samples or equivalently combining smaller samples into a larger sample

in Zohdi and Wriggers [233]. The proofs, which are constructive, and illustrative of the main ideas, are provided next. They result from a direct manipulation of classical energy minimization principles.

6.6.1 A Proof Based on Partitioning Results

Consider a large sample whose domain consists of the union of many smaller domains, $\overline{\cup_{K=1}^{N}\Omega_K} = \overline{\Omega}$, as depicted in Fig. 6.9. The corresponding solution $\mathbf{u}$, denoted the "globally exact" solution (for the whole sample), is characterized by the weak boundary value problem stated in Box 6.1 Specific choices for the prescribed displacements, tractions and body forces will be given momentarily. Now consider an individual subdomain Ω_K, $1 \leq K \leq N$ (N is the total number of subdomains) with the following weak form

Find $\tilde{\mathbf{u}}_K \in \mathbf{H}^1(\Omega_K), \tilde{\mathbf{u}}_K|_{\partial\Omega_K\cap(\Omega\cup\Gamma_u)} = \mathbf{U} \in \mathbf{H}^1(\Omega)$, such that

$$\int_{\Omega_K} \nabla \mathbf{v}_K : \tilde{\boldsymbol{\sigma}}_K \, d\Omega = \int_{\Omega_K} \mathbf{f}\cdot\mathbf{v}_K \, d\Omega + \int_{\partial\Omega_K\cap\Gamma_t} \mathbf{t}\cdot\mathbf{v}_K \, dA$$

$$\forall \mathbf{v}_K \in \mathbf{H}^1(\Omega_K), \mathbf{v}_K|_{\partial\Omega_K\cap(\Omega\cup\Gamma_u)} = \mathbf{0}. \tag{6.11}$$

A specific choice for $\mathbf{U}$ will also be given momentarily. The individual subdomain solutions form an approximate solution to the globally exact problem, $\mathbf{u}$. This approximate solution is constructed by a direct assembly process

$$\tilde{\mathbf{u}} \stackrel{\text{def}}{=} \mathbf{U} + (\tilde{\mathbf{u}}_1 - \mathbf{U})|_{\Omega_1} + (\tilde{\mathbf{u}}_2 - \mathbf{U})|_{\Omega_2} + \cdots + (\tilde{\mathbf{u}}_N - \mathbf{U})|_{\Omega_N}. \tag{6.12}$$

The approximate displacement field is continuous, however, the approximate traction field is (usually) discontinuous. Clearly, if there are no jumps in the tractions, the solution is exact.

6.6.2 Relation to the Material Tests

We have for any kinematically admissible function $\mathbf{w}$, using the (bilinear operator) notation,

$$0 \leq ||\mathbf{u}-\mathbf{w}||^2_{E(\Omega)} = \mathscr{B}(\mathbf{u}-\mathbf{w},\mathbf{u}-\mathbf{w}) = 2\mathscr{J}(\mathbf{w}) - 2\mathscr{J}(\mathbf{u}) \Rightarrow \mathscr{J}(\mathbf{u}) \leq \mathscr{J}(\mathbf{w}), \tag{6.13}$$

where $\mathscr{J}(\mathbf{w}) \stackrel{\text{def}}{=} \frac{1}{2}\mathscr{B}(\mathbf{w},\mathbf{w}) - \mathscr{F}(\mathbf{w})$. In other words, the true solution has a minimum potential (PMPE). By applying the PMPE, for the test loading $\mathbf{u}|_{\partial\Omega} = \mathscr{E}\cdot\mathbf{x} = \mathbf{d}$, $\mathbf{f} = \mathbf{0}$, with the specific (kinematically admissible) choice $\mathbf{w} = \mathbf{U}$, we obtain $||\mathbf{u}-\mathbf{U}||^2_{E(\Omega)} = 2(\mathscr{J}(\mathbf{U}) - \mathscr{J}(\mathbf{u})) \Rightarrow \mathscr{J}(\mathbf{u}) = \mathscr{J}(\mathbf{U}) - \frac{1}{2}||\mathbf{u}-\mathbf{U}||^2_{E(\Omega)}$. *The critical observation is that when choosing* $\mathbf{w} = \mathbf{U} \stackrel{\text{def}}{=} \mathscr{E}\cdot\mathbf{x}$ *in* Box 6.11, *then* $\tilde{\mathbf{u}}$, *as defined in* (6.12), *is also kinematically admissible.* Therefore, we have by direct expansion $||\mathbf{u}-\tilde{\mathbf{u}}||^2_{E(\Omega)} = 2(\mathscr{J}(\tilde{\mathbf{u}}) - \mathscr{J}(\mathbf{u})) = 2(\mathscr{J}(\tilde{\mathbf{u}}) - \mathscr{J}(\mathbf{U})) + ||\mathbf{u}-\mathbf{U}||^2_{E(\Omega)}$. Since $\tilde{\mathbf{u}}_K$ is a solution to a subdomain boundary value problem posed over Ω_K, it minimizes the corresponding subdomain potential energy function $\mathscr{J}_K(\cdot)$. Therefore, $\mathscr{J}(\mathbf{U}) = \Sigma_{K=1}^{S}\mathscr{J}_K(\mathbf{U}) \geq \Sigma_{K=1}^{S}\mathscr{J}_K(\tilde{\mathbf{u}}_K) = \mathscr{J}(\tilde{\mathbf{u}})$. Consequently,

$$||\mathbf{u}-\tilde{\mathbf{u}}||^2_{E(\Omega)} = \underbrace{2(\mathscr{J}(\tilde{\mathbf{u}}) - \mathscr{J}(\mathbf{U}))}_{\text{negative}} + ||\mathbf{u}-\mathbf{U}||^2_{E(\Omega)}. \tag{6.14}$$

By direct expansion we have $||\mathbf{u}-\mathbf{U}||^2_{E(\Omega)} = \mathscr{E} : (\langle\mathbb{E}\rangle - \mathbb{E}^*) : \mathscr{E}|\Omega| \leq \mathscr{E} : (\langle\mathbb{E}\rangle - \langle\mathbb{E}^{-1}\rangle^{-1}) : \mathscr{E}|\Omega|$. We have by definition, $2\mathscr{J}(\mathbf{u}) = \mathscr{E} : \mathbb{E}^* : \mathscr{E}|\Omega|$, $2\mathscr{J}(\mathbf{U}) = \mathscr{E} : \langle\mathbb{E}\rangle : \mathscr{E}|\Omega|$, and $2\mathscr{J}_K(\tilde{\mathbf{u}}) = \mathscr{E} : \tilde{\mathbb{E}}^*_K : \mathscr{E}|\Omega_K|$, which implies $2\mathscr{J}(\tilde{\mathbf{u}}) = \mathscr{E} : \tilde{\mathbb{E}}^* : \mathscr{E}|\Omega|$. By direct substitution this completes the first of the assertions in Box 6.10. We can directly interpret the testing of many smaller samples as simply the partitioning of a very large one which we cannot easily test (Fig. 6.9). Therefore, for displacement tests, the averaged effective responses generated will always bound the response of the very large sample from above. Therefore, the average of the 100 sample tests provide us with tighter upper bounds on the response of a very large sample. Therefore, by the previously proven results in (6.10), the Reuss-Voigt bounds in (6.8) are tightened by the following factors

$$\frac{96.17-91.37}{111.79-91.37} = 0.2747 \quad \text{and} \quad \frac{42.35-30.76}{57.68-30.76} = 0.4305. \tag{6.15}$$

Also, we may interpret the bounds as providing an error estimate

$$SUBSAMPLING\ ERROR = \frac{\mathscr{E} : (\tilde{\mathbb{E}}^* - \langle\mathbb{E}^{-1}\rangle^{-1}_{\Omega}) : \mathscr{E}|\Omega|}{\mathscr{E} : \langle\mathbb{E}\rangle_{\Omega} : \mathscr{E}|\Omega|} = 0.1073, \tag{6.16}$$

on the error in the internal fields induced by the partitioning or subsampling.

6.7 Dependency on Volume Fraction

We repeated the 100 sample tests procedure for samples containing significantly higher volume fraction, approximately 32%. Unlike fiber-reinforced composites, which can contain well over 50% volume fraction of fibers, particulate composites typically contain no more than approximately 25% volume fraction of particulates.

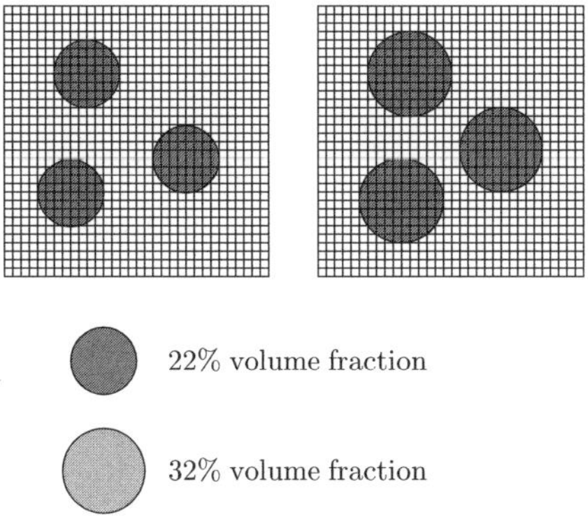

Fig. 6.10 A comparison of relative sizes of particles occupying approximately 22% ($\xi = 0.375$) and 32% ($\xi = 0.425$) volume fraction

Particulate volume fractions over 15% are already high in three dimensions. For the relevant examples, 22% particulate volume fraction in three dimensions corresponds roughly to 44% in two dimensions, while 32% in three dimension corresponds roughly to 57% in two dimensions. Cross-sections of these volume fractions are depicted in Fig. 6.10. The results for the 32% case are tabulated in Table 6.3. The averages for the effective properties of the samples were

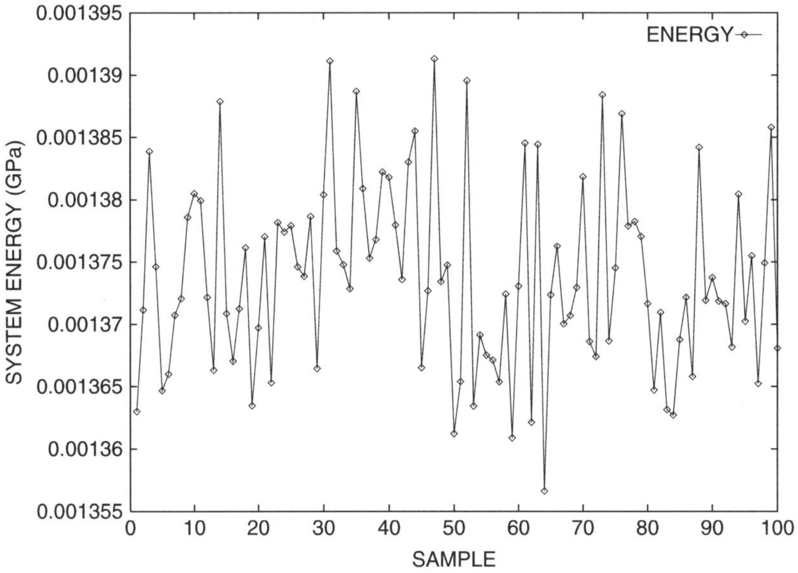

Fig. 6.11 100 SAMPLES: The energy responses, $W = \frac{1}{|\Omega|}\int_\Omega \nabla \mathbf{u} : \mathbb{E} : \nabla \mathbf{u}\, d\Omega$(GPa), of a block with 20 randomly distributed Boron spheres embedded in an Aluminum matrix. Each point represents the results of one test

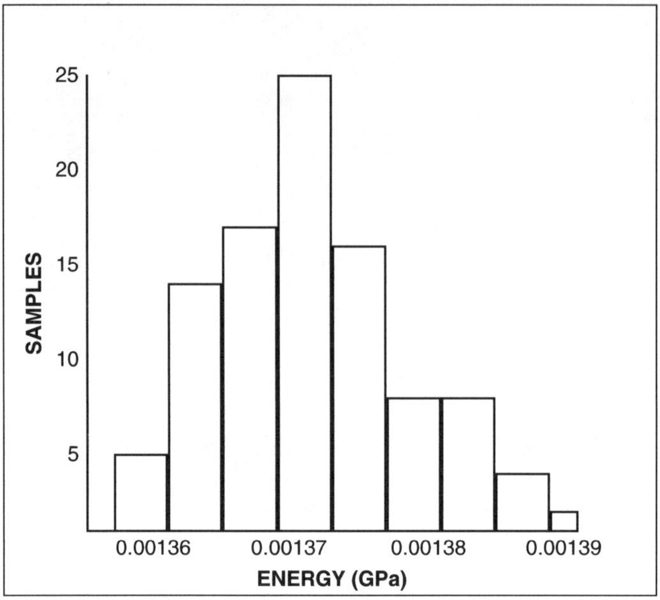

Fig. 6.12 100 SAMPLES: A histogram for the variations in energy, $W = \frac{1}{|\Omega|}\int_{\Omega} \nabla \mathbf{u} : \mathbb{E} : \nabla \mathbf{u}\, d\Omega$(GPa)

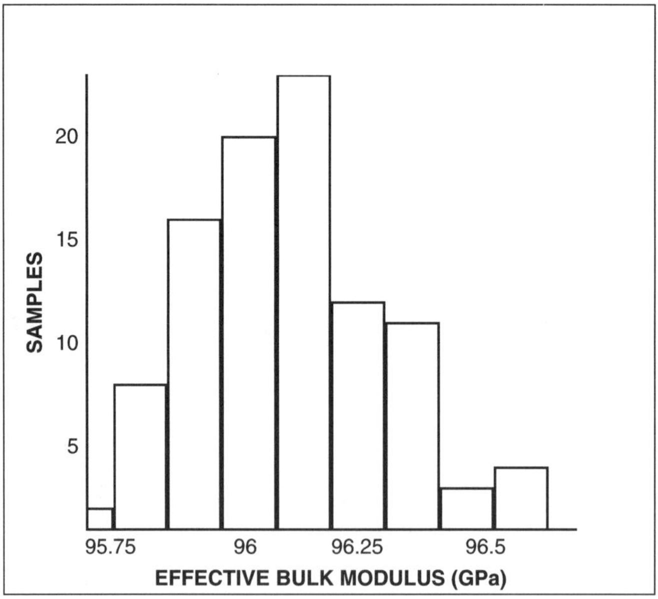

Fig. 6.13 100 SAMPLES: A histogram for the variations in the effective bulk responses, κ^* of a block with 20 randomly distributed Boron spheres embedded in an Aluminum matrix

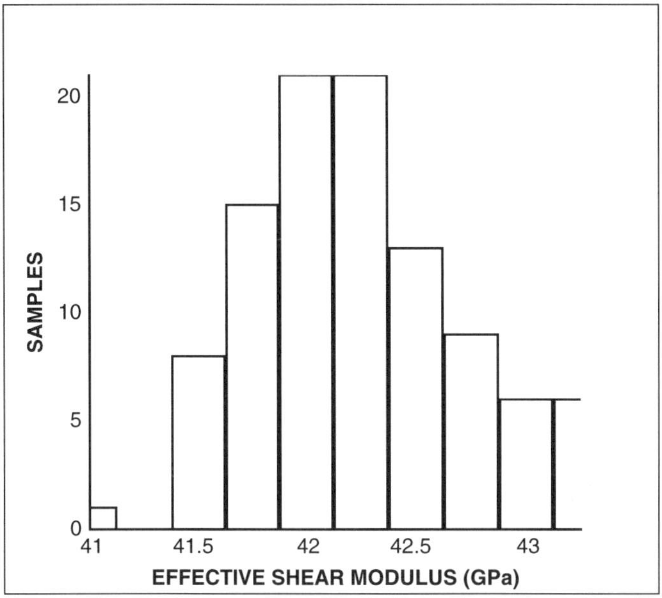

Fig. 6.14 100 SAMPLES: A histogram for the variations in the effective shear responses, μ^* of a block with 20 randomly distributed Boron spheres embedded in an Aluminum matrix

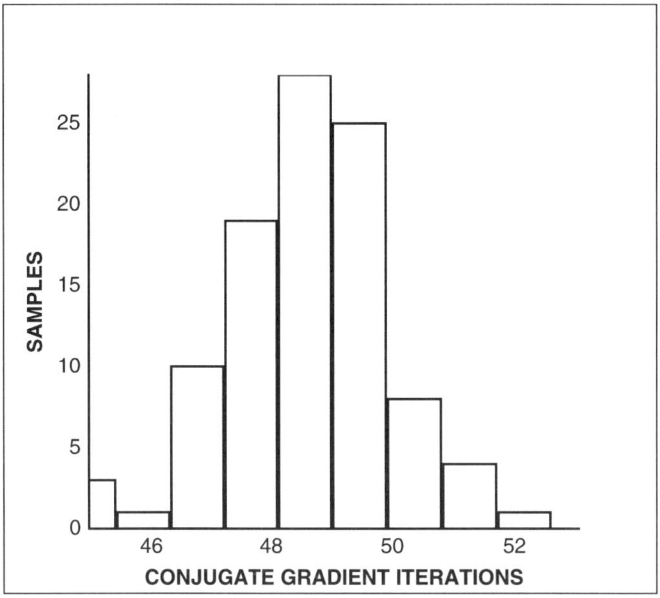

Fig. 6.15 100 SAMPLES: A histogram for the variations in Conjugate Gradient iterations needed for a solution of a block with 20 randomly distributed Boron spheres embedded in an Aluminum matrix

Table 6.3 Results of 100 material tests for randomly distributed particulate microstructures (20 spheres, ξ =0.425, or approximately 32% volume fraction)

Quan.	Aver.	Stan. Dev.	Max–Min
W(GPa)	0.001593	9.11×10^{-6}	5×10^{-5}
κ^*(GPa)	106.831	0.2505	1.350
μ^* (GPa)	52.635	0.5876	3.095
$CG-IT$	48.950	1.716	8

$$\begin{gathered} 99.04 = \langle\kappa^{-1}\rangle_\Omega^{-1} \le \tilde{\kappa}^* = 106.83 \le \langle\kappa\rangle_\Omega = 126.99, \\ 34.39 = \langle\mu^{-1}\rangle_\Omega^{-1} \le \tilde{\mu}^* = 52.63 \le \langle\mu\rangle_\Omega = 72.38, \end{gathered} \tag{6.17}$$

where $\tilde{\kappa}^*$ and $\tilde{\mu}^*$ are the averaged quantities from the 100 tests, and where the lower and upper bounds are, respectively, the classical Reuss [173] and Voigt [212] bounds. The Reuss-Voigt bounds in Box 6.17 are tightened by the following factors

$$\frac{106.831 - 99.043}{126.999 - 99.043} = 0.2785 \quad \text{and} \quad \frac{52.635 - 34.395}{72.385 - 34.395} = 0.4801. \tag{6.18}$$

The error estimate in this case is

$$\begin{aligned} SUBSAMPLING\ ERROR &\approx \frac{\mathscr{E} : (\tilde{\mathbb{E}}^* - \langle\mathbb{E}^{-1}\rangle_\Omega^{-1}) : \mathscr{E}|\Omega|}{\mathscr{E} : \langle\mathbb{E}\rangle_\Omega : \mathscr{E}|\Omega|} \\ &= 0.1436, \end{aligned} \tag{6.19}$$

which is approximately 3.5% higher error estimate than the case of 22% particulate volume fraction. Intuitively, one would expect such a result, since the particles interact more at such a high volume fraction (Fig. 6.10). In other words, they "feel" the presence of the other particles more at higher volume fractions than at lower volume fractions. As before, we also compared the computed results to the Hashin-Shtrikman bounds [74, 75] which are, strictly speaking, *only applicable* to asymptotic cases of an infinite (sample length)/(particulate length) ratio, and purely isotropic macroscopic responses

$$\begin{gathered} 103.378 = \kappa^{(-)} \le \tilde{\kappa}^* = 106.831 \le \kappa^{(+)} = 114.671, \\ 41.624 = \mu^{(-)} \le \tilde{\mu}^* = 52.635 \le \mu^{(-)} = 56.437, \end{gathered} \tag{6.20}$$

6.8 Increasing the Number of Samples

Suppose we increase the number of samples tested. For illustration purposes, let us test 512 samples of material. The number 512 is not accidental, since it is a common number of independent processors in modern parallel processing machines. Table 6.4 illustrates that the averaged results are virtually identical to the 100 sample tests for all the quantities. *Testing more and more samples will not help obtain better average results.* However, despite practically the same average values, one can observe from the Figs. 6.16 and 6.17 that the 512 sample tests have a more Gaussian distribution, relative to the 100 sample tests, for the responses. However, for even more accurate average responses, we must test larger samples of material. This is explored further in the next section.

Table 6.4 Results of 512 material subdomains, each containing 20 randomly distributed spheres for a total of 10240

Quan.	Average	Stan. Dev.	Max−Min
W (GPa)	0.001373	7.2×10^{-6}	7.5×10^{-5}
κ^*(GPa)	96.169	0.1967	1.203
μ^* (GPa)	42.353	0.4647	3.207
$CG-IT$	48.8984	1.7737	11

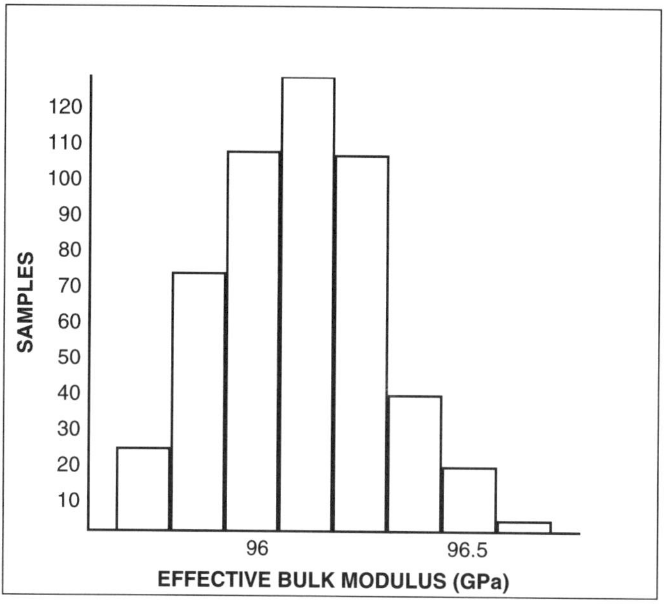

Fig. 6.16 512 SAMPLES: A histogram for the variations in the effective bulk responses, κ^* of a block with 20 randomly distributed boron spheres embedded in an aluminum matrix

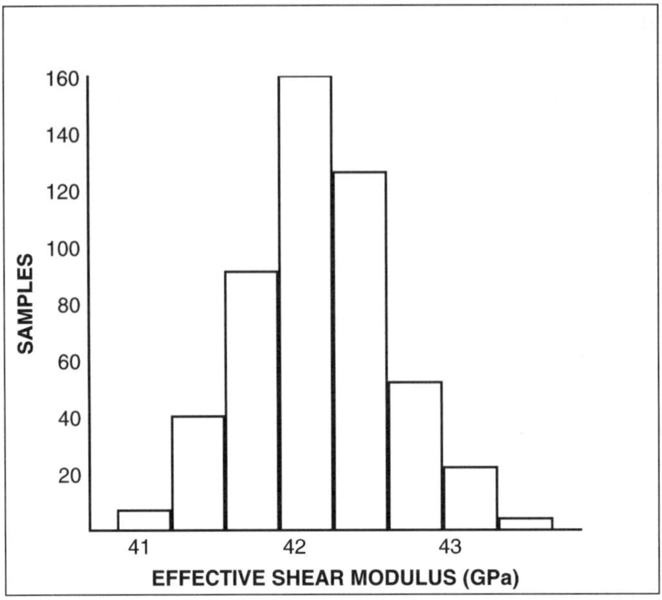

Fig. 6.17 512 SAMPLES: A histogram for the variations in the effective shear responses, μ^* of a block with 20 randomly distributed boron spheres embedded in an aluminum matrix

6.9 Increasing Sample Size

Beyond a certain threshold, it is simply impossible to obtain any more information by testing samples of a certain size. As we have shown the reason for the inability to obtain further information is the fact that the testing conditions are uniform on the "subsamples". This idealization is valid only for an infinitely large sample.

Table 6.5 Results of material tests for randomly distributed particulate microstructures for 100 ($\xi = 0.375$, approximately 22%) samples of 40 and 60 particles per sample

Part	$\frac{d}{L}$	DOF	Quan.	Aver.	Stan. Dev.	Max–Min
40	0.2193	98304	W(GPa)	0.0013617	5.0998×10^{-6}	2.35×10^{-5}
40	0.2193	98304	κ^*(GPa)	95.7900	0.1413	0.6600
40	0.2193	98304	μ^* (GPa)	41.6407	0.3245	1.590
40	0.2193	98304	$CG-IT$	60.3200	1.6302	7
60	0.1916	139968	W(GPa)	0.0013586	4.3963×10^{-6}	2.25×10^{-5}
60	0.1916	139968	κ^*(GPa)	95.6820	0.1197	0.6214
60	0.1916	139968	μ^* (GPa)	41.4621	0.2801	1.503
60	0.1916	139968	$CG-IT$	66.500	1.9974	10

However, suppose we increase the number of particles per subsample even further, from 20 to say 40 then 60, each time performing the 100 tests procedure. With this information one could then possibly extrapolate to a (giant) sample limit. The results for the 40 and 60 particle cases are shown in Table 6.5 for 22% boron volume fraction. Using these results, along with the 20 particle per sample tests, we have the following curve fits

$$\begin{aligned} W =& 0.0013205 + 0.0001895\frac{d}{L}, \quad R^2 = 0.9997, \\ \kappa^* =& 94.527 + 5.909\frac{d}{L}, \quad R^2 = 0.986, \\ \mu^* =& 39.345 + 10.775\frac{d}{L}, \quad R^2 = 0.986, \end{aligned} \tag{6.21}$$

where L is the sample size, d is the diameter of the particles. Thus as $\frac{d}{L} \to 0$, we obtain estimates of $W = 0.0013205$ GPa, $\kappa^* = 94.527$ GPa and $\mu^* = 39.345$ GPa as the asymptotic energy, effective bulk modulus, and effective shear modulus, respectively. Indeed, judging from the degree of accuracy of the curve-fit ($R^2 = 1.0$ is perfect), it appears to be nearly perfect for W. For κ^* and μ^*, the slightly less accurate reliability (regression values of $R^2 = 0.986$) is attributed to the fact that absolute perfect isotropy is impossible to achieve with finite sized samples. In other words the extrapolations using various samples exhibit slight isotropic inconsistencies. However, the energy W, has no built-in assumptions whatsoever, thus leading to the nearly perfect curve-fit. The monotonicity of the testing curves is to be expected, and is explained further in the next chapter, using minimum energy principles.

Chapter 7
Various Extensions and Further Interpretations of Partitioning

There are a variety of ways to interpret and extend the partitioning concept introduced in the previous chapter. This will be discussed presently.

7.1 Partitioning and Traction Test Cases

Suppose we repeat the partitioning process for an applied (internal) traction set of tests. As will be shown, traction tests will bound the response of the very large sample from below. Employing the *Principle of Complementary Potential Energy*, we obtain for the test loading $\mathbf{t}|_{\partial\Omega_K} = \mathscr{L}\cdot\mathbf{n}$, with $\gamma = \mathscr{L}$ (a statically admissible trial field),

$$\begin{aligned} ||\sigma-\tilde{\sigma}||^2_{E^{-1}(\Omega)} &= 2(\mathscr{K}(\tilde{\sigma})-\mathscr{K}(\sigma)) \\ &= \underbrace{2(\mathscr{K}(\tilde{\sigma})-\mathscr{K}(\gamma))}_{\text{negative}} + ||\sigma-\gamma||^2_{E^{-1}(\Omega)}, \end{aligned} \tag{7.1}$$

where $\mathscr{K}(\sigma) \stackrel{\text{def}}{=} \frac{1}{2}\int_\Omega \sigma : \mathbb{E}^{-1} : \sigma\, d\Omega - \int_{\Gamma_u} \sigma\cdot\mathbf{n}\cdot\mathbf{d}\, dA$, $\tilde{\sigma}$ being the stress field produced by solving each subsample problem with the uniform stress boundary conditions and forming $\tilde{\sigma} \stackrel{\text{def}}{=} \tilde{\sigma}_1|_{\overline{\Omega}_1} + \tilde{\sigma}_2|_{\overline{\Omega}_2} \ldots \tilde{\sigma}_N|_{\overline{\Omega}_N}$. By direct expansion we have

$$||\sigma-\gamma||^2_{E^{-1}(\Omega)} = \mathscr{L} : (\langle \mathbb{E}^{-1}\rangle - \mathbb{E}^{-1*}) : \mathscr{L}|\Omega| \leq \mathscr{L} : (\langle \mathbb{E}^{-1}\rangle - \langle \mathbb{E}\rangle^{-1}) : \mathscr{L}|\Omega|. \tag{7.2}$$

The complementary forms collapse to $2\mathscr{K}(\sigma) = \mathscr{L} : \mathbb{E}^{-1*} : \mathscr{L}|\Omega|$, $2\mathscr{K}(\gamma) = \mathscr{L} : \langle \mathbb{E}^{-1}\rangle : \mathscr{L}|\Omega|$ and $2\mathscr{K}_K(\tilde{\sigma}) = \mathscr{L} : \tilde{\mathbb{E}}_K^{-1*} : \mathscr{L}|\Omega_K| \Rightarrow 2\mathscr{K}(\tilde{\sigma}) = \mathscr{L} : \tilde{\mathbb{E}}^{-1*} : \mathscr{L}|\Omega|$, where $\mathscr{K}_K(\cdot)$ is the subdomain complementary potential energy function for subdomain K. By direct substitution this yields

$$
\begin{aligned}
&\langle \tilde{\varepsilon} \rangle_{\Omega_K} = \tilde{\mathbb{E}}_K^{-1*} : \langle \tilde{\sigma} \rangle_{\Omega_K}, \qquad \tilde{\mathbb{E}}^{-1*} \stackrel{\text{def}}{=} \textstyle\sum_{K=1}^{N} \tilde{\mathbb{E}}_K^{-1*} \frac{|\Omega_K|}{|\Omega|}, \\
&||\sigma - \tilde{\sigma}||^2_{E^{-1}(\Omega)} = \mathscr{L} : (\tilde{\mathbb{E}}^{-1*} - \mathbb{E}^{*-1}) : \mathscr{L}|\Omega|, \\
&\qquad\qquad\qquad \leq \mathscr{L} : (\tilde{\mathbb{E}}^{-1*} - \langle \mathbb{E} \rangle_\Omega^{-1}) : \mathscr{L}|\Omega|, \\
&\langle \mathbb{E}^{-1} \rangle_\Omega \geq \tilde{\mathbb{E}}^{-1*} \geq \mathbb{E}^{-1*} \geq \langle \mathbb{E} \rangle_\Omega^{-1}.
\end{aligned}
\tag{7.3}
$$

As for the primal (displacement) case, the second line of Box 7.3, and generalizations to nonuniform loading, were developed in Zohdi and Wriggers [225]. The fourth line has been derived in Huet [92] by other means.

Consequences/Difficulties with Traction Tests

If the sample were an RVE, we have $\mathbb{E}^{-1*} = \mathbb{E}^{*-1}$, then the preceding analysis yields the following two sided ordering of approximate effective material responses,

$$
\langle \mathbb{E}^{-1} \rangle_\Omega^{-1} \leq (\tilde{\mathbb{E}}^{-1*})^{-1} \leq \mathbb{E}^* \leq \tilde{\mathbb{E}}^* \leq \langle \mathbb{E} \rangle_\Omega. \tag{7.4}
$$

We emphasize that $\mathbb{E}^{-1*} = \mathbb{E}^{*-1}$ is an assumption which may not be true for a finite sized sample. Therefore, in theory, under the RVE assumption, traction tests form lower bounds on the effective responses. However, traction tests pose difficulties, which are as follows:

1. Numerically pure traction boundary data cause rigid motions (singular FEM stiffness matrices), however this can be circumvented by extracting the rigid body modes (three translations and three rotations) using a deflation of the stiffness matrix by these modes.
2. The FEM is a method based upon generating kinematically admissible solutions $\tilde{\sigma}$. The traction tests result is based upon the assumption that statically admissible trial field are generated. Statically admissible fields cannot be achieved by a standard FEM approach
3. True laboratory tests specifying the force on a sample are far more difficult than specifying the displacements.

7.1.1 Isolating the Subsampling Error/Numerical Error Orthogonality

According to the previous results, we have the following normalized estimate:

$$
\frac{||\mathbf{u} - \tilde{\mathbf{u}}||^2_{E(\Omega)}}{\mathscr{E} : \langle \mathbb{E} \rangle_\Omega : \mathscr{E}|\Omega|} = \frac{\mathscr{E} : (\tilde{\mathbb{E}}^* - \mathbb{E}^*) : \mathscr{E}|\Omega|}{\mathscr{E} : \langle \mathbb{E} \rangle_\Omega : \mathscr{E}|\Omega|} \leq \frac{\mathscr{E} : (\tilde{\mathbb{E}}^* - \langle \mathbb{E}^{-1} \rangle_\Omega^{-1}) : \mathscr{E}|\Omega|}{\mathscr{E} : \langle \mathbb{E} \rangle_\Omega : \mathscr{E}|\Omega|}, \tag{7.5}
$$

where, since we have used displacement controlled tests, we have used the Voigt material's energy, which corresponds to assuming a constant strain throughout the material equal to $\mathscr{E}$, to normalize the results.

It is important to note that the numerical error, which is implicitly included in the estimate in (7.5), is orthogonal to the "partitioning" or "subsampling" error. In other words, while the numerical error is known, it can be directly filtered out of the sampling error estimates. To see this, consider that the boundary value formulations in Boxes 6.1 and 6.11 directly imply, for any type of loading $\int_\Omega \nabla \mathbf{v} : \mathbb{E} : \nabla(\mathbf{u}-\tilde{\mathbf{u}})\, d\Omega = 0$, $\forall \mathbf{v} \in \mathbf{H}^1(\Omega), \mathbf{v}|_{\Omega\cap\partial\Omega_K} = \mathbf{0}$, $K = 1,2,3...N$. This directly implies that $\int_\Omega \nabla(\tilde{\mathbf{u}}-\tilde{\mathbf{u}}^h) : \mathbb{E} : \nabla(\mathbf{u}-\tilde{\mathbf{u}})\, d\Omega = 0$. Therefore,

$$
\begin{aligned}
||\mathbf{u}-\tilde{\mathbf{u}}^h||^2_{E(\Omega)} &= ||\mathbf{u}-\tilde{\mathbf{u}}+\tilde{\mathbf{u}}-\tilde{\mathbf{u}}^h||^2_{E(\Omega)} \\
&= \int_\Omega \nabla(\mathbf{u}-\tilde{\mathbf{u}}) : \mathbb{E} : \nabla(\mathbf{u}-\tilde{\mathbf{u}})\, d\Omega \\
&\quad \underbrace{-2\int_\Omega \nabla(\tilde{\mathbf{u}}-\tilde{\mathbf{u}}^h) : \mathbb{E} : \nabla(\mathbf{u}-\tilde{\mathbf{u}})\, d\Omega}_{=0 \text{ by the above}} \\
&\quad +\int_\Omega \nabla(\tilde{\mathbf{u}}-\tilde{\mathbf{u}}^h) : \mathbb{E} : \nabla(\tilde{\mathbf{u}}-\tilde{\mathbf{u}}^h)\, d\Omega \\
&= ||\mathbf{u}-\tilde{\mathbf{u}}||^2_{E(\Omega)} + ||\tilde{\mathbf{u}}-\tilde{\mathbf{u}}^h||^2_{E(\Omega)}.
\end{aligned} \tag{7.6}
$$

Under the special case that $\mathbf{U} \stackrel{\text{def}}{=} \mathscr{E}\cdot\mathbf{x}$, we have the following orthogonal decomposition of sampling and numerical error:

$$
\underbrace{\mathscr{E} : (\tilde{\mathbb{E}}^{*,h} - \mathbb{E}^*) : \mathscr{E}|\Omega|}_{||\mathbf{u}-\tilde{\mathbf{u}}^h||^2_{E(\Omega)}} = \underbrace{\mathscr{E} : (\tilde{\mathbb{E}}^{*} - \mathbb{E}^*) : \mathscr{E}|\Omega|}_{||\mathbf{u}-\tilde{\mathbf{u}}||^2_{E(\Omega)}} + \underbrace{\sum_{K=1}^{N} ||\tilde{\mathbf{u}}^h_K - \tilde{\mathbf{u}}_K||^2_{E(\Omega)}}_{\mathscr{E}:(\tilde{\mathbb{E}}^{*,h}-\tilde{\mathbb{E}}^*):\mathscr{E}|\Omega|}, \tag{7.7}
$$

where $\langle\tilde{\sigma}^h\rangle_{\Omega_K} \stackrel{\text{def}}{=} \tilde{\mathbb{E}}^{*,h}_K : \langle\tilde{\varepsilon}^h\rangle_{\Omega_K}$ and $\tilde{\mathbb{E}}^{*,h} \stackrel{\text{def}}{=} \sum_{K=1}^N \tilde{\mathbb{E}}^{*,h}_K \frac{|\Omega_K|}{|\Omega|}$. Therefore, the expression in (7.7) indicates that the estimates made on the effective response is an orthogonal sum of both the numerical and sampling error. Therefore, one can isolate either the numerical or sampling error, if the other is known, or estimated, by simply subtracting it from the total expression (the lefthand side of (7.7)).

7.2 An Ergodic Interpretation of the Results

It is possible to interpret the results in ergodic manner, which we now briefly discuss. In statistical mechanics a frequently invoked assumption is that of statistical ergodicity. Importantly, in the context of material testing, Hill's condition can be interpreted as such an assumption. To see this, consider several bodies with the same

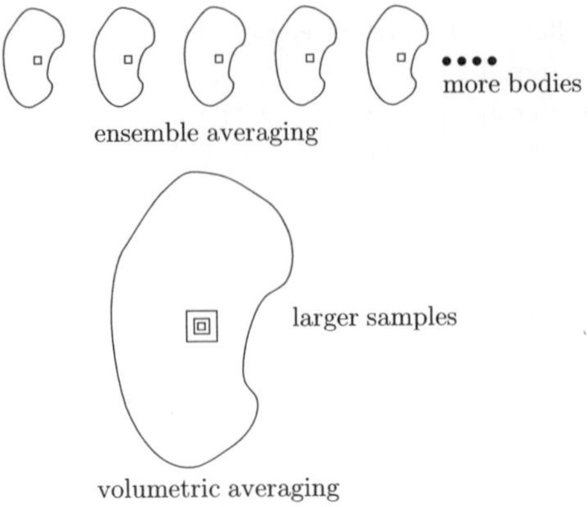

Fig. 7.1 Ensemble and volumetric averaging processes

external geometry composed of the same material volume fractions but with different random microstructure. Let us focus on the same spot on each body (indexed by $i = 1,2,...N$) and compute $\frac{1}{N}\sum_{i=1}^{N}\sigma^{(i)} \stackrel{\text{def}}{=} \mathbf{EI}^{*(i)} : (\frac{1}{N}\sum_{i=1}^{N}\varepsilon^{(i)})$, where N is the number of macroscopic structures. The tensorial quantity, $\mathbf{EI}^{*}$, is called an ensemble average (Fig. 7.1). Now we perform another experiment. We have a single body, and focus on one spot, but repeatedly compute and enlarge the sample size (indexed by $j = 1,2,...N$), $\langle\sigma^{(j)}\rangle_{\Omega_j} = \mathbb{E}^{*(j)} : \langle\varepsilon^{(j)}\rangle_{\Omega_j}$. This is our familiar (volumetric averaged) effective property. A classical ergodicity assumption is that as $i, j \to \infty$ that $\mathbf{EI}^{*} = \mathbb{E}^{*}$. Therefore, volumetric averaging and ensemble averaging yield the same result. In other words, *an infinitely large sample's volumetric average must equal the ensemble average of infinitely many finite samples at a point, from different bodies.*

If we split the stress and strain fields into a purely fluctuating (zero mean) $(\tilde{\sigma}, \tilde{\varepsilon})$ and average parts $(\langle\sigma\rangle_\Omega, \langle\varepsilon\rangle_\Omega)$ we see by direct expansion:

$$\langle((\langle\sigma\rangle_\Omega + \tilde{\sigma}) : ((\langle\varepsilon\rangle_\Omega + \tilde{\varepsilon}))_\Omega = \langle\sigma\rangle_\Omega : \langle\varepsilon\rangle_\Omega + \langle\tilde{\sigma} : \tilde{\varepsilon}\rangle_\Omega. \tag{7.8}$$

since $\langle\tilde{\sigma}\rangle_\Omega = \mathbf{0}$ and $\langle\tilde{\varepsilon}\rangle_\Omega = \mathbf{0}$. The ergodicity assumption is that $\langle\tilde{\sigma} : \tilde{\varepsilon}\rangle_\Omega = 0$. In other words, the product of two purely fluctuating random fields is again purely fluctuating. This implies that, as the sample becomes infinitely large, $\tilde{\sigma} : \tilde{\varepsilon}$ is purely fluctuating. This is essentially the implication of the Hill condition. It motivates the use of exterior uniform loading for an infinitely large sample. *Clearly, the partitioning, and the error estimates, can be interpreted testing an ergodic assumption.* However, the loadings are still not "rough" enough, since it is clear that uniform loading is an idealization and thus will only be present within the microstructure of a finite sized engineering (macro) structure to an error. For more on ergodic hypotheses see the classical work of Kröner [111].

7.3 Statistical Shifting Theorems

Consider any tested quantity, Q, with a distribution of values (Q_i, i=1, 2, ...N = samples) about an arbitrary reference point, denoted $Q^{\star}$, as follows:

$$\mathbf{M}_r^{Q_i - Q^\star} \overset{\text{def}}{=} \frac{\sum_{i=1}^{N}(Q_i - Q^\star)^r}{N} \overset{\text{def}}{=} \overline{(Q_i - Q^\star)^r}, \tag{7.9}$$

where $\frac{\sum_{i=1}^{N}(\cdot)}{N} \overset{\text{def}}{=} \overline{(\cdot)}$ and $A \overset{\text{def}}{=} \overline{Q_i}$. The various moments characterize the distribution, for example:

1. $\mathbf{M}_1^{Q_i - A}$ measures the first deviation from the average, which equals zero,
2. $\mathbf{M}_1^{Q_i - 0} \overset{\text{def}}{=} \frac{\sum_{i=1}^{N}(Q_i - 0)}{N} \overset{\text{def}}{=} \overline{(Q_i - 0)} = A$,
3. $\mathbf{M}_2^{Q_i - A}$ is the standard deviation, and
4. $\mathbf{M}_3^{Q_i - A}$ is the skewness. The skewness measures the bias, or asymmetry of the distribution of data.

The moments of the data can be expressed about the average and related to any other reference point, denoted here as Q^*, using parallel axis type theorems. Using the notation in (7.9), let us define $q_i = Q_i - A \Rightarrow \overline{Q_i} = \overline{A + q_i} \Rightarrow \overline{Q_i} = A + \overline{q_i} \Rightarrow Q_i - \overline{Q_i} = q_i - \overline{q_i}$. There are useful properties associated with these relations, in particular a shifting property for the standard deviation

$$\begin{aligned}
\mathbf{M}_2^{Q_i - A} &\overset{\text{def}}{=} \overline{(Q_i - A)^2} = \overline{(q_i - \overline{q_i})^2} = \overline{q_i^2 - 2\overline{q_i}q_i + \overline{q_i}^2}, \\
&= \overline{q_i^2} - 2\overline{q_i}^2 + \overline{q_i}^2, \\
&= \overline{q_i^2} - \overline{q_i}^2, \\
&= \mathbf{M}_2^{Q_i - Q^*} - (\mathbf{M}_1^{Q_i - Q^*})^2,
\end{aligned} \tag{7.10}$$

as well as for the skewness

$$\begin{aligned}
\mathbf{M}_3^{Q_i - A} &\overset{\text{def}}{=} \overline{(Q_i - A)^3} = \overline{(q_i - \overline{q_i})^3} = \overline{q_i^3 - 3q_i^2\overline{q_i} + 3\overline{q_i}^2 q_i - \overline{q_i}^3}, \\
&= \overline{q_i^3} - 3\overline{q_i^2}\overline{q_i} + 2\overline{q_i}^3, \\
&= \mathbf{M}_3^{Q_i - Q^*} - 3(\mathbf{M}_1^{Q_i - Q^*})(\mathbf{M}_2^{Q_i - Q^*}) + 2(\mathbf{M}_1^{Q_i - Q^*})^3.
\end{aligned} \tag{7.11}$$

From (7.10) we may write

$$\begin{aligned}
3(\mathbf{M}_1^{Q_i - Q^*})(\mathbf{M}_2^{Q_i - Q^*}) - 2(\mathbf{M}_1^{Q_i - Q^*})^3 &= \mathbf{M}_1^{Q_i - Q^*}\left(3(\mathbf{M}_2^{Q_i - Q^*}) - 2(\mathbf{M}_1^{Q_i - Q^*})^2\right), \\
&= \mathbf{M}_1^{Q_i - Q^*}\left(\mathbf{M}_2^{Q_i - Q^*} + 2(\mathbf{M}_2^{Q_i - Q^*}) - 2(\mathbf{M}_1^{Q_i - Q^*})^2\right), \\
&= \mathbf{M}_1^{Q_i - Q^*}\left(\mathbf{M}_2^{Q_i - Q^*} + 2\mathbf{M}_2^{Q_i - A}\right),
\end{aligned}$$

therefore,

$$\mathbf{M}_3^{Q_i-A} + \mathbf{M}_1^{Q_i-Q^*}\left(\mathbf{M}_2^{Q_i-Q^*} + 2\mathbf{M}_2^{Q_i-A}\right) = \mathbf{M}_3^{Q_i-Q^*}. \tag{7.12}$$

We now proceed to bound the average, standard deviation and skewness associated with the ensemble averaging of a population of samples.

7.4 Partitioning and Ensemble Averaging

Large samples can be divided into sub-domains. Several method which yield bounds for the constitutive parameters in a homogenization process can be considered.

7.4.1 Primal Partitioning

Consider a (large) sample of material with the following general boundary value representation

Find $\mathbf{u} \in \mathbf{H}^1(\Omega), \mathbf{u}|_{\Gamma_u} = \mathbf{d}$, such that

$$\int_\Omega \nabla\mathbf{v} : \underbrace{\mathbb{E} : \nabla\mathbf{u}}_{\sigma}\, d\Omega = \int_\Omega \mathbf{f}\cdot\mathbf{v}\, d\Omega + \int_{\Gamma_t} \mathbf{t}\cdot\mathbf{v}\, dA, \quad \forall\mathbf{v} \in \mathbf{H}^1(\Omega), \mathbf{v}|_{\Gamma_u} = \mathbf{0}. \tag{7.13}$$

Now partition the domain into S subdomains, $\Omega = \cup_{K=1}^{S}\Omega_K$. The pieces do not have to be the same size or shape, although for illustration purposes it is convenient to take a uniform (regular) partitioning (Fig. 7.2). Consider a kinematically admissible function, $\mathbf{U} \in \mathbf{H}^1(\Omega)$ and $\mathbf{U}|_{\Gamma_u} = \mathbf{d}$, which is projected onto the *internal boundaries* ($\partial\Omega_K$) of the subdomains. Any subdomain boundaries coinciding with the exterior surface retain their original boundary conditions (Fig. 7.2). Accordingly, we have the following virtual work formulation, for each subdomain, $1 \le K \le S$:

Find $\tilde{\mathbf{u}}_K \in \mathbf{H}^1(\Omega_K), \tilde{\mathbf{u}}_K|_{\partial\Omega_K \cap (\Omega\cup\Gamma_u)} = \mathbf{U} \in \mathbf{H}^1(\Omega)$, such that

$$\int_{\Omega_K} \nabla\mathbf{v}_K : \underbrace{\mathbb{E} : \nabla\tilde{\mathbf{u}}_K}_{\tilde{\sigma}_K}\, d\Omega = \int_{\Omega_K} \mathbf{f}\cdot\mathbf{v}_K\, d\Omega + \int_{\partial\Omega_K\cap\Gamma_t} \mathbf{t}\cdot\mathbf{v}_K\, dA \tag{7.14}$$

$\forall\mathbf{v}_K \in \mathbf{H}^1(\Omega_K), \mathbf{v}_K|_{\partial\Omega_K\cap(\Omega\cup\Gamma_u)} = \mathbf{0}.$

Fig. 7.2 A regular partitioning of a large sample of material

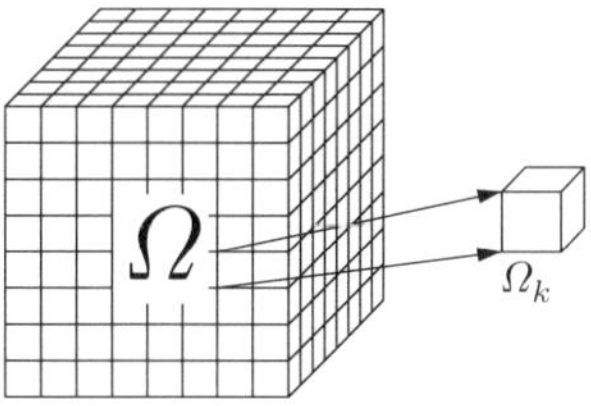

The individual subdomain solutions, $\tilde{\mathbf{u}}_K$, are zero outside of the corresponding subdomain $\overline{\Omega}_K$. In this case the approximate solution is constructed by a direct assembly process, $\tilde{\mathbf{u}} \stackrel{\text{def}}{=} \mathbf{U} + (\tilde{\mathbf{u}}_1 - \mathbf{U})|_{\Omega_1} + (\tilde{\mathbf{u}}_2 - \mathbf{U})|_{\Omega_2} + ... + (\tilde{\mathbf{u}}_S - \mathbf{U})|_{\Omega_S}$. The approximate displacement field is in $\mathbf{H}^1(\Omega)$, however, the approximate traction field is possibly discontinuous. Logical choices of $\mathbf{U}$, i.e. $\mathbf{U} = \mathscr{E} \cdot \mathbf{x}$, will be given momentarily. It should be clear that if $\tilde{\mathbf{u}} = \mathbf{u}$ on the internal partition boundaries, then the approximate solution is exact. Since we employ energy type principles to generate approximate solutions, we use an induced energy norm to measure the solution differences

$$0 \leq ||\mathbf{u} - \tilde{\mathbf{u}}||^2_{E(\Omega)} \stackrel{\text{def}}{=} \int_\Omega \nabla(\mathbf{u} - \tilde{\mathbf{u}}) : \mathbb{E} : \nabla(\mathbf{u} - \tilde{\mathbf{u}})\, d\Omega\,. \tag{7.15}$$

It is convenient to cast the error in terms of the potential energy,

$$\mathscr{J}(\mathbf{w}) \stackrel{\text{def}}{=} \frac{1}{2}\int_\Omega \nabla\mathbf{w} : \mathbb{E} : \nabla\mathbf{w}\, d\Omega - \int_\Omega \mathbf{f} \cdot \mathbf{w}\, d\Omega - \int_{\Gamma_t} \mathbf{t} \cdot \mathbf{w}\, dA\,, \tag{7.16}$$

where $\mathbf{w}$ is any kinematically admissible function. This leads to

$$||\mathbf{u} - \mathbf{w}||^2_{E(\Omega)} = 2(\mathscr{J}(\mathbf{w}) - \mathscr{J}(\mathbf{u})) \quad \text{or} \quad \mathscr{J}(\mathbf{u}) \leq \mathscr{J}(\mathbf{w}) \tag{7.17}$$

which is a form of the Principle of Minimum Potential Energy. In other words, the true solution possesses a minimum potential. By direct substitution we have $||\mathbf{u} - \tilde{\mathbf{u}}||^2_{E(\Omega)} = 2(\mathscr{J}(\tilde{\mathbf{u}}) - \mathscr{J}(\mathbf{u}))$.

In the special case that $\mathbf{u}|_{\partial\Omega} = \mathscr{E} \cdot \mathbf{x}$, which is equivalent to testing each subsample with $\mathbf{u}|_{\partial\Omega_K} = \mathscr{E} \cdot \mathbf{x}$ and $\mathbf{f} = \mathbf{0}$, then

$$||\mathbf{u} - \tilde{\mathbf{u}}||^2_{E(\Omega)} = \mathscr{E} : (\tilde{\mathbb{E}}^* - \mathbb{E}^*) : \mathscr{E}|\Omega|\,, \tag{7.18}$$

where $\langle\tilde{\sigma}\rangle_{\Omega_K} \stackrel{\text{def}}{=} \tilde{\mathbb{E}}^*_K : \langle\tilde{\varepsilon}\rangle_{\Omega_K}$ and $\tilde{\mathbb{E}}^* \stackrel{\text{def}}{=} \sum_{K=1}^S \tilde{\mathbb{E}}^*_K \frac{|\Omega_K|}{|\Omega|}$.

7.4.2 Complementary Partitioning

We can repeat the partitioning process for an applied internal traction set of tests. The equivalent complementary form for the exact (undecomposed) problem is

Find $\sigma, \nabla \cdot \sigma + \mathbf{f} = \mathbf{0}, \sigma \cdot \mathbf{n}|_{\Gamma_t} = \mathbf{t}$ such that

$$\int_{\Omega} \tau : \mathbb{E}^{-1} : \sigma \, d\Omega = \int_{\Gamma_u} \tau \cdot \mathbf{n} \cdot \mathbf{d} \, dA, \ \forall \tau, \nabla \cdot \tau = \mathbf{0}, \tau \cdot \mathbf{n}|_{\Gamma_t} = \mathbf{0}. \tag{7.19}$$

For the complementary problem, similar restrictions are placed on the solution and test fields to force the integrals to make sense. In other words, we assume that solutions produce finite global energy. When employing the applied internal traction approach, in order to construct approximate solutions, a statically admissible function, Σ, with the property that $\Sigma \cdot \mathbf{n}|_{\Gamma_t} = \mathbf{t}$, is projected onto the *internal boundaries* of the subdomain partitions. As in the applied displacement case, any subdomain boundaries coinciding with the exterior surface retain their original boundary conditions. Accordingly, we have the following complementary virtual work formulation, for each subdomain, $1 \leq K \leq S$:

Find $\hat{\sigma}_K, \nabla \cdot \hat{\sigma}_K + \mathbf{f} = \mathbf{0}, \hat{\sigma}_K \cdot \mathbf{n}|_{\partial\Omega_K \cap (\Omega \cup \Gamma_t)} = \Sigma \cdot \mathbf{n}|_{\partial\Omega_K \cap (\Omega \cup \Gamma_t)}$ such that

$$\int_{\Omega_K} \tau_K : \mathbb{E}^{-1} : \hat{\sigma}_K \, d\Omega = \int_{\Gamma_u} \tau_K \cdot \mathbf{n} \cdot \mathbf{d} \, dA$$

$$\forall \tau_K, \nabla \cdot \tau_K = \mathbf{0}, \tau_K \cdot \mathbf{n}|_{\partial\Omega_K \cap (\Omega \cup \Gamma_t)} = \mathbf{0}. \tag{7.20}$$

The individual subdomain solutions, $\hat{\sigma}_K$, are zero outside of the corresponding subdomain $\overline{\Omega}_K$. In this case the approximate solution is constructed by a direct assembly process $\hat{\sigma} \stackrel{\text{def}}{=} \Sigma + (\hat{\sigma}_1 - \Sigma)|_{\Omega_1} + (\hat{\sigma}_2 - \Sigma)|_{\Omega_2} + \ldots + (\hat{\sigma}_S - \Sigma)|_{\Omega_S}$. The stress field is statically admissible, however, the approximate displacement field is possibly discontinuous. Logical choices of Σ, $\Sigma = \mathscr{L} = constant$, will be given momentarily. It should be clear that if $\Sigma = \sigma$ on the internal partition boundaries, then the approximate solution is exact. We define the complementary norm

$$0 \leq ||\sigma - \hat{\sigma}||^2_{E^{-1}(\Omega)} \stackrel{\text{def}}{=} \int_{\Omega} (\sigma - \hat{\sigma}) : \mathbb{E}^{-1} : (\sigma - \hat{\sigma}) \, d\Omega. \tag{7.21}$$

As in the primal case, it is convenient to cast the error in terms of the potential complementary energy for the case of linear elasticity, where

$$\mathscr{K}(\gamma) \stackrel{\text{def}}{=} \frac{1}{2} \int_{\Omega} \gamma : \mathbb{E}^{-1} : \gamma \, d\Omega - \int_{\Gamma_u} \gamma \cdot \mathbf{n} \cdot \mathbf{u} \, dA, \tag{7.22}$$

where γ is any statically admissible function. The well known relationship, for a statically admissible function γ, is

$$||\sigma-\gamma||^2_{E^{-1}(\Omega)}=2(\mathscr{K}(\sigma)-\mathscr{K}(\sigma)) \quad \text{or } \mathscr{K}(\sigma)\leq\mathscr{K}(\gamma). \tag{7.23}$$

This is a form of the Principle of Minimum Complementary Potential Energy. In other words, the true solution possesses a minimum complementary potential. Choosing $\gamma=\hat{\sigma}$, we have $||\sigma-\hat{\sigma}||^2_{E(\Omega)}=2(\mathscr{K}(\hat{\sigma})-\mathscr{K}(\sigma))$.

In the special case that $\mathbf{t}|_{\partial\Omega}=\mathscr{L}\cdot\mathbf{n}$, which is equivalent to testing each subsample with $\mathbf{t}|_{\partial\Omega_K}=\mathscr{L}\cdot\mathbf{n}$,

$$||\sigma-\hat{\sigma}||^2_{E^{-1}(\Omega)}=\mathscr{L}:(\hat{\mathbb{E}}^{-1*}-\mathbb{E}^{-1*}):\mathscr{L}|\Omega|, \tag{7.24}$$

where $\langle\hat{\varepsilon}\rangle_{\Omega_K}\overset{\text{def}}{=}\hat{\mathbb{E}}_K^{-1*}:\langle\hat{\sigma}\rangle_{\Omega_K}$ and $\hat{\mathbb{E}}^{-1*}\overset{\text{def}}{=}\sum_{K=1}^{S}\hat{\mathbb{E}}_K^{-1*}\frac{|\Omega_K|}{|\Omega|}$.

7.4.3 *Homogenized Material Orderings*

If the sample is statistically representative, we have $\mathbb{E}^{-1*}=\mathbb{E}^{*-1}$, then the previous results imply, under the assumption that the uniform loadings are arbitrary, the following two sided ordering of approximate effective material responses,

$$\langle\mathbb{E}^{-1}\rangle_{\Omega}^{-1}\leq\hat{\mathbb{E}}^*\leq\mathbb{E}^*\leq\tilde{\mathbb{E}}^*\leq\langle\mathbb{E}\rangle_{\Omega}, \tag{7.25}$$

where the tensor inequality notation means, for example, that the difference tensor $(\tilde{\mathbb{E}}^*-\mathbb{E}^*)$ is positive definite, etc. Since $\mathscr{J}(\tilde{\mathbf{u}})\leq\mathscr{J}(\mathbf{U})$, we also have $\tilde{\mathbb{E}}^*\leq\langle\mathbb{E}\rangle_{\Omega}$. Alternatively, since $\mathscr{K}(\hat{\sigma})\leq\mathscr{K}(\Sigma)$, then $\hat{\mathbb{E}}^{-1*}\leq\langle\mathbb{E}^{-1}\rangle_{\Omega}$. To the knowledge of the author, the result in Box 7.25 was first derived in Huet [92], however by other techniques.

Remark 7.1. For isotropic responses, we have

$$\langle\kappa^{-1}\rangle_{\Omega}^{-1}\leq\hat{\kappa}^*\leq\kappa^*\leq\tilde{\kappa}^*\leq\langle\kappa\rangle_{\Omega} \tag{7.26}$$

and

$$\langle\mu^{-1}\rangle_{\Omega}^{-1}\leq\hat{\mu}^*\leq\mu^*\leq\tilde{\mu}^*\leq\langle\mu\rangle_{\Omega}. \tag{7.27}$$

7.4.4 *Embedded Orthogonal Monotonicities*

Since $\tilde{\mathbf{u}}$ is kinematically admissible, we have $||\mathbf{u}-\tilde{\mathbf{u}}||^2_{E(\Omega)}=2(\mathscr{J}(\tilde{\mathbf{u}})-\mathscr{J}(\mathbf{u}))$. If we repartition the existing subdomains into more subdomains (Fig. 7.3), and use

$\mathbf{U}$ for the local boundary conditions on the finer partition, upon solving the local boundary value problems and assembling the local solutions together (just as before for $\tilde{\mathbf{u}}$), we have, denoting the solution by $\tilde{\tilde{\mathbf{u}}}$, $||\tilde{\mathbf{u}} - \tilde{\tilde{\mathbf{u}}}||^2_{E(\Omega)} = 2(\mathscr{J}(\tilde{\tilde{\mathbf{u}}}) - \mathscr{J}(\tilde{\mathbf{u}}))$. Adding the two previous relations together yields an orthogonal decomposition

$$||\mathbf{u} - \tilde{\tilde{\mathbf{u}}}||^2_{E(\Omega)} = ||\tilde{\mathbf{u}} - \tilde{\tilde{\mathbf{u}}}||^2_{E(\Omega)} + ||\mathbf{u} - \tilde{\mathbf{u}}||^2_{E(\Omega)}. \tag{7.28}$$

This implies that the error monotonically grows for successively finer embedded partitions. Intuitively one expects this type of growth in the error, since one is projecting more inaccurate data onto the interfaces. *Simply stated, more embedded subdomains, more error. Furthermore, the relationship is monotone.* As in the displacement controlled tests, for traction controlled tests we have

$$||\sigma - \hat{\hat{\sigma}}||^2_{E^{-1}(\Omega)} = ||\hat{\sigma} - \hat{\hat{\sigma}}||^2_{E^{-1}(\Omega)} + ||\sigma - \hat{\sigma}||^2_{E^{-1}(\Omega)}. \tag{7.29}$$

In terms of effective properties, (7.28) implies $\mathscr{E} : (\tilde{\mathbb{E}}^* - \mathbb{E}^*) : \mathscr{E} = \mathscr{E} : (\tilde{\tilde{\mathbb{E}}}^* - \tilde{\mathbb{E}}^*) : \mathscr{E} + \mathscr{E} : (\tilde{\mathbb{E}}^* - \mathbb{E}^*) : \mathscr{E}$, while (7.29) implies $\mathscr{L} : (\hat{\mathbb{E}}^{-1*} - \mathbb{E}^{-1*}) : \mathscr{L} = \mathscr{L} : (\hat{\hat{\mathbb{E}}}^{-1*} - \hat{\mathbb{E}}^{-1*}) : \mathscr{L} + \mathscr{L} : (\hat{\mathbb{E}}^{-1*} - \mathbb{E}^{-1*}) : \mathscr{L}$. To streamline the notation we employ the notation $\overset{\tilde{Z}}{\mathbb{E}^*}$ to denote Z embedded partitions, etc. The higher the number the more embedded partitions exist (Fig. 7.3). Therefore, for $1 \leq M \leq Z$ and $1 \leq P \leq L$, we have

$$\langle \mathbb{E}^{-1} \rangle^{-1}_{\Omega} \leq \ldots \overset{\hat{L}}{\mathbb{E}^*} \leq \overset{\hat{P}}{\mathbb{E}^*} \ldots \leq \hat{\mathbb{E}}^* \leq \mathbb{E}^* \leq \tilde{\mathbb{E}}^* \leq \ldots \overset{\tilde{M}}{\mathbb{E}^*} \leq \overset{\tilde{Z}}{\mathbb{E}^*} \ldots \leq \langle \mathbb{E} \rangle_{\Omega}. \tag{7.30}$$

The results of this section generalize and extend relations found in Huet [92], Hazanov and Huet [77], Huet [100], Zohdi and Wriggers [225], Zohdi [230] and Zohdi et al. [236].

Remark. For isotropic material responses we have

$$\begin{aligned} &\langle \kappa^{-1} \rangle^{-1}_{\Omega} \leq \ldots \overset{\hat{L}}{\kappa^*} \leq \overset{\hat{P}}{\kappa^*} \ldots \leq \hat{\kappa}^* \leq \kappa^* \leq \tilde{\kappa}^* \leq \ldots \overset{\tilde{M}}{\kappa^*} \leq \overset{\tilde{Z}}{\kappa^*} \ldots \leq \langle \kappa \rangle_{\Omega} \\ &\langle \mu^{-1} \rangle^{-1}_{\Omega} \leq \ldots \overset{\hat{L}}{\mu^*} \leq \overset{\hat{P}}{\mu^*} \ldots \leq \hat{\mu}^* \leq \mu^* \leq \tilde{\mu}^* \leq \ldots \overset{\tilde{M}}{\mu^*} \leq \overset{\tilde{Z}}{\mu^*} \ldots \leq \langle \mu \rangle_{\Omega}. \end{aligned} \tag{7.31}$$

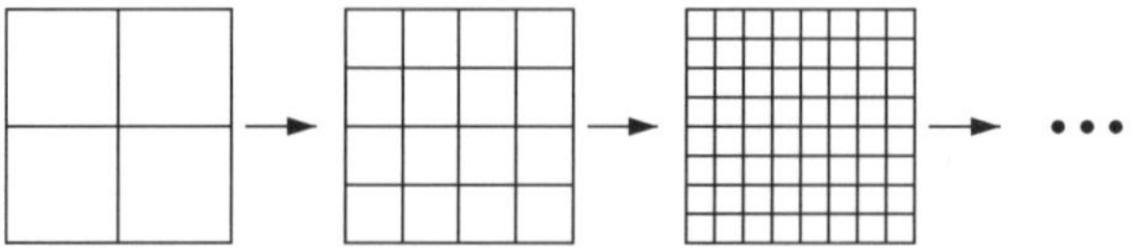

Fig. 7.3 Successively embedded partitions

7.5 Moment Bounds on Population Responses

Let us take any effective property, averaged over the *ith* subsample, denoted by Q_i, and correspondingly then A is the average of all of the samples. Let Q^* denote the true effective property. Employing the notation $[\tilde{\mathbb{E}}^* - \mathbb{E}^*] \stackrel{\text{def}}{=} \mathscr{T} : (\tilde{\mathbb{E}}^* - \mathbb{E}^*) : \mathscr{T}$, where $\mathscr{T}$ is any arbitrary second order (strain) tensor.

7.5.1 First Order (Average) Bounds

Equation 7.30 implies

$$\begin{aligned}
&\mathbf{M}_1^{[\overset{\tilde{Z}}{\mathbb{E}^*}_i - \overset{\tilde{M}}{\mathbb{E}^*}]} \leq \mathbf{M}_1^{[\overset{\tilde{Z}}{\mathbb{E}^*}_i - \mathbb{E}^*]} \leq \mathbf{M}_1^{[\overset{\tilde{Z}}{\mathbb{E}^*}_i - \overset{\hat{P}}{\mathbb{E}^*}]} \leq \mathbf{M}_1^{[\overset{\tilde{Z}}{\mathbb{E}^*}_i - \overset{\hat{L}}{\mathbb{E}^*}]},\\
&\mathbf{M}_1^{[\overset{\hat{L}}{\mathbb{E}^*}_i - \overset{\hat{P}}{\mathbb{E}^*}]} \geq \mathbf{M}_1^{[\overset{\hat{L}}{\mathbb{E}^*}_i - \mathbb{E}^*]} \geq \mathbf{M}_1^{[\overset{\hat{L}}{\mathbb{E}^*}_i - \overset{\tilde{M}}{\mathbb{E}^*}]} \geq \mathbf{M}_1^{[\overset{\hat{L}}{\mathbb{E}^*}_i - \overset{\tilde{Z}}{\mathbb{E}^*}]}.
\end{aligned} \tag{7.32}$$

7.5.2 Second Order (Standard Deviation) Bounds

For the second order moments, the shifting theorems imply

$$\begin{aligned}
\mathbf{M}_2^{[\overset{\tilde{Z}}{\mathbb{E}^*}_i - \mathbb{E}^*]} &= \mathbf{M}_2^{[\overset{\tilde{Z}}{\mathbb{E}^*}_i - \overset{\tilde{Z}}{\mathbb{E}^*}]} + \left(\mathbf{M}_1^{[\overset{\tilde{Z}}{\mathbb{E}^*}_i - \mathbb{E}^*]}\right)^2,\\
\mathbf{M}_2^{[\overset{\tilde{Z}}{\mathbb{E}^*}_i - \overset{\tilde{M}}{\mathbb{E}^*}]} &= \mathbf{M}_2^{[\overset{\tilde{Z}}{\mathbb{E}^*}_i - \overset{\tilde{Z}}{\mathbb{E}^*}]} + \left(\mathbf{M}_1^{[\overset{\tilde{Z}}{\mathbb{E}^*}_i - \overset{\tilde{M}}{\mathbb{E}^*}]}\right)^2,\\
\mathbf{M}_2^{[\overset{\tilde{Z}}{\mathbb{E}^*}_i - \overset{\hat{L}}{\mathbb{E}^*}]} &= \mathbf{M}_2^{[\overset{\tilde{Z}}{\mathbb{E}^*}_i - \overset{\tilde{Z}}{\mathbb{E}^*}]} + \left(\mathbf{M}_1^{[\overset{\tilde{Z}}{\mathbb{E}^*}_i - \overset{\hat{L}}{\mathbb{E}^*}]}\right)^2,\\
\mathbf{M}_2^{[\overset{\tilde{Z}}{\mathbb{E}^*}_i - \overset{\hat{P}}{\mathbb{E}^*}]} &= \mathbf{M}_2^{[\overset{\tilde{Z}}{\mathbb{E}^*}_i - \overset{\tilde{Z}}{\mathbb{E}^*}]} + \left(\mathbf{M}_1^{[\overset{\tilde{Z}}{\mathbb{E}^*}_i - \overset{\hat{P}}{\mathbb{E}^*}]}\right)^2.
\end{aligned} \tag{7.33}$$

and

$$
\begin{aligned}
\mathbf{M}_2^{[\overset{\hat{L}}{\mathbb{E}^*_i}-\mathbb{E}^*]} &= \mathbf{M}_2^{[\overset{\hat{L}}{\mathbb{E}^*_i}-\overset{\hat{L}}{\mathbb{E}^*}]} + \left(\mathbf{M}_1^{[\overset{\hat{L}}{\mathbb{E}^*_i}-\mathbb{E}^*]}\right)^2, \\
\mathbf{M}_2^{[\overset{\hat{L}}{\mathbb{E}^*_i}-\overset{\hat{P}}{\mathbb{E}^*}]} &= \mathbf{M}_2^{[\overset{\hat{L}}{\mathbb{E}^*_i}-\overset{\hat{L}}{\mathbb{E}^*}]} + \left(\mathbf{M}_1^{[\overset{\hat{L}}{\mathbb{E}^*_i}-\overset{\hat{P}}{\mathbb{E}^*}]}\right)^2, \\
\mathbf{M}_2^{[\overset{\hat{L}}{\mathbb{E}^*_i}-\overset{\tilde{Z}}{\mathbb{E}^*}]} &= \mathbf{M}_2^{[\overset{\hat{L}}{\mathbb{E}^*_i}-\overset{\hat{L}}{\mathbb{E}^*}]} + \left(\mathbf{M}_1^{[\overset{\hat{L}}{\mathbb{E}^*_i}-\overset{\tilde{Z}}{\mathbb{E}^*}]}\right)^2, \\
\mathbf{M}_2^{[\overset{\hat{L}}{\mathbb{E}^*_i}-\overset{\tilde{M}}{\mathbb{E}^*}]} &= \mathbf{M}_2^{[\overset{\hat{L}}{\mathbb{E}^*_i}-\overset{\hat{L}}{\mathbb{E}^*}]} + \left(\mathbf{M}_1^{[\overset{\hat{L}}{\mathbb{E}^*_i}-\overset{\tilde{M}}{\mathbb{E}^*}]}\right)^2.
\end{aligned}
\tag{7.34}
$$

The results in (7.32–7.34) imply

$$
\begin{aligned}
\mathbf{M}_2^{[\overset{\tilde{Z}}{\mathbb{E}^*_i}-\overset{\tilde{M}}{\mathbb{E}^*}]} \leq \mathbf{M}_2^{[\overset{\tilde{Z}}{\mathbb{E}^*_i}-\mathbb{E}^*]} \leq \mathbf{M}_2^{[\overset{\tilde{Z}}{\mathbb{E}^*_i}-\overset{\hat{P}}{\mathbb{E}^*}]} \leq \mathbf{M}_2^{[\overset{\tilde{Z}}{\mathbb{E}^*_i}-\overset{\hat{L}}{\mathbb{E}^*}]} \\
\mathbf{M}_2^{[\overset{\hat{L}}{\mathbb{E}^*_i}-\overset{\hat{P}}{\mathbb{E}^*}]} \leq \mathbf{M}_2^{[\overset{\hat{L}}{\mathbb{E}^*_i}-\mathbb{E}^*]} \leq \mathbf{M}_2^{[\overset{\hat{L}}{\mathbb{E}^*_i}-\overset{\tilde{M}}{\mathbb{E}^*}]} \leq \mathbf{M}_2^{[\overset{\hat{L}}{\mathbb{E}^*_i}-\overset{\tilde{Z}}{\mathbb{E}^*}]}
\end{aligned}
\tag{7.35}
$$

Remark. In the case of isotropy, the moment results hold for the effective shear and bulk moduli individually.

7.5.3 Third Order (Skewness) Bounds

From (7.12), we may write the skewness as

$$
\begin{aligned}
\mathbf{M}_3^{[\overset{\tilde{Z}}{\mathbb{E}^*_i}-\mathbb{E}^*]} &= \mathbf{M}_3^{[\overset{\tilde{Z}}{\mathbb{E}^*_i}-\overset{\tilde{Z}}{\mathbb{E}^*}]} + \mathbf{M}_1^{[\overset{\tilde{Z}}{\mathbb{E}^*_i}-\mathbb{E}^*]}\mathbf{M}_2^{[\overset{\tilde{Z}}{\mathbb{E}^*_i}-\mathbb{E}^*]} + 2\mathbf{M}_1^{[\overset{\tilde{Z}}{\mathbb{E}^*_i}-\mathbb{E}^*]}\mathbf{M}_2^{[\overset{\tilde{Z}}{\mathbb{E}^*_i}-\overset{\tilde{Z}}{\mathbb{E}^*}]}, \\
\mathbf{M}_3^{[\overset{\tilde{Z}}{\mathbb{E}^*_i}-\overset{\tilde{M}}{\mathbb{E}^*}]} &= \mathbf{M}_3^{[\overset{\tilde{Z}}{\mathbb{E}^*_i}-\overset{\tilde{Z}}{\mathbb{E}^*}]} + \mathbf{M}_1^{[\overset{\tilde{Z}}{\mathbb{E}^*_i}-\overset{\tilde{M}}{\mathbb{E}^*}]}\mathbf{M}_2^{[\overset{\tilde{Z}}{\mathbb{E}^*_i}-\overset{\tilde{M}}{\mathbb{E}^*}]} + 2\mathbf{M}_1^{[\overset{\tilde{Z}}{\mathbb{E}^*_i}-\overset{\tilde{M}}{\mathbb{E}^*}]}\mathbf{M}_2^{[\overset{\tilde{Z}}{\mathbb{E}^*_i}-\overset{\tilde{Z}}{\mathbb{E}^*}]}, \\
\mathbf{M}_3^{[\overset{\tilde{Z}}{\mathbb{E}^*_i}-\overset{\hat{L}}{\mathbb{E}^*}]} &= \mathbf{M}_3^{[\overset{\tilde{Z}}{\mathbb{E}^*_i}-\overset{\tilde{Z}}{\mathbb{E}^*}]} + \mathbf{M}_1^{[\overset{\tilde{Z}}{\mathbb{E}^*_i}-\overset{\hat{L}}{\mathbb{E}^*}]}\mathbf{M}_2^{[\overset{\tilde{Z}}{\mathbb{E}^*_i}-\overset{\hat{L}}{\mathbb{E}^*}]} + 2\mathbf{M}_1^{[\overset{\tilde{Z}}{\mathbb{E}^*_i}-\overset{\hat{L}}{\mathbb{E}^*}]}\mathbf{M}_2^{[\overset{\tilde{Z}}{\mathbb{E}^*_i}-\overset{\tilde{Z}}{\mathbb{E}^*}]}, \\
\mathbf{M}_3^{[\overset{\tilde{Z}}{\mathbb{E}^*_i}-\overset{\hat{P}}{\mathbb{E}^*}]} &= \mathbf{M}_3^{[\overset{\tilde{Z}}{\mathbb{E}^*_i}-\overset{\tilde{Z}}{\mathbb{E}^*}]} + \mathbf{M}_1^{[\overset{\tilde{Z}}{\mathbb{E}^*_i}-\overset{\hat{P}}{\mathbb{E}^*}]}\mathbf{M}_2^{[\overset{\tilde{Z}}{\mathbb{E}^*_i}-\overset{\hat{P}}{\mathbb{E}^*}]} + 2\mathbf{M}_1^{[\overset{\tilde{Z}}{\mathbb{E}^*_i}-\overset{\hat{P}}{\mathbb{E}^*}]}\mathbf{M}_2^{[\overset{\tilde{Z}}{\mathbb{E}^*_i}-\overset{\tilde{Z}}{\mathbb{E}^*}]},
\end{aligned}
\tag{7.36}
$$

and as

$$
\begin{aligned}
\mathbf{M}_3^{[\overset{\hat{L}}{\mathbb{E}}{}^*_i-\mathbb{E}^*]} &= \mathbf{M}_3^{[\overset{\hat{L}}{\mathbb{E}}{}^*_i-\overset{\hat{L}}{\mathbb{E}}{}^*]} + \mathbf{M}_1^{[\overset{\hat{L}}{\mathbb{E}}{}^*_i-\mathbb{E}^*]}\mathbf{M}_2^{[\overset{\hat{L}}{\mathbb{E}}{}^*_i-\mathbb{E}^*]} + 2\mathbf{M}_1^{[\overset{\hat{L}}{\mathbb{E}}{}^*_i-\mathbb{E}^*]}\mathbf{M}_2^{[\overset{\hat{L}}{\mathbb{E}}{}^*_i-\overset{\hat{L}}{\mathbb{E}}{}^*]},\\
\mathbf{M}_3^{[\overset{\hat{L}}{\mathbb{E}}{}^*_i-\overset{\hat{P}}{\mathbb{E}}{}^*]} &= \mathbf{M}_3^{[\overset{\hat{L}}{\mathbb{E}}{}^*_i-\overset{\hat{L}}{\mathbb{E}}{}^*]} + \mathbf{M}_1^{[\overset{\hat{L}}{\mathbb{E}}{}^*_i-\overset{\hat{P}}{\mathbb{E}}{}^*]}\mathbf{M}_2^{[\overset{\hat{L}}{\mathbb{E}}{}^*_i-\overset{\hat{P}}{\mathbb{E}}{}^*]} + 2\mathbf{M}_1^{[\overset{\hat{L}}{\mathbb{E}}{}^*_i-\overset{\hat{P}}{\mathbb{E}}{}^*]}\mathbf{M}_2^{[\overset{\hat{L}}{\mathbb{E}}{}^*_i-\overset{\hat{L}}{\mathbb{E}}{}^*]},\\
\mathbf{M}_3^{[\overset{\hat{L}}{\mathbb{E}}{}^*_i-\overset{\tilde{Z}}{\mathbb{E}}{}^*]} &= \mathbf{M}_3^{[\overset{\hat{L}}{\mathbb{E}}{}^*_i-\overset{\hat{L}}{\mathbb{E}}{}^*]} + \mathbf{M}_1^{[\overset{\hat{L}}{\mathbb{E}}{}^*_i-\overset{\tilde{Z}}{\mathbb{E}}{}^*]}\mathbf{M}_2^{[\overset{\hat{L}}{\mathbb{E}}{}^*_i-\overset{\tilde{Z}}{\mathbb{E}}{}^*]} + 2\mathbf{M}_1^{[\overset{\hat{L}}{\mathbb{E}}{}^*_i-\overset{\tilde{Z}}{\mathbb{E}}{}^*]}\mathbf{M}_2^{[\overset{\hat{L}}{\mathbb{E}}{}^*_i-\overset{\hat{L}}{\mathbb{E}}{}^*]},\\
\mathbf{M}_3^{[\overset{\hat{L}}{\mathbb{E}}{}^*_i-\overset{\tilde{M}}{\mathbb{E}}{}^*]} &= \mathbf{M}_3^{[\overset{\hat{L}}{\mathbb{E}}{}^*_i-\overset{\hat{L}}{\mathbb{E}}{}^*]} + \mathbf{M}_1^{[\overset{\hat{L}}{\mathbb{E}}{}^*_i-\overset{\tilde{M}}{\mathbb{E}}{}^*]}\mathbf{M}_2^{[\overset{\hat{L}}{\mathbb{E}}{}^*_i-\overset{\tilde{M}}{\mathbb{E}}{}^*]} + 2\mathbf{M}_1^{[\overset{\hat{L}}{\mathbb{E}}{}^*_i-\overset{\tilde{M}}{\mathbb{E}}{}^*]}\mathbf{M}_2^{[\overset{\hat{L}}{\mathbb{E}}{}^*_i-\overset{\hat{L}}{\mathbb{E}}{}^*]},
\end{aligned}
\tag{7.37}
$$

The above expressions imply

$$
\begin{aligned}
&\mathbf{M}_3^{[\overset{\tilde{Z}}{\mathbb{E}}{}^*_i-\overset{\tilde{M}}{\mathbb{E}}{}^*]} \le \mathbf{M}_3^{[\overset{\tilde{Z}}{\mathbb{E}}{}^*_i-\mathbb{E}^*]} \le \mathbf{M}_3^{[\overset{\tilde{Z}}{\mathbb{E}}{}^*_i-\overset{\hat{P}}{\mathbb{E}}{}^*]} \le \mathbf{M}_3^{[\overset{\tilde{Z}}{\mathbb{E}}{}^*_i-\overset{\hat{L}}{\mathbb{E}}{}^*]}\\
&\mathbf{M}_3^{[\overset{\hat{L}}{\mathbb{E}}{}^*_i-\overset{\hat{P}}{\mathbb{E}}{}^*]} \ge \mathbf{M}_3^{[\overset{\hat{L}}{\mathbb{E}}{}^*_i-\mathbb{E}^*]} \ge \mathbf{M}_3^{[\overset{\hat{L}}{\mathbb{E}}{}^*_i-\overset{\tilde{M}}{\mathbb{E}}{}^*]} \ge \mathbf{M}_3^{[\overset{\hat{L}}{\mathbb{E}}{}^*_i-\overset{\tilde{Z}}{\mathbb{E}}{}^*]}.
\end{aligned}
\tag{7.38}
$$

The derived results allow one to bound, above and below, the unknown SRVE response in terms of the ensembles averages. Related forms of the first order bounds have been derived in various forms dating back to Huet [88, 89], [92], Hazanov and Huet [77], Hazanov and Amieur [78], Huet [99, 100], Zohdi et al. [223], Oden and Zohdi [156], Zohdi and Wriggers [225], [233], [235], Zohdi [230] and Zohdi et al. [236]. The second and third order bounds appear to have been unknown. Using similar techniques, bounds on even higher order moments, such as the kurtosis (fourth moment), which measures the "tightness" of the distribution, are possible.

7.6 Remarks

The results derived here can be used in conjunction with a variety of methods to perform large-scale micro-macro simulations. Noteworthy are the *multiscale methods*: Fish and Wagiman [39], Fish et al. [40], Fish and Belsky [41, 42, 43], Fish et al. [44, 45], Fish and Shek [46], Fish and Ghouli [47], Fish and Yu [48], Fish and Chen [49], Chen and Fish [25], Wentorf et al. [213], Ladeveze and Leguillon [112], Ladeveze and Dureisseix [113], [115], Ladeveze [114] and Champaney et al. [23]; *Voronoi cell methods*: Ghosh and Mukhopadhyay [54], Ghosh and Moorthy [55], Ghosh et al. [56], Ghosh and Moorthy [57], Ghosh et al. [58, 59], Lee et al. [116], Li et al. [123], Moorthy and Ghosh [146] and

Raghavan et al. [171], *transformation methods*: Moulinec et al. [147] and Michel et al. [144], *partitioning methods*: Huet [88, 89], [92], Hazanov and Huet [77], Hazanov and Amieur [78] and Huet [99, 100] and the *adaptive hierarchical modeling methods*: Zohdi et al. [223], Oden and Zohdi [156], Moes et al. [145], Oden and Vemaganti [157], Oden et al. [158] and Vemaganti and Oden [209] and finally *multipole methods* adapted to such problems by Fu et al. [51]. Particularly, attractive are iterative domain decomposition type strategies, whereby a global domain is divided into nonoverlapping subdomains. On the interior subdomain partitions an approximate globally kinematically admissible displacement is projected. This allows the subdomains to be mutually decoupled, and therefore separately solvable. The subdomain boundary value problems are solved with the exact microstructural representation contained within their respective boundaries, but with approximate displacement boundary data. The resulting microstructural solution is the assembly of the subdomain solutions, each restricted to its corresponding subdomain. As in the ensemble testing, the approximate solution is far more inexpensive to compute than the direct problem. Numerical and theoretical studies of such approaches have been studied in Huet [88], Hazanov and Huet [78], Zohdi et al. [223], Oden and Zohdi [156], Zohdi and Wriggers [225], [233], [235], Zohdi [230] and Zohdi et al. [236]. Clearly, when decomposing the structure by a projection of a kinematically admissible function onto the partitioning interfaces, regardless of the constitutive law, the error is due to the jumps in tractions at the interfaces (statical inadmissibility). If the interfaces would be in equilibrium, then there would be no traction jumps. Therefore, if the resulting approximate solution is deemed not accurate enough, via a-posteriori error estimation techniques, the decoupling function on the boundaries of the subdomain is updated using information from the previously computed solution, and the subdomains are resolved. Methods for updating subdomain boundaries can be found in Zohdi et al. [236]. They bear a strong relation to alternating Schwarz methods (see Le Tallec [119] for reviews) and methods of equilibration (see Ainsworth and Oden [2]).

Chapter 8
Domain Decomposition Analogies and Extensions

One can consider the multiple sample testing introduced in the last chapter as a type of domain decomposition process. In order to see this, we consider an approach whereby a global domain, under arbitrary loading, is divided into nonoverlapping subdomains. On the interior subdomain partitions an approximate globally kinematically admissible solution is projected. This allows the subdomains to be mutually decoupled, and therefore separately solvable. The subdomain boundary value problems are solved with the exact microstructural representation contained within their respective boundaries, but with approximate displacement boundary data. The resulting microstructural solution is the assembly of the subdomain solutions, each restricted to its corresponding subdomain. The approximate solution is far more inexpensive to compute than the direct problem. The work follows from results found in Zohdi et al. [236].

8.1 Boundary Value Problem Formulations

The globally exact solution, $\mathbf{u}$, is characterized by the following virtual work formulation:

$$\begin{aligned}&\text{Find } \mathbf{u} \in \mathbf{H}^1(\Omega), \mathbf{u}|_{\Gamma_u} = \mathbf{d}, \text{ such that}\\ &\int_\Omega \nabla \mathbf{v} : \sigma \, d\Omega = \int_\Omega \mathbf{f} \cdot \mathbf{v} \, d\Omega + \int_{\Gamma_t} \mathbf{t} \cdot \mathbf{v} \, dA \qquad \forall \mathbf{v} \in \mathbf{H}^1(\Omega),\ \mathbf{v}|_{\Gamma_u} = \mathbf{0},\end{aligned} \tag{8.1}$$

where σ is the Cauchy stress. In the infinitesimal strain, linearly elastic, case $\sigma = \mathbb{E} : \nabla \mathbf{u}$. In order to construct approximate solutions, next we consider the subdomain boundary value problems, which have the exact microstructural representation contained within the domain, but approximate *displacement* boundary data on the interior subdomain boundaries.

To construct the approximate microstructural solutions we first partition the domain, Ω, into N nonintersecting open subdomains $\overline{\cup_{K=1}^{N}\Omega_K} = \overline{\Omega}$. We define the boundary of an individual subdomain Ω_K, as $\partial\Omega_K$. When employing the applied internal displacement approach, a kinematically admissible function, $\mathbf{U} \in \mathbf{H}^1(\Omega)$ and $\mathbf{U}|_{\Gamma_u} = \mathbf{d}$, is projected onto the *internal boundaries* of the subdomain partitions. Any subdomain boundaries coinciding with the exterior surface retain their original boundary conditions (Fig. 8.1). Accordingly, we have the following virtual work formulation, for each subdomain, $1 \leq K \leq N$:

$$
\begin{aligned}
&\text{Find } \tilde{\mathbf{u}}_K \in \mathbf{H}^1(\Omega_K), \tilde{\mathbf{u}}_K|_{\partial\Omega_K\cap(\Omega\cup\Gamma_u)} = \mathbf{U} \in \mathbf{H}^1(\Omega), \text{ such that} \\
&\int_{\Omega_K} \nabla\mathbf{v}_K : \tilde{\sigma}_K \, d\Omega = \int_{\Omega_K} \mathbf{f}\cdot\mathbf{v}_K \, d\Omega + \int_{\partial\Omega_K\cap\Gamma_t} \mathbf{t}\cdot\mathbf{v}_K \, dA \qquad (8.2) \\
&\forall \mathbf{v}_K \in \mathbf{H}^1(\Omega_K), \mathbf{v}_K|_{\partial\Omega_K\cap(\Omega\cup\Gamma_u)} = \mathbf{0}.
\end{aligned}
$$

The constitutive law and microstructure are identical to that of the globally exact problem. In the infinitesimal strain, linearly elastic, case $\tilde{\sigma} = \mathbb{E} : \nabla\tilde{\mathbf{u}}_K$. The individual subdomain solutions, $\tilde{\mathbf{u}}_K$, are zero outside of the corresponding subdomain $\overline{\Omega}_K$. In this case the approximate solution is constructed by a direct assembly process

$$
\tilde{\mathbf{u}} \stackrel{\text{def}}{=} \mathbf{U} + (\tilde{\mathbf{u}}_1 - \mathbf{U})|_{\Omega_1} + (\tilde{\mathbf{u}}_2 - \mathbf{U})|_{\Omega_2} + \cdots + (\tilde{\mathbf{u}}_N - \mathbf{U})|_{\Omega_N}. \qquad (8.3)
$$

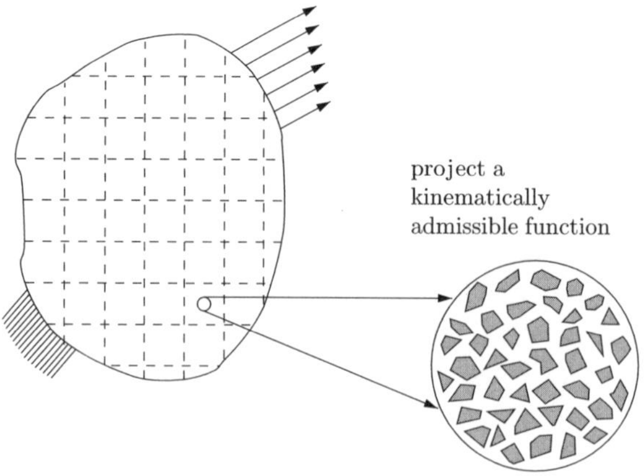

Fig. 8.1 Construction of the approximate solution

The approximate displacement field is in $\mathbf{H}^1(\Omega)$, however, the approximate traction field is possibly discontinuous. Logical choices of $\mathbf{U}$ will be given later in the work. It should be clear that if $\tilde{\mathbf{u}} = \mathbf{u}$ on the internal partition boundaries, then the approximate solution is exact.

Remark I. In theory, an approach of projecting the tractions could be developed. However, such an approach has a number of difficulties, in particular the pure traction subdomain problems must have equilibrated tractions to be well posed. To impose this on the numerical implementation level is not a trivial task. Primarily for this reason the projected displacement approach is preferable.

Remark II. For example, if one considers a cubical domain, whose response is simulated using n numerical unknowns, for example employing a finite element discretization, approximately $\mathcal{O}(n^\gamma)$ floating point operations are needed to solve the system. The value of γ, typically $2 \leq \gamma \leq 3$, depends on the type of algebraic system solver used. If the cube was divided into N equal subdomains (subcubes), the number of floating point operations needed would be approximately on the order of $\left(\frac{n}{N}\right)^\gamma N$, and thus

$$DIRECT\ COSTS/DECOMPOSED\ COSTS \approx N^{\gamma-1}. \tag{8.4}$$

For example, if one had 1000 subdomains, the "broken" solution costs between 1000 and 1000000 times less to compute than the globally exact solution. The advantages are not limited to the possible reduction of operation counts, since the subdomain problems can be solved separately, and trivially in parallel. If parallel processing is used for the decomposed problem, which by construction has a perfect speedup, then the ratio of costs becomes $PN^{\gamma-1}$, where P denotes the number of processors.

8.2 Error in the "Broken" Problems

Since we employ energy type principles to generate approximate solutions, we use energy type measures for the error. Directly by use of the divergence theorem one has

$$\begin{aligned}\int_\Omega \nabla(\mathbf{u}-\tilde{\mathbf{u}}) : (\sigma - \tilde{\sigma})\, d\Omega &= \sum_{K=1}^{N} \int_{\Omega_K} \nabla(\mathbf{u}-\tilde{\mathbf{u}}_K) : (\sigma - \tilde{\sigma}_K)\, d\Omega_K, \\ &= \sum_{K=1}^{N} \int_{\Omega_K} \nabla \cdot ((\mathbf{u}-\tilde{\mathbf{u}}_K) \cdot (\sigma - \tilde{\sigma}_K))\, d\Omega_K, \\ &\quad - \sum_{K=1}^{N} \int_{\Omega_K} (\mathbf{u}-\tilde{\mathbf{u}}_K) \cdot \underbrace{\nabla \cdot (\sigma - \tilde{\sigma}_K)}_{=\mathbf{0}}\, d\Omega_K,\end{aligned}$$

$$= \sum_{K=1}^{N} \int_{\partial\Omega_K \cap \Omega} \underbrace{(\tilde{\boldsymbol{\sigma}}_K \cdot \mathbf{n}_K)}_{\text{not continuous}} \cdot \underbrace{(\tilde{\mathbf{u}}_K - \mathbf{u})}_{\text{error}} \, dA_K,$$

$$= \sum_{\mathscr{I}=1}^{N_{\mathscr{I}}} \int_{\Gamma_{\mathscr{I}}} (\text{traction jump}) \cdot (\text{displacement error}) \, dA_K, \tag{8.5}$$

where $\Gamma_{\mathscr{I}}$ is an interior subdomain interface, and $\mathscr{I} = 1, 2, \ldots N_{\mathscr{I}}$ = number of interior subdomain interfaces. For the applied internal displacement case, the tractions may suffer discontinuities at the interior subdomain boundaries (see Fig. 8.2). In the case of infinitesimal strain, linear elasticity, we have

$$||\mathbf{u} - \tilde{\mathbf{u}}||^2_{E(\Omega)} = \sum_{K=1}^{N} \int_{\partial\Omega_K \cap \Omega} \underbrace{(\mathbb{E} : \nabla\tilde{\mathbf{u}}_K \cdot \mathbf{n}_K)}_{\text{not continuous}} \cdot \underbrace{(\tilde{\mathbf{u}}_K - \mathbf{u})}_{\text{error}} \, dA_K,$$
$$||\mathbf{u} - \tilde{\mathbf{u}}||^2_{E(\Omega)} \stackrel{\text{def}}{=} \int_{\Omega} \nabla(\mathbf{u} - \tilde{\mathbf{u}}) : \mathbb{E} : \nabla(\mathbf{u} - \tilde{\mathbf{u}}) \, d\Omega. \tag{8.6}$$

It is convenient to cast the error in terms of the potential energy for the case of linear elasticity, $\mathscr{J}(\mathbf{w}) \stackrel{\text{def}}{=} \frac{1}{2}\int_{\Omega} \nabla\mathbf{w} : \mathbb{E} : \nabla\mathbf{w} \, d\Omega - \int_{\Omega} \mathbf{f} \cdot \mathbf{w} \, d\Omega - \int_{\Gamma_t} \mathbf{t} \cdot \mathbf{w} \, dA$, where $\mathbf{w}$ is any kinematically admissible function. The well-known relationship for any kinematically admissible function $\mathbf{w}$ is

$$||\mathbf{u} - \mathbf{w}||^2_{E(\Omega)} = 2(\mathscr{J}(\mathbf{w}) - \mathscr{J}(\mathbf{u})) \; or \; \mathscr{J}(\mathbf{u}) \leq \mathscr{J}(\mathbf{w}), \tag{8.7}$$

which is the Principle of Minimum Potential Energy (PMPE). In other words, the true solution possesses a minimum potential. Therefore, since $\tilde{\mathbf{u}}$ is kinematically admissible, we immediately have $||\mathbf{u} - \tilde{\mathbf{u}}||^2_{E(\Omega)} = 2(\mathscr{J}(\tilde{\mathbf{u}}) - \mathscr{J}(\mathbf{u}))$. The critical observation is that if we can bound $\mathscr{J}(\mathbf{u})$ from below, then we can bound the error from above. In other words, what we seek is $\mathscr{J}^- \leq \mathscr{J}(\mathbf{u})$.

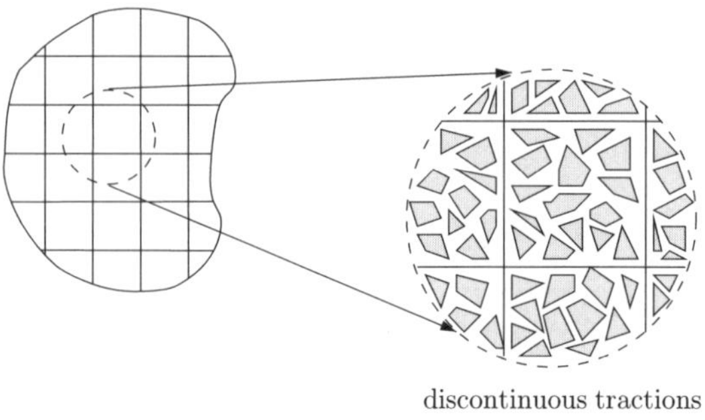

Fig. 8.2 The consequences of the projection approach

To bound $\mathscr{J}(\mathbf{u})$ in terms of easily accessible quantities, we first calculate a computationally inexpensive, kinematically admissible, regularized test solution. The regularized test solution, $\mathbf{u}^R$, is characterized by a virtual work formulation:

Find a $\mathbf{u}^R \in \mathbf{H}^1(\Omega), \mathbf{u}^R|_{\Gamma_u} = \mathbf{d}$, such that

$$\int_\Omega \nabla \mathbf{v} : \sigma^R \, d\Omega = \int_\Omega \mathbf{f} \cdot \mathbf{v} \, d\Omega + \int_{\Gamma_t} \mathbf{t} \cdot \mathbf{v} \, dA \qquad \forall \mathbf{v} \in \mathbf{H}^1(\Omega), \mathbf{v}|_{\Gamma_u} = \mathbf{0}. \tag{8.8}$$

Here the constitutive law for σ^R should be taken to be as simple as a possible, i.e. a constant (regularized) fourth order symmetric positive definite linear elasticity tensor $\mathbb{R}$, defined via $\sigma^R = \mathbb{R} : \nabla \mathbf{u}^R$. *In general, the regularized solution $(\sigma^R, \mathbf{u}^R)$ does not coincide with the field associated with the decoupling solution* $\mathbf{U}$. If we choose $\mathbf{w} = \mathbf{u}^R$ in (8.7), which is an admissible function, we obtain $||\mathbf{u} - \mathbf{u}^R||^2_{E(\Omega)} = 2(\mathscr{J}(\mathbf{u}^R) - \mathscr{J}(\mathbf{u}))$, which implies $\mathscr{J}(\mathbf{u}) = \mathscr{J}(\mathbf{u}^R) - \frac{1}{2}||\mathbf{u} - \mathbf{u}^R||^2_{E(\Omega)}$. Our objective is to form an *upper bound* on $||\mathbf{u} - \mathbf{u}^R||_{E(\Omega)}$ in terms of $\mathbb{R}$, $\mathbb{E}$ and $\nabla \mathbf{u}^R$, to obtain a *lower bound* on $\mathscr{J}(\mathbf{u})$. For the bound to be useful, it should contain no unknown constants, and should be solely in terms of the regularized solution and the material data. In other words we seek $||\mathbf{u} - \mathbf{u}^R||_{E(\Omega)} \leq \mathscr{H}^{+(u)}$, which will lead to

$$\mathscr{J}^- \stackrel{\text{def}}{=} \mathscr{J}(\mathbf{u}^R) - \frac{1}{2}(\mathscr{H}^{+(u)})^2 \leq \mathscr{J}(\mathbf{u}), \tag{8.9}$$

thus supplying an upper bound for the quantities in Box 8.6.

8.2.1 Multiscale Proximity Bounds

The solution corresponding to a material with microstructure is $\mathbf{u}$, and is characterized by the following virtual work formulation:

Find $\mathbf{u} \in \mathbf{H}^1(\Omega), \mathbf{u}|_{\Gamma_u} = \mathbf{d}$, such that

$$\underbrace{\int_\Omega \nabla \mathbf{v} : \sigma \, d\Omega}_{\stackrel{\text{def}}{=} \mathscr{B}(\mathbf{u},\mathbf{v})} = \underbrace{\int_\Omega \mathbf{f} \cdot \mathbf{v} \, d\Omega + \int_{\Gamma_t} \mathbf{t} \cdot \mathbf{v} \, dA}_{\stackrel{\text{def}}{=} \mathscr{F}(\mathbf{v})} \qquad \forall \mathbf{v} \in \mathbf{H}^1(\Omega), \mathbf{v}|_{\Gamma_u} = \mathbf{0}. \tag{8.10}$$

In the infinitesimal strain linearly elastic case, then $\sigma = \mathbb{E} : \nabla \mathbf{u}$. The equivalent complementary form is:

Find $\sigma, \nabla \cdot \sigma + \mathbf{f} = \mathbf{0}, \sigma \cdot \mathbf{n}|_{\Gamma_t} = \mathbf{t}$ such that

$$\underbrace{\int_{\Omega} \tau : \mathbb{E}^{-1} : \sigma \, d\Omega}_{\stackrel{\text{def}}{=} \mathscr{A}(\sigma,\tau)} = \underbrace{\int_{\Gamma_u} \tau \cdot \mathbf{n} \cdot \mathbf{d} \, dA}_{\stackrel{\text{def}}{=} \mathscr{G}(\tau)} \qquad \forall \tau, \nabla \cdot \tau = \mathbf{0}, \tau \cdot \mathbf{n}|_{\Gamma_t} = \mathbf{0}. \tag{8.11}$$

The solution to the constant coefficient problem, denoted the *regular* solution, $\mathbf{u}^R$, is characterized by a virtual work formulation:

Find a $\mathbf{u}^R \in \mathbf{H}^1(\Omega), \mathbf{u}^R|_{\Gamma_u} = \mathbf{d}$, such that

$$\underbrace{\int_{\Omega} \nabla \mathbf{v} : \sigma^R \, d\Omega}_{\stackrel{\text{def}}{=} \mathscr{B}^R(\mathbf{u}^R,\mathbf{v})} = \underbrace{\int_{\Omega} \mathbf{f} \cdot \mathbf{v} \, d\Omega + \int_{\Gamma_t} \mathbf{t} \cdot \mathbf{v} \, dA}_{\stackrel{\text{def}}{=} \mathscr{F}(\mathbf{v})} \qquad \forall \mathbf{v} \in \mathbf{H}^1(\Omega), \mathbf{v}|_{\Gamma_u} = \mathbf{0}, \tag{8.12}$$

where $\sigma^R = \mathbb{R} : \nabla \mathbf{u}^R$. The equivalent complementary form is:

Find $\sigma^R, \nabla \cdot \sigma^R + \mathbf{f} = \mathbf{0}, \sigma^R \cdot \mathbf{n}|_{\Gamma_t} = \mathbf{t}$ such that

$$\underbrace{\int_{\Omega} \tau : \mathbb{R}^{-1} : \sigma^R \, d\Omega}_{\stackrel{\text{def}}{=} \mathscr{A}^R(\sigma^R,\tau)} = \underbrace{\int_{\Gamma_u} \tau \cdot \mathbf{n} \cdot \mathbf{d} \, dA}_{\stackrel{\text{def}}{=} \mathscr{G}(\tau)} \qquad \forall \tau, \nabla \cdot \tau = \mathbf{0}, \tau \cdot \mathbf{n}|_{\Gamma_t} = \mathbf{0}. \tag{8.13}$$

For the complementary problem, similar restrictions are placed on the solution and test fields to force the integrals to make sense. In other words, we assume that solutions produce finite global energy.

We have for any kinematically admissible function $\mathbf{w}$, a definition of the primal energy norm

$$0 \leq ||\mathbf{u}-\mathbf{w}||^2_{E(\Omega)} \stackrel{\text{def}}{=} \underbrace{\int_\Omega (\nabla \mathbf{u} - \nabla \mathbf{w}) : \mathbb{E} : (\nabla \mathbf{u} - \nabla \mathbf{w})\, d\Omega}_{\mathscr{B}(\mathbf{u}-\mathbf{w},\mathbf{u}-\mathbf{w})} = 2\mathscr{J}(\mathbf{w}) - 2\mathscr{J}(\mathbf{u}), \tag{8.14}$$

where we define the "elastic potential" by

$$\mathscr{J}(\mathbf{w}) \stackrel{\text{def}}{=} \frac{1}{2}\mathscr{B}(\mathbf{w},\mathbf{w}) - \mathscr{F}(\mathbf{w}) = \frac{1}{2}\int_\Omega \nabla \mathbf{w} : \mathbb{E} : \nabla \mathbf{w}\, d\Omega - \int_\Omega \mathbf{f}\cdot \mathbf{w}\, d\Omega - \int_{\Gamma_t} \mathbf{t}\cdot \mathbf{w}\, dA. \tag{8.15}$$

Equation (8.14) is a form of the Principle of Minimum Potential Energy. In other words, the true solution possesses a minimum potential. Similarly, for the complementary formulation, for any statically admissible function (γ), one has

$$0 \leq ||\sigma - \gamma||^2_{E^{-1}(\Omega)} \stackrel{\text{def}}{=} \underbrace{\int_\Omega (\sigma - \gamma) : \mathbb{E}^{-1} : (\sigma - \gamma)\, d\Omega}_{\mathscr{A}(\sigma-\gamma,\sigma-\gamma)} = 2\mathscr{K}(\gamma) - 2\mathscr{K}(\sigma), \tag{8.16}$$

where

$$\mathscr{K}(\gamma) \stackrel{\text{def}}{=} \frac{1}{2}\mathscr{A}(\gamma,\gamma) - \mathscr{G}(\gamma) = \frac{1}{2}\int_\Omega \gamma : \mathbb{E}^{-1} : \gamma\, d\Omega - \int_{\Gamma_u} \gamma \cdot \mathbf{n} \cdot \mathbf{d}\, dA. \tag{8.17}$$

This is a form of the Principle of Minimum Complementary Potential Energy. By directly adding together the potential energy and the complementary energy we obtain an equation of balance:

$$\begin{aligned} \mathscr{J}(\mathbf{u}) + \mathscr{K}(\sigma) &= \frac{1}{2}\int_\Omega \nabla \mathbf{u} : \mathbb{E} : \nabla \mathbf{u}\, d\Omega - \int_\Omega \mathbf{f}\cdot \mathbf{u}\, d\Omega - \int_{\Gamma_t} \mathbf{t}\cdot \mathbf{u}\, dA \\ &\quad + \frac{1}{2}\int_\Omega \sigma : \mathbb{E}^{-1} : \sigma\, d\Omega - \int_{\Gamma_u} (\sigma \cdot \mathbf{n}) \cdot \mathbf{d}\, dA = 0. \end{aligned} \tag{8.18}$$

If we choose $\mathbf{w} = \mathbf{u}^R$, which is a kinematically admissible function, we obtain $||\mathbf{u} - \mathbf{u}^R||^2_{E(\Omega)} = 2(\mathscr{J}(\mathbf{u}^R) - \mathscr{J}(\mathbf{u}))$ which implies $\mathscr{J}(\mathbf{u}) = \mathscr{J}(\mathbf{u}^R) - \frac{1}{2}||\mathbf{u} - \mathbf{u}^R||^2_{E(\Omega)}$. Also, choosing $\gamma = \sigma^R$, which is statically admissible, we have $||\sigma - \sigma^R||^2_{E^{-1}(\Omega)} = 2(\mathscr{K}(\sigma^R) - \mathscr{K}(\sigma))$ which implies $\mathscr{K}(\sigma) = \mathscr{K}(\sigma^R) - \frac{1}{2}||\sigma - \sigma^R||^2_{E^{-1}(\Omega)}$. Combining the two previous results yields

$$2(\mathscr{J}(\mathbf{u}^R) + \mathscr{K}(\sigma^R)) = ||\mathbf{u} - \mathbf{u}^R||^2_{E(\Omega)} + ||\sigma - \sigma^R||^2_{E^{-1}(\Omega)} + \underbrace{2(\mathscr{J}(\mathbf{u}) + \mathscr{K}(\sigma))}_{=0}. \tag{8.19}$$

This "estimate" is an exact measure of this particular norm, and requires *no computation of the exact microfield boundary value problem*. If one wishes to isolate either the primal or complementary norms, one must resort to two-sided bounding techniques. Clearly, the primal norm is desirable in order to determine the approximate solution's proximity to the exact fine-scale solution. Since $\mathbf{u}^R$ is kinematically admissible, from (8.14), we immediately have

$$\mathscr{J}(\mathbf{u}) = \mathscr{J}(\mathbf{u}^R) - \frac{1}{2}||\mathbf{u}-\mathbf{u}^R||^2_{E(\Omega)} \geq \mathscr{J}(\mathbf{u}^R) - \frac{1}{2}(\mathscr{H}^{+(u)})^2 \stackrel{\text{def}}{=} \mathscr{J}^-, \tag{8.20}$$

where the potential lower bound, $\mathscr{J}^-$, is sought-after. From (8.10) and (8.12), we have $\int_\Omega \nabla\mathbf{v} : \mathbb{E} : \nabla\mathbf{u}\, d\Omega = \int_\Omega \nabla\mathbf{v} : \mathbb{R} : \nabla\mathbf{u}^R\, d\Omega = \int_\Omega \mathbf{f}\cdot\mathbf{v}\, d\Omega + \int_{\Gamma_t} \mathbf{t}\cdot\mathbf{v}\, dA$. Subtracting $\int_\Omega \nabla\mathbf{v} : \mathbb{E} : \nabla\mathbf{u}^R\, d\Omega$ from both sides yields $\int_\Omega \nabla\mathbf{v} : \mathbb{E} : \nabla(\mathbf{u}-\mathbf{u}^R)\, d\Omega = \int_\Omega \nabla\mathbf{v} : (\mathbb{R} : \nabla\mathbf{u}^R - \mathbb{E} : \nabla\mathbf{u}^R)\, d\Omega$. Since $\mathbf{v}$ is an arbitrary virtual displacement, we may set $\mathbf{v} = \mathbf{u} - \mathbf{u}^R$ to yield

$$\begin{aligned}
||\mathbf{u}-\mathbf{u}^R||^2_{E(\Omega)} &= \int_\Omega \nabla(\mathbf{u}-\mathbf{u}^R) : (\mathbb{R} : \nabla\mathbf{u}^R - \mathbb{E} : \nabla\mathbf{u}^R)\, d\Omega \\
&= \int_\Omega (\mathbb{E}^{\frac{1}{2}} : \nabla(\mathbf{u}-\mathbf{u}^R)) : \mathbb{E}^{\frac{1}{2}} : \mathbb{E}^{-1} : ((\mathbb{R}-\mathbb{E}) : \nabla\mathbf{u}^R)\, d\Omega \\
&\leq \underbrace{\left(\int_\Omega \nabla(\mathbf{u}-\mathbf{u}^R) : \mathbb{E} : \nabla(\mathbf{u}-\mathbf{u}^R)\, d\Omega\right)^{\frac{1}{2}}}_{||\mathbf{u}-\mathbf{u}^R||_{E(\Omega)}} \\
&\quad \times \left(\int_\Omega ((\mathbb{R}-\mathbb{E}) : \nabla\mathbf{u}^R) : \mathbb{E}^{-1} : ((\mathbb{R}-\mathbb{E}) : \nabla\mathbf{u}^R)\, d\Omega\right)^{\frac{1}{2}}.
\end{aligned} \tag{8.21}$$

Therefore, a primal solution difference upper bound is

$$||\mathbf{u}-\mathbf{u}^R||^2_{E(\Omega)} \leq \underbrace{\int_\Omega ((\mathbb{R}-\mathbb{E}) : \nabla\mathbf{u}^R) : \mathbb{E}^{-1} : ((\mathbb{R}-\mathbb{E}) : \nabla\mathbf{u}^R)\, d\Omega.}_{\stackrel{\text{def}}{=}(\mathscr{H}^{+(u)})^2} \tag{8.22}$$

To derive this result we used the fact that a symmetric positive-definite matrix has a unique square root, $\mathbb{E} = \mathbb{E}^{\frac{1}{2}} : \mathbb{E}^{\frac{1}{2}}$ and the Cauchy-Schwarz inequality. Note that the $\mathscr{H}^{+(u)}$-bound depends only on $\mathbb{R}$, $\mathbb{E}$ and $\mathbf{u}^R$. The process is repeated in a similar manner to form a complementary upper bound:

$$||\sigma-\sigma^R||^2_{E^{-1}(\Omega)} \leq \underbrace{\int_\Omega \left((\mathbb{R}^{-1}-\mathbb{E}^{-1}) : \sigma^R\right) : \mathbb{E} : \left((\mathbb{R}^{-1}-\mathbb{E}^{-1}) : \sigma^R\right) d\Omega.}_{\stackrel{\text{def}}{=}\left(\overline{\mathscr{H}}^{+(u)}\right)^2} \tag{8.23}$$

Analogous to the $\mathscr{H}^{+(u)}$-bound, the $\overline{\mathscr{H}}^{+(u)}$-bound depends only on $\mathbb{R}$, $\mathbb{E}$ and σ^R. From (8.19), lower primal and complementary bounds follow,

$$
\begin{aligned}
||\mathbf{u}-\mathbf{u}^R||^2_{E(\Omega)} &= 2(\mathscr{J}(\mathbf{u}^R)+\mathscr{K}(\sigma^R)) - ||\sigma-\sigma^R||^2_{E^{-1}(\Omega)},\\
&\geq 2(\mathscr{J}(\mathbf{u}^R)+\mathscr{K}(\sigma^R)) - (\overline{\mathscr{H}}^{+(u)})^2\\
&= 2(\mathscr{J}(\mathbf{u}^R)+\mathscr{K}^-)\stackrel{\text{def}}{=}(\mathscr{H}^{-(u)})^2,\\
||\sigma-\sigma^R||^2_{E^{-1}(\Omega)} &= 2(\mathscr{J}(\mathbf{u}^R)+\mathscr{K}(\sigma^R)) - ||\mathbf{u}-\mathbf{u}^R||^2_{E(\Omega)},\\
&\geq 2(\mathscr{J}(\mathbf{u}^R)+\mathscr{K}(\sigma^R)) - (\mathscr{H}^{+(u)})^2\\
&= 2(\mathscr{K}(\sigma^R)-\mathscr{J}^-)\stackrel{\text{def}}{=}(\overline{\mathscr{H}}^{-(u)})^2.
\end{aligned} \tag{8.24}
$$

Therefore, in summary

$$
\begin{aligned}
&2(\mathscr{J}(\mathbf{u}^R)+\mathscr{K}^-) = (\mathscr{H}^{-(u)})^2 \leq ||\mathbf{u}-\mathbf{u}^R||^2_{E(\Omega)} \leq (\mathscr{H}^{+(u)})^2\\
&= 2(\mathscr{J}(\mathbf{u}^R)-\mathscr{J}^-),\\
&2(\mathscr{K}(\sigma^R)+\mathscr{J}^-) = (\overline{\mathscr{H}}^{-(u)})^2 \leq ||\sigma-\sigma^R||^2_{E^{-1}(\Omega)} \leq (\overline{\mathscr{H}}^{+(u)})^2\\
&= 2(\mathscr{K}(\sigma^R)-\mathscr{K}^-).
\end{aligned} \tag{8.25}
$$

These two sided bounds hold for any loading, external geometry and pointwise positive-definite microstructure. Related forms of such estimates can be found in Zohdi [230].

8.2.2 Uses of the Bounds for Domain Decomposition

From the previous results one immediately has

$$
||\mathbf{u}-\mathbf{u}^R||^2_{E(\Omega)} \leq \int_\Omega ((\mathbb{E}-\mathbb{R}):\nabla\mathbf{u}^R):\mathbb{E}^{-1}:((\mathbb{E}-\mathbb{R}):\nabla\mathbf{u}^R)\,d\Omega \stackrel{\text{def}}{=} \mathscr{H}^{+(u)^2}. \tag{8.26}
$$

To derive this result we used the fact that a symmetric positive definite matrix has a unique positive square root, $\mathbb{E} = \mathbb{E}^{\frac{1}{2}} : \mathbb{E}^{\frac{1}{2}}$ and the Cauchy-Schwarz inequality.

Note that the $\mathscr{H}^{+(u)}$-bound depends only on $\mathbf{I\!R}$, $\mathbf{I\!E}$ and $\nabla \mathbf{u}^R$. Therefore, using (8.22) we have $\mathscr{J}^- \stackrel{\text{def}}{=} \mathscr{J}(\mathbf{u}^R) - \frac{1}{2}(\mathscr{H}^{+(u)})^2 \leq \mathscr{J}(\mathbf{u})$, and thus

$$\underbrace{||\mathbf{u}-\tilde{\mathbf{u}}||^2_{E(\Omega)}}_{\text{ERROR}=2(\mathscr{J}(\tilde{\mathbf{u}})-\mathscr{J}(\mathbf{u}))} \leq \underbrace{2(\mathscr{J}(\tilde{\mathbf{u}}) - \mathscr{J}^-)}_{\text{UPPER BOUND}}. \tag{8.27}$$

Remark I. Bounds of such type have been studied in Zohdi et al. [223], Oden and Zohdi [156] and Zohdi and Wriggers [225], however, they required that the decoupling solution $\mathbf{U}$ be identical to $\mathbf{u}^R$. *Clearly this is unnecessary.* In the preceding bound, the tensor $\mathbf{I\!R}$ did not have to be constant, and this has been exploited to generate hierarchical micromechanical models in Zohdi et al. [223]. More general theoretical aspects of such bounds, including their complementary (dual) formulation counterparts, can be found in Zohdi and Wriggers [225] and Zohdi [230]. Under certain conditions, such bounds coincide with results found in Huet [87, 88].

Remark II. It is important to realize that the only approximation in this entire estimate is $||\mathbf{u}-\mathbf{u}^R||_{E(\Omega)} \leq \mathscr{H}^{+(u)}$. Thus the ratio $\rho \stackrel{\text{def}}{=} \frac{\mathscr{H}^{+(u)}}{||\mathbf{u}-\mathbf{u}^R||_{E(\Omega)}}$ is an extremely good indicator of the quality of the estimate. Numerical and theoretical studies of the proximity of ρ to unity has been studied in Zohdi et al. [223], Oden and Zohdi [156], Zohdi and Wriggers [225, 229] and Zohdi [230]. It is clear that the tensorial parameter $\mathbf{I\!R}$ is free to choose, provided it is symmetric and positive definite. Thus, ideally, one would want to minimize $\rho - 1$, in other words $\frac{\partial(\rho-1)}{\partial \mathbf{I\!R}} = \mathbf{0}$, $\frac{\partial^2(\rho-1)}{\partial \mathbf{I\!R}^2} > \mathbf{0} \Rightarrow \mathbf{I\!R}^\star$. *Unfortunately, minimization of* $\rho - 1$ *is usually impossible, since* $||\mathbf{u}-\mathbf{u}^R||_{E(\Omega)}$ *is unknown.* However, another avenue to optimize the bound is possible. Formally, the process to obtain the optimal choice, $\mathbf{I\!R}^\star$, *is to maximize the lower bound* $\mathscr{J}^-$, in other words $\frac{\partial \mathscr{J}^-}{\partial \mathbf{I\!R}} = \mathbf{0}$, $\frac{\partial^2 \mathscr{J}^-}{\partial \mathbf{I\!R}^2} < \mathbf{0} \Rightarrow \mathbf{I\!R}^\star$. For example, if we consider a one-dimensional displacement controlled structure

$$0 < \mathrm{x} < \mathrm{L} : \frac{d}{dx}(E(x)\frac{du}{dx}) = 0 \qquad u(0) = \Delta_0 \qquad u(L) = \Delta_L, \tag{8.28}$$

it is easy to show that

$$\rho = \frac{\mathscr{H}^{+(u)}}{||u-u^R||_{E(\Omega)}} = \sqrt{\frac{\int_0^1 \left(\frac{1}{R} - \frac{1}{E}\right)^2 E\,dx}{\int_0^1 \left(\frac{1}{R} - \frac{1}{E}\beta\right)^2 E\,dx}} \qquad \beta \stackrel{\text{def}}{=} \frac{\langle \frac{1}{R} \rangle_L}{\langle \frac{1}{E} \rangle_L}, \tag{8.29}$$

where the estimate is exact ($\rho = 1$) for $R = \langle \frac{1}{E} \rangle_L^{-1} \stackrel{\text{def}}{=} \left(\frac{1}{L}\int_0^L \frac{1}{E}\,dx\right)^{-1}$. This value of R also maximizes $\mathscr{J}^-$.

We remark that for traction controlled loading $u(0) = 0$, $E\frac{du}{dx}(L) = G$, for a given G, the error estimate is exact for any $\mathscr{R}$. For three dimensional problems, the optimal tensor $\mathbf{I\!R}^\star$, which is problem dependent, is easily computable, since it requires *no* microscale computations.

8.3 The Connection to Material Testing

As an example, we apply the decomposition approach to a sample of material containing a significant amount of microstructure (Fig. 8.3). We simulated a rectangular block composed of an Aluminum matrix containing 10240 Boron particles. The partitioning consisted of 512 subblocks (subdomains), each containing a random distribution of 20 nonintersecting particles occupying approximately 22% of volume. We considered external boundary loading of the form $\mathscr{E}_{ij} = 0.001, i, j = 1,2,3$:

$$\left\{ \begin{array}{c} u_1|_{\partial\Omega} \\ u_2|_{\partial\Omega} \\ u_3|_{\partial\Omega} \end{array} \right\} = \underbrace{\left[\begin{array}{ccc} \mathscr{E}_{11} & \mathscr{E}_{12} & \mathscr{E}_{13} \\ \mathscr{E}_{12} & \mathscr{E}_{22} & \mathscr{E}_{23} \\ \mathscr{E}_{31} & \mathscr{E}_{32} & \mathscr{E}_{33} \end{array} \right]}_{\mathscr{E}} \left\{ \begin{array}{c} x_1 \\ x_2 \\ x_3 \end{array} \right\}, \tag{8.30}$$

which is relatively standard in micromechanical analyses. As mentioned previously, the usual motivation for such a loading is that the sample is considered so large that it experiences nearly uniform strain (linear displacements) on its boundary. For this loading, a straightforward kinematically admissible choice for $\mathbf{U}$ (in Box 8.2) is $\mathbf{U} = \mathscr{E} \cdot \mathbf{x}$. Before we ran the full tests, we repeatedly refined the finite element mesh for several subdomains (subblocks), each containing the same number of spheres, but with different random distributions, until we obtained invariant total strain energy responses. The result was that beyond typically a $24 \times 24 \times 24$ mesh (46875 dof per test) per subdomain, roughly 2344 dof per particle, delivered macroscopically (volumetrically averaged) mesh invariant responses. *The resulting number of total*

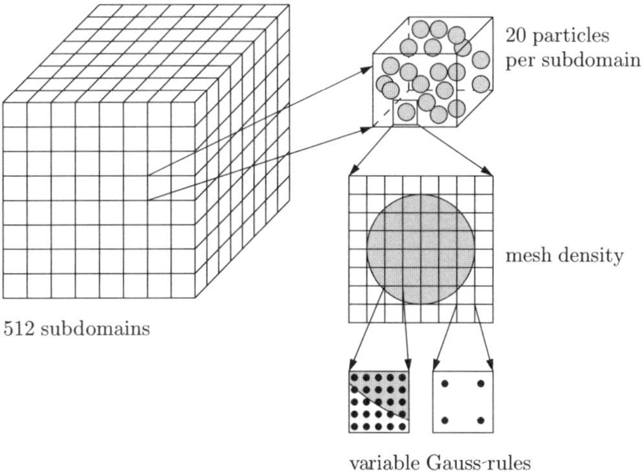

Fig. 8.3 A large sample of Aluminum with 10240 embedded Boron particles. In reality the sample is approximately $0.1\,\text{mm} \times 0.1\,\text{mm} \times 0.1\,\text{mm}$, while the particles are approximately 0.0035 mm in diameter

degrees of freedom (n), if the entire domain was directly simulated with this mesh density, would be $n = 21,567,171$.

Important: Under these special loading conditions, this decomposition process can be considered as equivalent to the statistical testing of 512 samples of heterogeneous materials in the previous chapter, since the loading on each subdomain will be uniform. It is important to realize that since all subdomains are the same size, that the projection of $\mathbf{U} = \mathscr{E} \cdot \mathbf{x}$ on the interfacial boundaries would produce the same strain energy in each subdomain if the material had been uniform.[1] Therefore, it is logical to (statistically) compare the solution behavior between the subdomains. Throughout the tests a standard preconditioned Conjugate Gradient (CG) solver was used. The primary premise for its (wide) use is that a solution, to tolerable accuracy, can be achieved in much less than $\mathscr{O}(n^3)$ operations, which is required with most Gaussian-type techniques. For overviews of a variety of such methods see Axelsson [9]. The CG method, is guaranteed to converge in n iterations, provided the algebra is performed exactly. At each iteration the computational cost is $\mathscr{O}(n^2)$ operations, and if the number of CG iterations is C, then the total cost is Cn^2 operations, where typically $C << n$. Use of an iterative solver allowed the use of the vector $\mathbf{U}$ as a starting vector to speed up the solution process. In other words, a starting vector that captures a-priori the "long wave" components (low frequency eigenmodes) of the solution is advantageous. This increases the effectiveness of the CG searches for this class of problems. A well known fact with regard to iterative solvers is the fact that they are quite adept at capturing high-frequency responses. However, they typically may require many iterations to capture "long-wave" modes. For a discussion see Briggs [18]. For this decomposition, since the number of CG iterations to solve each subdomain problem was $m_{loc} \approx 49$ (Table 6.4). Therefore, the cost to solve all of the subdomain problems was $m_{loc}(\frac{n}{N})^2 N \approx 49\frac{n^2}{512}$, as opposed to the direct cost of approximately $M_{glob}n^2$, where $M_{glob} > m_{loc}$. In this case the approximate cost savings are approximately $DIRECT\ COSTS/DECOMPOSED\ COSTS \approx \frac{NM_{glob}}{m_{loc}} \geq 512$. The time to preprocess, solve and postprocess each 20 particle subdomain took no more than one minute on a single RISC 6000 workstation. Clearly, the overall solution process is trivially parallelizable.

We computed the following upper bound, which coincided with the previous chapter's results, normalized by the energy of $\mathbf{U} = \mathscr{E} \cdot \mathbf{x}$, on the error

$$0 \leq \frac{||\mathbf{u} - \tilde{\mathbf{u}}||^2_{E(\Omega)}}{||\mathbf{U}||^2_{E(\Omega)}} \leq \frac{2(\mathscr{J}(\tilde{\mathbf{u}}) - \mathscr{J}^-)}{||\mathbf{U}||^2_{E(\Omega)}} = 0.1073, \tag{8.31}$$

where the maximized lower bound on the potential is

$$\mathscr{J}^- = \frac{1}{2}\mathscr{E} : \langle \mathbb{E}^{-1} \rangle_\Omega^{-1} : \mathscr{E}|\Omega|, \tag{8.32}$$

[1] Clearly, it is a homogeneous deformation in this case.

where $\mathbb{R}^{\star} = \langle \mathbb{E}^{-1} \rangle_{\Omega}^{-1} = \left(\frac{1}{|\Omega|} \int_{\Omega} \mathbb{E}^{-1} \, d\Omega \right)^{-1}$, has been used. The optimal $\mathbb{R}^{\star}$ stems from computing $\frac{\partial \mathscr{J}^{-}}{\partial \mathbb{R}} = \mathbf{0}$, $\frac{\partial^2 \mathscr{J}^{-}}{\partial \mathbb{R}^2} < \mathbf{0} \Rightarrow \mathbb{R}^{\star} = \langle \mathbb{E}^{-1} \rangle_{\Omega}^{-1}$.

Remarks. In the general case, in order to determine the optimal regularization for a given boundary value problem, one must search for the components of $\mathbb{R}$ which minimize any one of the estimates derived earlier, which, for the sake of generality, we denote by Π. This entails that boundary value problems must be solved repeatedly to form, for example, a multivariate Newton-type search for the components of $\mathbb{R}$ denoted R_i. Explicitly written, a Newton type system is $[\mathbb{H}]\{\Delta \mathbb{R}\} = -\{\mathbf{g}\}$, where $[\mathbb{H}]$ is the Hessian matrix $(N \times N)$, with components $H_{ij} = \frac{\partial^2 \Pi(\mathbb{R})}{\partial R_i \partial R_j}$, $\{\mathbf{g}\}$ is the gradient $(N \times 1)$, with components $g_i = \frac{\partial \Pi(\mathbb{R})}{\partial R_i}$ and where $\{\Delta \mathbb{R}\}$ is the increment $(N \times 1)$, with components ΔR_i, needed to update the components of $\mathbb{R}$ (which are initially guessed). To construct the system $[\mathbb{H}]\{\Delta \mathbb{R}\} = -\{\mathbf{g}\}$, $\mathbf{u}^R$ and σ^R must be determined numerically, for example using the finite element method, to determine Π and its derivatives. As with most complicated systems, finite difference approximations of the gradient and Hessian components are constructed with respect to the search parameters by perturbing the search variables around a base point. The reader is referred to Gill et al. [61] for details. The number of objective function evaluations to construct the derivatives of Π numerically with respect to the design variables can be large. For example, if $\mathbb{R}$ is triclinic, then there are up to 21 free constants, while for an isotropic material there are only two free constants. The construction of the discrete Hessian is the main expense, since the number of objective functions to form it scales with the number of search variables as $\mathscr{O}(N^2)$. There exist a variety of *quasi-Newton* methods, which approximate the Hessian in order to reduce computational effort (see Gill et al. [61]). It is important to realize that these problems may possess multiple local minima with respect to $\mathbb{R}$, because of the possibly nonconvex dependence of Π on $\mathbb{R}$. To deal with this one should restart the search from different starting guesses, to locate other potential minima, if they exist. Alternatively, so-called *GO* (Global Optimization) techniques based upon rapid, non-derivative stochastic function evaluation, may be useful alternatives. We refer the reader to Horst and Tuy [85], Zhigljavsky [219] or Davis [32] for reviews. Finally, we note that regardless of the search method, a restriction is that the tensor $\mathbb{R}$ remain positive definite. Clearly, for the class of problems under consideration, solutions must be generated numerically, for example by the finite element method. We now consider the effect of the use of the finite element method to generate a subspatial approximation to $\mathbf{u}^R$, denoted $\mathbf{u}^{R,H}$

Find $\mathbf{u}^{R,H} \in \mathbf{H}^H_{u^R}(\Omega) \subset \mathbf{H}^1(\Omega), \mathbf{u}^{R,H}|_{\Gamma_u} = \mathbf{d}$, such that

$$\int_{\Omega} \nabla \mathbf{v} : \mathbb{E} : \nabla \mathbf{u}^{R,H} \, d\Omega = \int_{\Omega} \mathbf{f} \cdot \mathbf{v} \, d\Omega + \int_{\Gamma_t} \mathbf{t} \cdot \mathbf{v} \, dA \tag{8.33}$$

$\forall \mathbf{v}^H \in \mathbf{H}^H_v(\Omega) \subset \mathbf{H}^1(\Omega), \mathbf{v}^H|_{\Gamma_u} = \mathbf{0}$.

A critical point is that $\mathbf{H}^H_{u^R}(\Omega), \mathbf{H}^H_v(\Omega) \subset \mathbf{H}^1(\Omega)$, and thus $\mathbf{u}^{R,H}$ is kinematically admissible with respect to the true fine-scale problem. For the regularized and discrete regularized solutions we have

$$||\mathbf{u}-\mathbf{u}^R||^2_{E(\Omega)} = 2(\mathscr{J}(\mathbf{u}^R) - \mathscr{J}(\mathbf{u})), \tag{8.34}$$

and

$$||\mathbf{u}-\mathbf{u}^{R,H}||^2_{E(\Omega)} = 2(\mathscr{J}(\mathbf{u}^{R,H}) - \mathscr{J}(\mathbf{u})). \tag{8.35}$$

Subtracting the two results yields

$$||\mathbf{u}-\mathbf{u}^{R,H}||^2_{E(\Omega)} = \underbrace{||\mathbf{u}-\mathbf{u}^R||^2_{E(\Omega)}}_{\text{difference in scales}} + \underbrace{2(\mathscr{J}(\mathbf{u}^{R,H}) - \mathscr{J}(\mathbf{u}^R))}_{\text{coarse-scale discretization error}}. \tag{8.36}$$

Clearly, the total difference is composed of two mutually orthogonal parts: (1) a difference in scales and (2) a numerical error, introduced by spatial discretization of the coarse scale problem. Formally speaking, estimates can be derived for the complementary norm in a similar manner. However, there is a fundamental difficulty with such an approach since the space of admissible trial stress functions must satisfy equilibrium, $\nabla \cdot \sigma^{R,h} + \mathbf{f} = \mathbf{0}$, and the test functions must be divergence free, $\nabla \cdot \gamma^h = \mathbf{0}$. However, formally proceeding we have

$$||\sigma-\sigma^R||^2_{E^{-1}(\Omega)} = 2(\mathscr{K}(\sigma^R) - \mathscr{K}(\sigma)), \tag{8.37}$$

and

$$||\sigma-\sigma^{R,h}||^2_{E^{-1}(\Omega)} = 2(\mathscr{K}(\sigma^{R,h}) - \mathscr{K}(\sigma)), \tag{8.38}$$

leading to

$$||\sigma-\sigma^{R,h}||^2_{E^{-1}(\Omega)} = \underbrace{||\sigma-\sigma^R||^2_{E^{-1}(\Omega)}}_{\text{difference in scales}} + \underbrace{2(\mathscr{K}(\sigma^{R,h}) - \mathscr{K}(\sigma^R))}_{\text{coarse-scale discretization error}}. \tag{8.39}$$

Finally, by adding (8.35) and (8.38) we obtain an exact "mixed estimate" for a solution generated by a standard finite element method (H) and a solution generated by a method which generates a statically admissible solution (h)

$$||\mathbf{u}-\mathbf{u}^{R,H}||^2_{E(\Omega)} + ||\sigma-\sigma^{R,h}||^2_{E^{-1}(\Omega)} = 2(\mathscr{J}(\mathbf{u}^{R,H}) + \mathscr{K}(\sigma^{R,h})). \tag{8.40}$$

8.4 A "total" Orthogonal Sum

Clearly, for the class of problems under consideration, solutions must be generated numerically, for example, as in the last section, by the finite element method. We now consider the effect of the use of the finite element method to generate a subspatial approximation to $\tilde{\mathbf{u}}$, denoted $\tilde{\mathbf{u}}^h$, governed per subdomain by

Find $\tilde{\mathbf{u}}_K^h \in \mathbf{H}_{\tilde{u}}^h(\Omega_K) \subset \mathbf{H}^1(\Omega), \tilde{\mathbf{u}}_K^h|_{\partial\Omega_K \cap (\Omega \cup \Gamma_u)} = \mathbf{U} \in \mathbf{H}^1(\Omega)$, such that

$$\int_{\Omega_K} \nabla \mathbf{v}_K : \tilde{\boldsymbol{\sigma}}_K^h \, d\Omega = \int_{\Omega_K} \mathbf{f} \cdot \mathbf{v}_K \, d\Omega + \int_{\partial\Omega_K \cap \Gamma_t} \mathbf{t} \cdot \mathbf{v}_K \, dA \tag{8.41}$$

$\forall \mathbf{v}_K^h \in \mathbf{H}_v^h(\Omega_K) \subset \mathbf{H}^1(\Omega_K), \mathbf{v}_K^h|_{\partial\Omega_K \cap (\Omega \cup \Gamma_u)} = \mathbf{0}$.

A critical point is that $\mathbf{H}_{\tilde{u}}^h(\Omega), \mathbf{H}_v^h(\Omega) \subset \mathbf{H}^1(\Omega)$. This "inner" approximation allows the development of a straightforward decomposition. Using a subspatial approximation, by direct expansion we have

$$\begin{aligned}
\mathscr{J}(\tilde{\mathbf{u}}^h) &= \sum_{K=1}^N \mathscr{J}_K(\tilde{\mathbf{u}}_K^h) = \sum_{K=1}^N \mathscr{J}_K(\tilde{\mathbf{u}}_K + (\tilde{\mathbf{u}}_K^h - \tilde{\mathbf{u}}_K)), \\
&= \sum_{K=1}^N \mathscr{J}_K(\tilde{\mathbf{u}}_K) + \sum_{K=1}^N \int_{\Omega_K} \nabla(\tilde{\mathbf{u}}_K^h - \tilde{\mathbf{u}}_K) : \mathbb{E} : \nabla \tilde{\mathbf{u}}_K \, d\Omega, \\
&\quad - \sum_{K=1}^N \int_{\Omega_K} \mathbf{f} \cdot (\tilde{\mathbf{u}}_K^h - \tilde{\mathbf{u}}_K) \, d\Omega - \sum_{K=1}^N \int_{\partial\Omega_K \cap \Gamma_t} \mathbf{t} \cdot (\tilde{\mathbf{u}}_K^h - \tilde{\mathbf{u}}_K) \, dA, \\
&\quad + \sum_{K=1}^N \frac{1}{2} \int_{\Omega_K} \nabla(\tilde{\mathbf{u}}_K^h - \tilde{\mathbf{u}}_K) : \mathbb{E} : \nabla(\tilde{\mathbf{u}}_K^h - \tilde{\mathbf{u}}_K) \, d\Omega, \\
&= \sum_{K=1}^N \mathscr{J}(\tilde{\mathbf{u}}_K) + \sum_{K=1}^N \frac{1}{2} \int_{\Omega_K} \nabla(\tilde{\mathbf{u}}_K^h - \tilde{\mathbf{u}}_K) : \mathbb{E} : \nabla(\tilde{\mathbf{u}}_K^h - \tilde{\mathbf{u}}_K) \, d\Omega,
\end{aligned} \tag{8.42}$$

where, by Box 8.41, for $K = 1, 2, \ldots, N$

$$\int_{\Omega_K} \nabla(\tilde{\mathbf{u}}_K^h - \tilde{\mathbf{u}}_K) : \mathbb{E} : \nabla \tilde{\mathbf{u}}_K \, d\Omega - \int_{\Omega_K} \mathbf{f} \cdot (\tilde{\mathbf{u}}_K^h - \tilde{\mathbf{u}}_K) \, d\Omega - \int_{\partial\Omega_K \cap \Gamma_t} \mathbf{t} \cdot (\tilde{\mathbf{u}}_K^h - \tilde{\mathbf{u}}_K) \, dA = 0. \tag{8.43}$$

Subtracting $\mathscr{J}(\mathbf{u})$ from both sides of (8.42) leads to

$$\mathscr{J}(\tilde{\mathbf{u}}^h) - \mathscr{J}(\mathbf{u}) = \mathscr{J}(\tilde{\mathbf{u}}) - \mathscr{J}(\mathbf{u}) + \frac{1}{2} \int_{\Omega} \nabla(\tilde{\mathbf{u}} - \tilde{\mathbf{u}}^h) : \mathbb{E} : \nabla(\tilde{\mathbf{u}} - \tilde{\mathbf{u}}^h) \, d\Omega, \tag{8.44}$$

which immediately implies

$$||\mathbf{u} - \tilde{\mathbf{u}}^h||_{E(\Omega)}^2 = ||\mathbf{u} - \tilde{\mathbf{u}}||_{E(\Omega)}^2 + ||\tilde{\mathbf{u}} - \tilde{\mathbf{u}}^h||_{E(\Omega)}^2. \tag{8.45}$$

Therefore the estimate of the error, for example in Box 8.25 *and* (7.5), *implicitly contains the numerical discretization error.* Clearly, the total error is composed of two mutually orthogonal parts: (1) a partitioning error introduced by subdividing the body and applying inexact boundary data on the interfaces and (2) a numerical error, introduced by spatial discretization. Therefore, if one has a numerical error estimator, one can isolate the pure decomposition error. Such "filtering" processes have been performed in Zohdi and Wriggers [229]. For details on elementary a priori numerical error estimates, see the text of Hughes [101], while for a posteriori estimates see Ainsworth and Oden [2].

8.5 Iterative Extensions

In order to achieve further reduction of the global/local error, the subdomain interfacial boundary conditions must be improved. Clearly, the error bound in Box 8.25, and the orthogonality expression in Box 8.45 still hold, provided that the modified, hopefully improved, local boundary data deliver a globally kinematically admissible solution. Thus, we now discuss a few methods which consist of updating the numerical degrees of freedom on the interface of a finite element discretization. Without loss of generality, and for simplicity of presentation, we consider that we have meshed each subdomain in the substructuring process so that a global nodally-conforming trilinear hexahedral finite element mesh results.

8.5.1 Method I: Global/Local CG Iterations

One can consider a process of updating interfacial nodal values as having as its goal the minimization of the discrete analog of $\mathscr{J}(\tilde{\mathbf{u}})$, namely $\mathscr{J}(\tilde{\mathbf{u}}^h)$, which is convex in terms of the finite element nodal displacements. There exist a variety of classical techniques which could drive a process to minimize potential energies, the most applicable in the linearly elastic case being the CG method, which, for the sake of completeness, we briefly discuss in the context of a global/local strategy.

Suppose we wish to solve the discrete system $[\mathbf{K}]\{\mathbf{u}^h\} = [\mathbf{R}]$. $[\mathbf{K}]$ is a symmetric positive definite $n \times n$ matrix, $\{\mathbf{u}^h\}$ is the $n \times 1$ solution vector, $[\mathbf{R}]$ is the $n \times 1$ righthand side, and n is the number of discrete unknowns. We define a discrete potential $\Pi \overset{\text{def}}{=} \frac{1}{2}\{\mathbf{u}^h\}^T[\mathbf{K}]\{\mathbf{u}^h\} - \{\mathbf{u}^h\}^T[\mathbf{R}]$. Correspondingly, from basic calculus we have $\nabla\Pi \overset{\text{def}}{=} \{\frac{\partial \Pi}{\partial u_1}, \frac{\partial \Pi}{\partial u_2}, \ldots \frac{\partial \Pi}{\partial u_n}\}^T = \mathbf{0} \Rightarrow [\mathbf{K}]\{\mathbf{u}^h\} - [\mathbf{R}] = \mathbf{0}$. Therefore the minimizer of the potential Π is also the solution to the discrete system. The CG method is based upon minimizing Π by successively updating a starting vector. Therefore, the substructuring method described in this work can be considered as a construction of an advanced initial starting guess, $\tilde{\mathbf{u}}^h$, to the solution of the global problem formed by directly discretizing the entire body with no decomposition. If we are to directly apply the CG method, one will reduce $\mathscr{J}(\tilde{\mathbf{u}}^h)$, to the minimum value for the corresponding global mesh discretization. Since a CG approach updates all values in the entire domain simultaneously, via a global matrix-vector multiply, each iteration can be thought of as providing global transfer of information, i.e. microstructural corrections to the initial solution $\tilde{\mathbf{u}}^h$, and thus the interfacial boundary conditions. The minimum of the discrete potential is the global solution for the given finite element discretization. However, instead of a direct application of the CG method, one can perform a set number of global CG iterations, M_{glob}, then reevaluate the local boundary value problems with the new local boundary data. Per subdomain we have $m_{loc}(\frac{n}{N})^2$ operation counts, where m_{loc} is the number of local CG iterations for subdomain solution convergence. For simplicity, we have assumed that one would use the CG method to solve each subdomain problem, and that m_{loc} is approximately the same for each subdomain. Therefore, for N total subdomains we have $m_{loc}(\frac{n}{N})^2 N$ operation counts. Let us denote the number of times that we globally iterate and resolve the local problems as, $\mathscr{S}$. Thus, in total, we have $\mathscr{S}(M_{glob}n^2 + Nm_{loc}(\frac{n}{N})^2) = \mathscr{S}(M_{glob} + \frac{m_{loc}}{N})n^2$. Therefore, a direct CG approach and a projecting-global/local resolve approach compare as follows

$$\underbrace{\hat{C}n^2 \stackrel{\text{def}}{=} \mathscr{S}(M_{glob} + \frac{m_{loc}}{N})n^2}_{\text{projecting and resolving}} \quad \text{and} \quad \underbrace{Cn^2}_{\text{Conjugate Gradient}} . \tag{8.46}$$

Clearly the magnitudes of $\hat{C}$ and C determine the relative performance of the two methods.

8.5.2 Method II: Iterative Equilibration

Clearly, when decomposing the structure by a projection of a kinematically admissible function onto the partitioning interfaces, regardless of the constitutive law, the error is due to the jumps in tractions at the interfaces (Box 8.5). If the interfaces would be in equilibrium, then there would be no traction jumps. Therefore, an approach which attempts to eliminate the jumps in tractions is sought next.

Global Equilibration

For global equilibration one treats the entire interface as a single subdomain, denoted $\omega \subset \Omega$, and solves the following (Fig. 8.4)

$$\text{Find } \mathbf{s} \in \mathbf{H}^1(\omega), \mathbf{s}|_{\partial\omega\cap(\omega\cup\Gamma_u)} = \tilde{\mathbf{u}} \in \mathbf{H}^1(\Omega), \text{ such that}$$

$$\int_\omega \nabla\mathbf{v} : \sigma(\mathbf{s})\, d\omega = \int_\omega \mathbf{f}\cdot\mathbf{v}\, d\omega + \int_{\partial\omega\cap\Gamma_t} \mathbf{t}\cdot\mathbf{v}\, da \tag{8.47}$$

$$\forall\mathbf{v} \in \mathbf{H}^1(\omega), \mathbf{v}|_{\partial\omega\cap(\omega\cup\Gamma_u)} = \mathbf{0}.$$

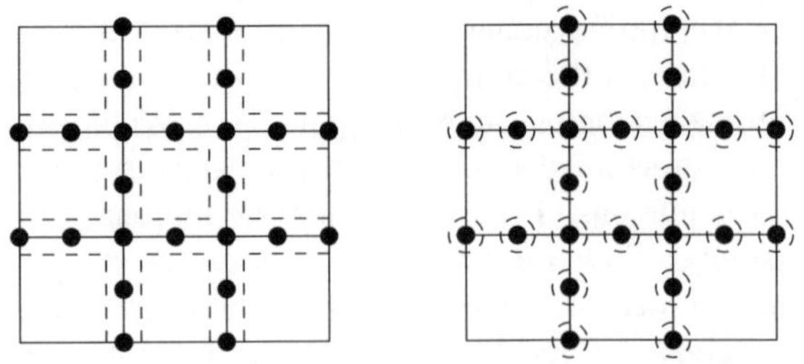
global interface
problem is solved
local interface (nodal)
problems are solved

Fig. 8.4 Global equilibration and local equilibration

The new interfacial displacements are then computed, and the subdomain problems are re-solved. The process is repeated until convergence. The entire interface problem can be quite large. A less expensive approach, amenable to trivial parallel processing, follows.

Local (Nodal) Equilibration

For local equilibration, one prescribes small non-intersecting "nodal domains", ω_K (Fig. 8.4), surrounding each interfacial node $1 \leq K \leq M$: (M= number of interfacial nodes)

$$\text{Find } \mathbf{s}_K \in \mathbf{H}^1(\omega_K), \mathbf{s}_K|_{\partial\omega_K \cap (\omega_K \cup \Gamma_u)} = \tilde{\mathbf{u}} \in \mathbf{H}^1(\Omega), \text{ such that}$$

$$\int_{\omega_K} \nabla \mathbf{v}_K : \sigma_K(\mathbf{s})\, d\omega_K = \int_{\omega_K} \mathbf{f} \cdot \mathbf{v}_K\, d\omega_K + \int_{\partial\omega_K \cap \Gamma_t} \mathbf{t} \cdot \mathbf{v}_K\, da_K \qquad (8.48)$$

$$\forall \mathbf{v}_K \in \mathbf{H}^1(\omega_K), \mathbf{v}_K|_{\partial\omega_K \cap (\omega_K \cup \Gamma_u)} = \mathbf{0}.$$

The new interfacial displacements are then computed, and the subdomain problems are resolved. The process is repeated until convergence. The advantage of such an approach is that the "nodal domains" can be processed independently of one another. The disadvantage is that global equilibrium is not satisfied throughout the entire interface. Either of the introduced approaches can be considered as a fixed-point iteration, which we discuss next.

General Fixed-Point Iterations

Consider a general system of coupled partial differential equations given by $\mathscr{A}(\mathbf{s}) = \mathscr{F}$, where $\mathbf{s}$ is a solution, and where it is assumed that the operator, $\mathscr{A}$, is invertible.

One desires that the sequence of iterated solutions, $\mathbf{s}^I$, $I = 1,2,...$, converge to $\mathscr{A}^{-1}(\mathscr{F})$ as $I \to \infty$. If $\mathbf{s}^I$ is a function of $\mathscr{A}, \mathscr{F}, \mathbf{s}^I, ...\mathbf{s}^{I-K}$ one says that K is the order of iteration. It is assumed that the Ith iterate can be represented by some arbitrary function $\mathbf{s}^I = \mathscr{T}^I(\mathscr{A}, \mathscr{F}, \mathbf{s}^{I-1})$. One makes the following split $\mathbf{s}^I = \mathscr{G}^I(\mathbf{s}^{I-1}) + \mathbf{r}^I$. For this method to be useful the exact solution should be reproduced. In other words, when $\mathbf{s} = \mathscr{A}^{-1}(\mathscr{F})$, then $\mathbf{s} = \mathscr{A}^{-1}(\mathscr{F}) = \mathscr{G}^I(\mathscr{A}^{-1}(\mathscr{F})) + \mathbf{r}^I$. Therefore, one has the following consistency condition $\mathbf{r}^I = \mathscr{A}^{-1}(\mathscr{F}) - \mathscr{G}^I(\mathscr{A}^{-1}(\mathscr{F}))$, and as a consequence, $\mathbf{s}^I = \mathscr{G}^I(\mathbf{s}^{I-1}) + \mathscr{A}^{-1}(\mathscr{F}) - \mathscr{G}^I(\mathscr{A}^{-1}(\mathscr{F}))$. Convergence of the iteration can be studied by defining the error vector:

$$
\begin{aligned}
\mathbf{e}^I = \mathbf{s}^I - \mathbf{s} = \mathbf{s}^I - \mathscr{A}^{-1}(\mathscr{F}) &= \mathscr{G}^I(\mathbf{s}^{I-1}) + \mathscr{A}^{-1}(\mathscr{F}) - \mathscr{G}^I(\mathscr{A}^{-1}(\mathscr{F})) - \mathscr{A}^{-1}(\mathscr{F}) \\
&= \mathscr{G}^I(\mathbf{s}^{I-1}) - \mathscr{G}^I(\mathscr{A}^{-1}(\mathscr{F})).
\end{aligned} \tag{8.49}
$$

One sees that, if $\mathscr{G}^I$ is linear and invertible, the above reduces to $\mathbf{e}^I = \mathscr{G}^I(\mathbf{s}^{I-1} - \mathscr{A}^{-1}(\mathscr{F})) = \mathscr{G}^I(\mathbf{e}^{I-1})$. Therefore, if the spectral radius of $\mathscr{G}^I$, i.e. the magnitude of its largest eigenvalue, is less than unity for each iteration I, then $\mathbf{e}^I \to \mathbf{0}$ for any arbitrary starting solution $\mathbf{s}^{I=0}$ as $I \to \infty$.

An Example

Again we use the one dimensional structure $0 < \mathrm{x} < \mathrm{L} : \frac{d}{dx}\left(E(x)\frac{du}{dx}\right) = 0$, $u(0) = \Delta_0$, $u(L) = \Delta_L$ depicted in Fig. 8.5. Consider two subdomains, and one interfacial problem, which are solved in an iterative manner $i = 1,2,...N$, until $|e^I(\frac{L}{2})| = |s^I(\frac{L}{2}) - u(\frac{L}{2})| \leq TOL$

$$
\begin{aligned}
&\text{SUB\#1 } (0 < \mathrm{x} < \tfrac{\mathrm{L}}{2}) : \frac{d}{dx}\left(E_1 \frac{d\tilde{u}_1^I}{dx}\right) = 0 && \tilde{u}_1^I(0) = \Delta_0 \quad \tilde{u}_1^I(\tfrac{L}{2}) = s^I(\tfrac{L}{2}) \\
&\text{SUB\#2 } (\tfrac{\mathrm{L}}{2} < \mathrm{x} < \mathrm{L}) : \frac{d}{dx}\left(E_2 \frac{d\tilde{u}_2^I}{dx}\right) = 0 && \tilde{u}_2^I(\tfrac{L}{2}) = s^I(\tfrac{L}{2}) \quad \tilde{u}_2^I(L) = \Delta_L \\
&\text{INT } (\tfrac{\mathrm{L}}{2} - \delta < \mathrm{x} < \tfrac{\mathrm{L}}{2} + \delta) : \frac{d}{dx}\left(E(x) \frac{ds^I}{dx}\right) = 0 && s^I(\tfrac{L}{2} - \delta) = \tilde{u}_1^I(\tfrac{L}{2} - \delta) \\
& && s^I(\tfrac{L}{2} + \delta) = \tilde{u}_2^I(\tfrac{L}{2} + \delta).
\end{aligned} \tag{8.50}
$$

Since there is only a single one-dimensional interface, the global and local equilibration methods are the same. A simple starting value, that of the regularized solution, $s^0(\frac{L}{2}) = \frac{u(L)+u(0)}{2}$ is chosen. After some algebra we have, $\mathscr{G} = 1 - \Theta$, where $\Theta = \frac{2\delta}{L}$, yielding

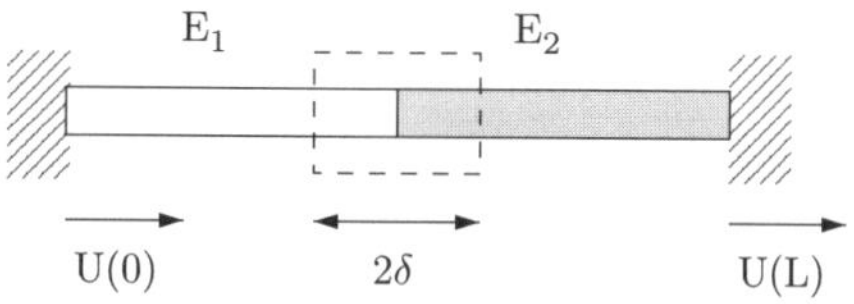

Fig. 8.5 A one-dimensional equilibration example

$$
\begin{aligned}
s^{I}(\tfrac{L}{2}) &= \underbrace{(1-\Theta)}_{\mathscr{G}} s^{I-1}(\tfrac{L}{2}) + \underbrace{\Theta\left(\left(\frac{E_2}{E_1+E_2}\right)(u(L)-u(0))+u(0)\right)}_{r},\\
e^{I}(\tfrac{L}{2}) &= s^{I}(\tfrac{L}{2}) - u(\tfrac{L}{2}) = (1-\Theta)(s^{I-1}(\tfrac{L}{2}) - u(\tfrac{L}{2}))\\
&= (1-\Theta)e^{I-1}(\tfrac{L}{2}) = (1-\Theta)^{I} e^{I=0}(\tfrac{L}{2}).
\end{aligned}
\tag{8.51}
$$

One sees that convergence occurs if $(1-\Theta) < 1$, which implies $\delta > 0$. The rate of convergence is linear and is controlled by the size of δ.

To illustrate the approach we compared it to classical a classical overlapping Schwarz method (see Le Tallec [119]), which for this simple structure could be written as

$$
\begin{aligned}
&\text{SUB\#1 } (0 < \mathrm{x} < \tfrac{\mathrm{L}}{2}+\delta) : \tfrac{d}{dx}(E\tfrac{d\tilde{u}_1^I}{dx}) = 0,\ \tilde{u}^I(0) = \Delta_0,\ \tilde{u}_1^I(\tfrac{L}{2}+\delta) = \tilde{u}_2^{I-1}(\tfrac{L}{2}+\delta),\\
&\text{SUB\#2 } (\tfrac{\mathrm{L}}{2}-\delta < \mathrm{x} < \mathrm{L}) : \tfrac{d}{dx}(E\tfrac{d\tilde{u}_2^I}{dx}) = 0,\ \tilde{u}_2^I(\tfrac{L}{2}-\delta) = \tilde{u}_1^I(\tfrac{L}{2}-\delta),\ \tilde{u}_2^I(L) = \Delta_L.
\end{aligned}
\tag{8.52}
$$

Depicted in Figs. 8.6, 8.7 and 8.8 are the relative performances of both approaches (equilibration and overlapping Schwarz). The parameters were $L = 1$, $E_1 = 10 \times 10^9$ GPa, $E_2 = 100 \times 10^9$ GPa. Each of the approaches iteratively improved the interfacial data by balancing the jumps in the tractions via information exchange between subdomains. Clearly the overlapping Schwarz method attained a faster rate of convergence since more information is provided via the fact that the entire extended subdomain problem is solved during the iterative process. Generally, the local equilibration method is less expensive to compute for three dimensional problems, however, as illustrated by the preceding one dimensional example, it will probably exhibit slower rates of convergence.

Remark I. Depending on the problem, one method may exhibit superiority over another in terms of the overall cost for a desired solution accuracy. It is important to remark that, in the general three dimensional case, for either the global/local CG approach or the iterative equilibration approach, one can determine whether the solution is improving by monitoring the potential of each iterate, $\mathscr{J}(\tilde{\mathbf{u}}^I)$. A decreasing potential indicates that the solution is improving since $||\mathbf{u}-\tilde{\mathbf{u}}^I||^2_{E(\Omega)} = 2(\mathscr{J}(\tilde{\mathbf{u}}^I) - \mathscr{J}(\mathbf{u}))$ and $\mathscr{J}(\mathbf{u})$ is fixed. Clearly the error bound in Box 8.25 holds for the iteratively generated kinematically admissible solutions, and is

$$
||\mathbf{u}-\tilde{\mathbf{u}}^I||^2_{E(\Omega)} = 2(\mathscr{J}(\tilde{\mathbf{u}}^I) - \mathscr{J}(\mathbf{u})) \le 2(\mathscr{J}(\tilde{\mathbf{u}}^I) - \mathscr{J}^-),
\tag{8.53}
$$

where $\mathscr{J}^- = \mathscr{J}(\mathbf{u}^R) - \frac{1}{2}\underbrace{\int_\Omega ((\mathbb{R}-\mathbb{E}) : \nabla\mathbf{u}^R) : \mathbb{E}^{-1} : ((\mathbb{R}-\mathbb{E}) : \nabla\mathbf{u}^R)\, d\Omega}_{=(\mathscr{H}^+(u))^2}.$

Thus one should maximize $\mathscr{J}^-$ over $\mathbf{u}^R$, via $\mathbb{R}$, and retain it for any subsequent iterations, since it is iteration independent. As we have indicated earlier, the only

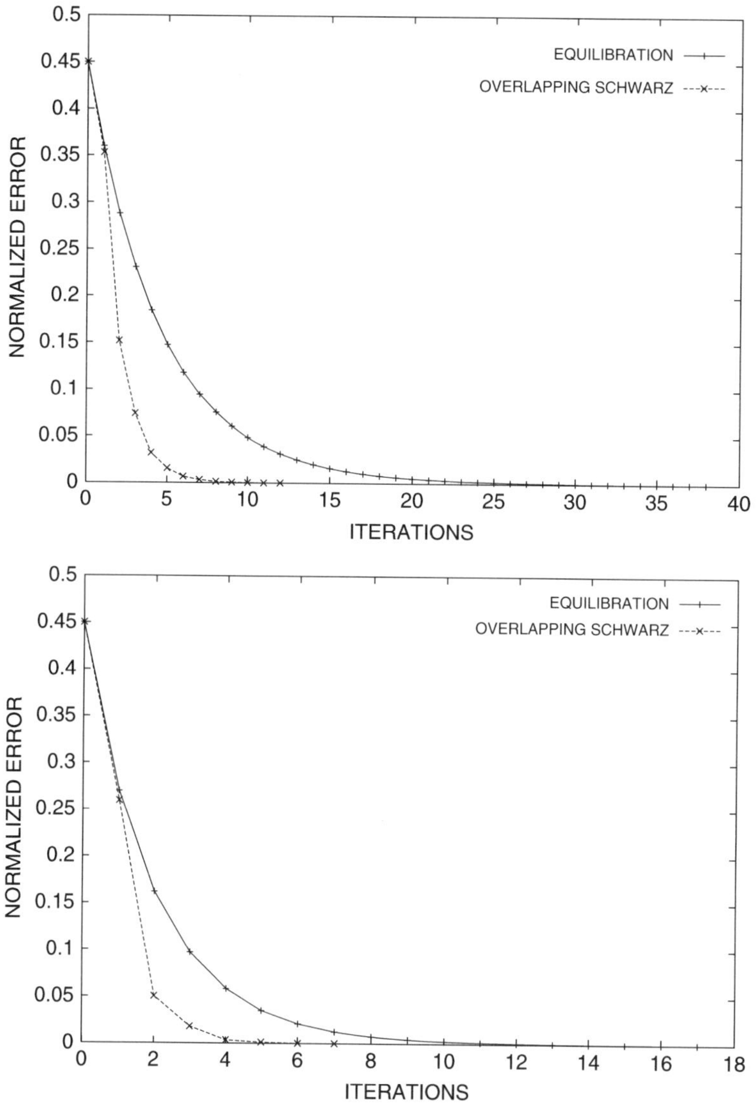

Fig. 8.6 NORMALIZED ERROR $\stackrel{\text{def}}{=} \left| \frac{\tilde{u}(\frac{L}{2}) - u(\frac{L}{2})}{u(\frac{L}{2})} \right|$. **Top** for an overlap of $\delta = 0.1$ and **Bottom** for $\delta = 0.2$

approximation in this entire estimate is $||\mathbf{u} - \mathbf{u}^R||_{E(\Omega)} \leq \mathscr{H}^{+(u)}$, *which is exact for the previous one dimensional example, using the regularized parameter* $\mathscr{R} = \left(\frac{1}{L}\int_0^L \frac{1}{E}\,dx\right)^{-1}$. Application of the global/local CG approach, iterative equilibration or classical overlapping methods to three dimensional problems is under current investigation.

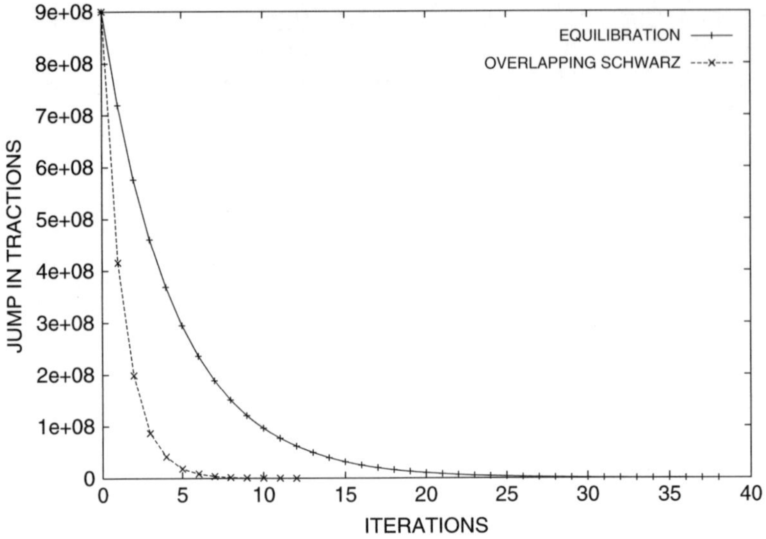

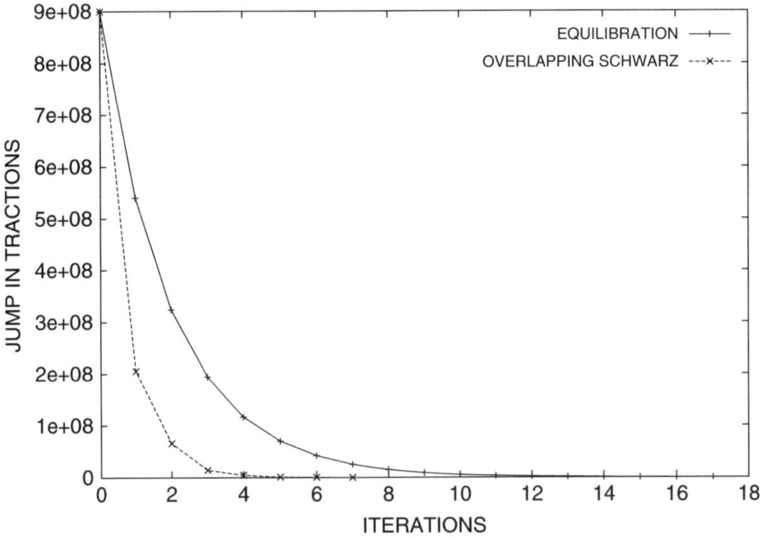

Fig. 8.7 JUMP IN TRACTIONS $\stackrel{\text{def}}{=} \left| E_1 \frac{d\tilde{u}_1(\frac{L}{2})}{dx} - E_2 \frac{d\tilde{u}_2(\frac{L}{2})}{dx} \right|$. **Top** for an overlap of $\delta = 0.1$ and **Bottom** for $\delta = 0.2$

Remark II. The rate of convergence of the CG method is

$$||\mathbf{u} - \mathbf{u}^i||_{E(\Omega)} \leq |\mathscr{S}|^i ||\mathbf{u} - \mathbf{u}^1||_{E(\Omega)}, \tag{8.54}$$

where $\mathscr{S}$ a functional of the condition number of the discretized stiffness matrix. Proofs of the various characteristics of the method can be found in Axelsson [9]. It is

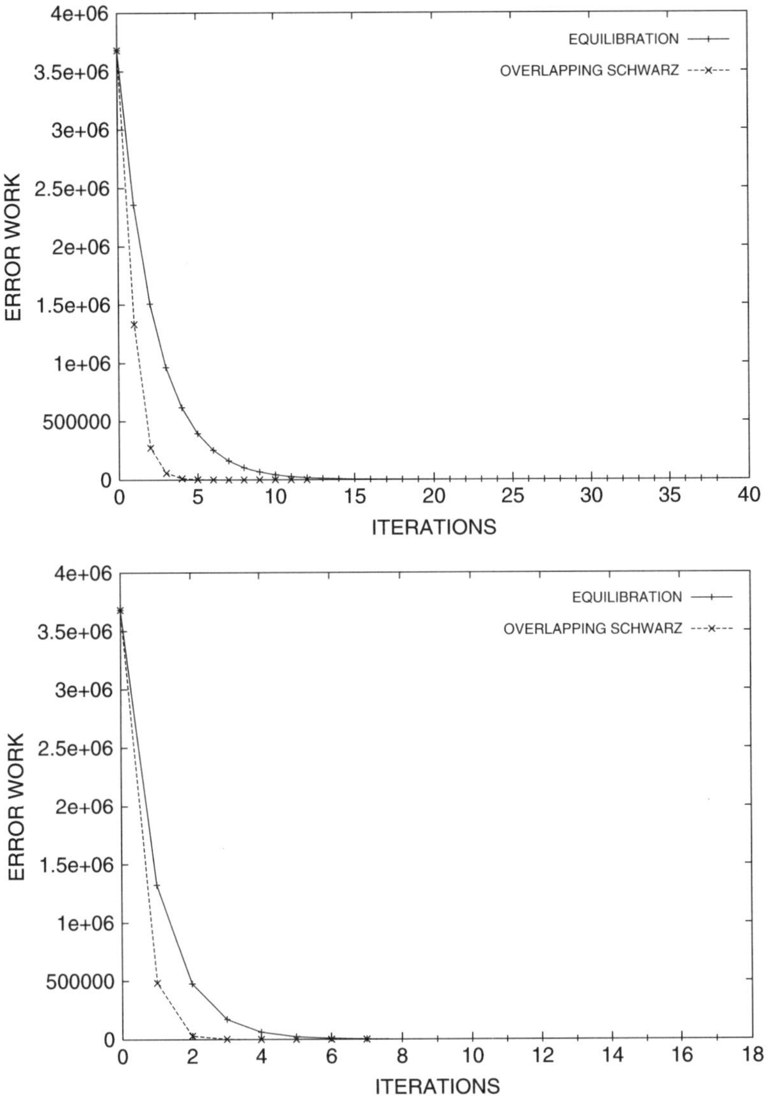

Fig. 8.8 ERROR WORK $\stackrel{\text{def}}{=} \left| (E_1 \frac{d\tilde{u}_1(\frac{L}{2})}{dx} - E_2 \frac{d\tilde{u}_2(\frac{L}{2})}{dx})(u(\frac{L}{2}) - \tilde{u}(\frac{L}{2})) \right|$. **Top** for an overlap of $\delta = 0.1$ and **Bottom** for $\delta = 0.2$

clear that the magnitude of $||\mathbf{u} - \mathbf{u}^1||_{E(\Omega)}$ controls most of the error during the initial stages of the CG process. Therefore, on the discrete level, a good initial choice for $\mathbf{u}^1$, which captures the long range features of the fine-scale solution is desired. As mentioned, the CG method will then successively endow the long range solution with local (high frequency) corrections. Therefore, it is clear that if $\mathbf{u}^R$ is available it should be projected onto the fine-grid to produce a starting vector $\mathbf{u}^1 = \mathbf{u}^R$.

Chapter 9
Nonconvex–Nonderivative Genetic Material Design

Until this point in the presentation, we have considered the microstructure of the material as given data. It is important to realize that microstructural selections, such as combinations of volume fraction, stiffness mismatches and particulate topology, are usually the result of tedious trial and error laboratory experiments. In this chapter, a computational strategy is developed to simulate and accelerate the associated trial and error development of tailored dispersed-type materials. The objective is to develop a procedure to determine multiple possible microstructures that can deliver the same prespecified effective linearly elastic response. The determination of multiple possible microstructural combinations gives an engineer a wide variety of design and manufacturing alternatives. However, there exist a variety of difficulties in the computational design of macroscopic solid material properties formed by doping a base matrix material with randomly distributed particles of different phases. Three primary problems are (1) the wide array of free microdesign variables, such as particle topology, volume fraction and mechanical property phase contrasts, which force associated objective functions to be highly nonconvex in those variables, (2) the associated objective functions are not continuously differentiable with respect to design space, primarily due to microscale design constraints, such as limits on the desired local stress field intensities, and (3) the effective responses of various finite sized samples, of equal volume but of different random distributions of the particulate matter, exhibit mutual fluctuations, leading to amplified noise in optimization strategies where objective function sensitivities or comparisons are needed. The effects in (3) become amplified when computing design sensitivities or comparisons needed in optimization strategies, thus relatively large samples must be used to eliminate detrimental size effects. The presented work concentrates on the parametrization of microscale parameters inherent in such materials, with the goal of computational optimization of material microstructure. The outline of the presentation is as follows. The fundamentals of computational material design formulations are given. Attention is drawn to some of the difficulties encountered, i.e. nonconvexity and nondifferentiability of objective functions, as well as size effects. Solution strategies based on genetic algorithms are then presented.

9.1 Computational Material Design

Presently, we are primarily concerned with the construction and solution of inverse problems, describing the design of materials composed of randomly dispersed particulates suspended in a homogeneous binding matrix. The design objectives are to find sets of microstructural parameters, such as the relative volume fractions of the constituents, the geometries of the particulates and their mechanical properties, which minimize

$$\Pi = \left(\frac{||\mathbb{E}^{*} - \mathbb{E}^{*,D}||}{||\mathbb{E}^{*,D}||} \right)^{q_0}, \tag{9.1}$$

where $0 < q_0 < \infty$, while simultaneously obeying constraints on the microscale stress field behavior. The constraints will be specified shortly. Here $\mathbb{E}^{*,D}$ is a prespecified desired effective response, $\mathbb{E}^{*}$ is the effective response described via $\langle \sigma \rangle_{\Omega} = \mathbb{E}^{*} : \langle \varepsilon \rangle_{\Omega}$, where $\langle \cdot \rangle_{\Omega} \stackrel{\text{def}}{=} \frac{1}{|\Omega|} \int_{\Omega} \cdot \, d\Omega$, and where σ and ε are the stress and strain tensor fields within a statistically representative volume element (RVE) of volume $|\Omega|$, produced by a trial microstructure.[1] If the effective response is assumed isotropic, which is the case if the particles are randomly distributed and randomly oriented (if nonspherical), then the effective bulk and shear moduli are given by $3\kappa^{*} \stackrel{\text{def}}{=} \frac{\langle \frac{tr\sigma}{3} \rangle_{\Omega}}{\langle \frac{tr\varepsilon}{3} \rangle_{\Omega}}$ and $2\mu^{*} \stackrel{\text{def}}{=} \sqrt{\frac{\langle \sigma' \rangle_{\Omega} : \langle \sigma' \rangle_{\Omega}}{\langle \varepsilon' \rangle_{\Omega} : \langle \varepsilon' \rangle_{\Omega}}}$. An extensive review of the state of the art in the analysis of random heterogeneous media can be found in the work of Torquato [201, 202, 203, 204, 205].

In order to systematize the computational minimization process, we characterize a microstructural design through an N-tuple design vector, denoted $\Lambda \stackrel{\text{def}}{=} (\Lambda_1, \Lambda_2, ..., \Lambda_N)$, representing the following components: (I) *Particulate mechanical properties:* for example, assuming local isotropy of the particles, the bulk and shear moduli, κ_2 and μ_2 (two variables), (II) *Particulate topology:* where we characterize the shape of the particulates by a generalized ellipsoidal equation (Fig. 9.1):

$$\left(\frac{|x - x_o|}{r_1} \right)^{s_1} + \left(\frac{|y - y_o|}{r_2} \right)^{s_2} + \left(\frac{|z - z_o|}{r_3} \right)^{s_3} = 1, \tag{9.2}$$

where the s's are exponents. Values of $s < 1$ produce nonconvex shapes, while $s > 2$ values produce "block-like" shapes (three design variables), (III) *Particulate aspect ratio:* for example, defined by $AR \stackrel{\text{def}}{=} \frac{r_1}{r_2} = \frac{r_1}{r_3}$, where $r_2 = r_3$, $AR > 1$ for prolate geometries and $AR < 1$ for oblate shapes (one variable) and (IV) *Particulate volume fraction:* for example, $v_2 \stackrel{\text{def}}{=} \frac{|\Delta|}{|\Omega|}$, where $|\Delta|$ is the volume occupied by the particles, and $|\Omega|$ is the total volume of the material (one variable). Therefore we have a total

[1] At the microscale, the mechanical properties of microheterogeneous materials are characterized by a spatially variable elasticity tensor $\mathbb{E}$. The symbol $||\cdot||$ indicates an appropriate admissible norm to be discussed later.

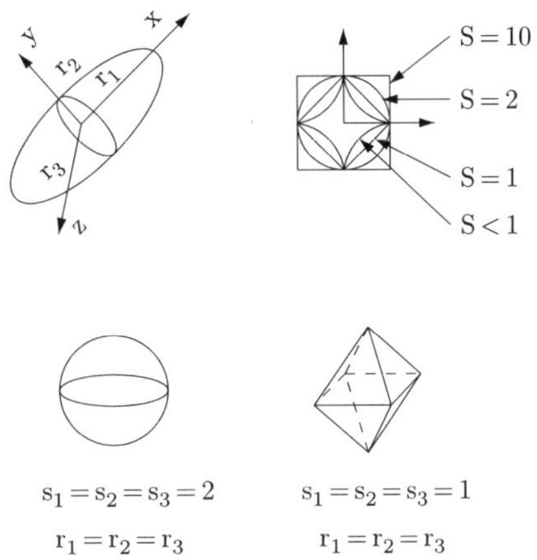

Fig. 9.1 Parametrization of a generalized ellipsoid

of seven design variables, $\Lambda = (\kappa_2, \mu_2, AR, v_2, s_1, s_2, s_3)$. Although we will not consider the matrix material's properties as being free variables, this poses no additional difficulties.

9.2 Characteristic of Such Objectives

Some key difficulties encountered in such problems, which we now discuss, are (1) Nonconvexity of objective functions such as in (9.1), due the variety of design variables, (2) Objective functions which are not continuously differentiable because of active-inactive constraints, and (3) Sample size effects, which induce a degree of stochastic behavior into the objective function.

9.2.1 Nonconvexity

Consider a one dimensional heterogeneous bar composed of two materials, E_1 and E_2 (transverse bonded strips), each occupying volume fractions v_1 and v_2 ($v_1 + v_2 = 1$), respectively. The effective mechanical response for such a system, $\mathbb{E}^*$, $\langle\sigma\rangle_\Omega \stackrel{\text{def}}{=} E^* \langle\varepsilon\rangle_\Omega$, is the harmonic average, which can be written as $E^* = \frac{\tau E_1}{(1-v_2)\tau + v_2}$, where $\tau = \frac{E_2}{E_1}$, Clearly, there is no unique combination of τ and v_2 to produce the same desired effective response. Consider the objective, $\Pi = \left(\frac{E^* - E^{*,D}}{E^{*,D}}\right)^2$, where $E^{*,D}$ is the desired effective response. If one were to pursue a standard Newton-type multivariate search for a new design increment one would construct the following Hessian system for the two design variables, $\Lambda = (\tau, v_2)$:

$$\begin{bmatrix} \dfrac{\partial^2 \Pi(\tau,v_2)}{\partial \tau^2} & \dfrac{\partial^2 \Pi(\tau,v_2)}{\partial \tau \partial v_2} \\ \dfrac{\partial^2 \Pi(\tau,v_2)}{\partial v_2 \partial \tau} & \dfrac{\partial^2 \Pi(\tau,v_2)}{\partial v_2^2} \end{bmatrix} \left\{ \begin{array}{c} \Delta \tau \\ \Delta v_2 \end{array} \right\} = - \left\{ \begin{array}{c} \dfrac{\partial \Pi(\tau,v_2)}{\partial \tau} \\ \dfrac{\partial \Pi(\tau,v_2)}{\partial v_2} \end{array} \right\}. \tag{9.3}$$

Unfortunately, this system becomes noninvertible throughout the design space, due to the nonconvexity of the objective function Π, as illustrated in Fig. 9.2. The behavior of such objectives becomes even worse when more design variables are present, such as in three dimensional cases involving topological variables discussed earlier. For more detailed discussions of such issues see Cherkaev and Gibiansky [26].

9.2.2 Size Effects

We remark that even if the issues of nonconvexity and nondifferentiability of the objective were not present, and one were to attempt to apply a gradient-type approach, the construction of numerical derivatives of the objective function can become highly unstable. This is due to the fact that effective responses of finite sized samples, of equal volume but of different random particle distributions, exhibit mutual fluctuations, leading to amplified noise in optimization strategies where objective function sensitivities (derivatives) are needed (Fig. 9.3). For example, referring to Fig. 9.3, the effects of fluctuations due sample size can be characterized by computing the difference between the smallest value on the right of an arbitrary design point Λ^o, due to size effects, and the largest value on the left, also due to size effects, resulting in

$$\delta^- = \Pi^-(\Lambda^o + \Delta\Lambda) - \Pi^+(\Lambda^o - \Delta\Lambda) \tag{9.4}$$

and computing the difference between the largest value on the right and smallest value on the left, resulting in

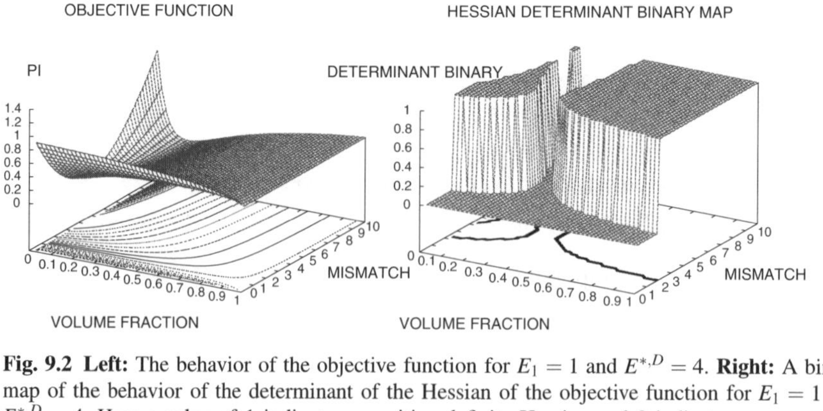

Fig. 9.2 Left: The behavior of the objective function for $E_1 = 1$ and $E^{*,D} = 4$. **Right:** A binary map of the behavior of the determinant of the Hessian of the objective function for $E_1 = 1$ and $E^{*,D} = 4$. Here a value of 1 indicates a positive definite Hessian and 0 indicates a nonpositive Hessian

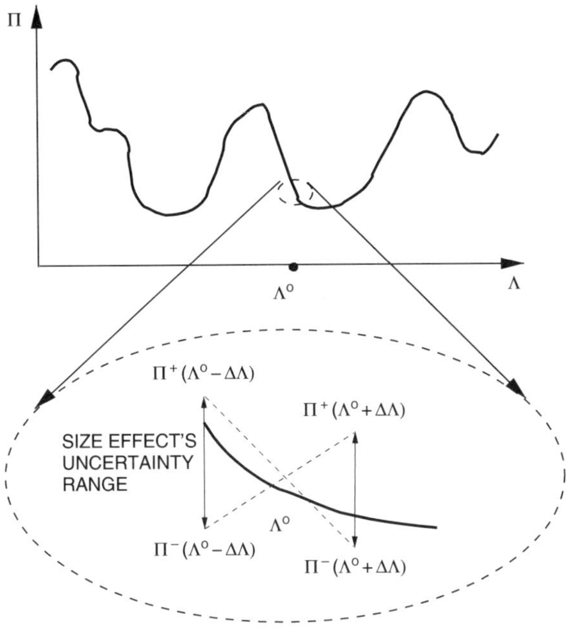

Fig. 9.3 An objective function suffering from size effects

$$\delta^+ = \Pi^+(\Lambda^o + \Delta\Lambda) - \Pi^-(\Lambda^o - \Delta\Lambda). \tag{9.5}$$

Denoting the uncertainty in the objective function's value at $\Lambda^o + \Delta\Lambda$ by $\mathscr{U}(\Lambda^o + \Delta\Lambda) \stackrel{\text{def}}{=} \Pi^+(\Lambda^o + \Delta\Lambda) - \Pi^-(\Lambda^o + \Delta\Lambda)$ and the uncertainty in the objective function's value at $\Lambda^o - \Delta\Lambda$ by $\mathscr{U}(\Lambda^o - \Delta\Lambda) \stackrel{\text{def}}{=} \Pi^+(\Lambda^o - \Delta\Lambda) - \Pi^-(\Lambda^o - \Delta\Lambda)$, we have an uncertainty in the sensitivity

$$0 \leq |\delta^+ - \delta^-| = \mathscr{U}(\Lambda^o + \Delta\Lambda) + \mathscr{U}(\Lambda^o - \Delta\Lambda). \tag{9.6}$$

Clearly, the effects of the fluctuations are amplified for the sensitivities. Therefore, even for large samples of randomly dispersed particulate material, which may exhibit slight fluctuations from one another, the resulting deviations in derivatives can be quite large. For further details see Zohdi [245] and [241]. Furthermore, when active-inactive constraints are introduced, as they later will be, the objective function will become not continuously differentiable throughout the domain, in locations that are indeterminant a priori. In summary, for the class of problems considered here, gradient methods should be avoided due to: (I) Nonconvexity leading to uninvertible Hessians and (II) Noise in the numerical derivatives of the objective function. In addition, a further issue arises, namely nondifferentiability of the constraints, in particular active-inactive constraints, as well as, in some cases, nondifferentiability of the unconstrained objective function itself. This further motivates, later in the analysis, a nonderivative (genetic) search procedure.

9.3 Introduction of Constraints

A drawback of adding particulate material to a homogeneous base matrix is that the presence of a second phase particles will perturb the otherwise smooth stress fields in the matrix, locally amplifying or reducing the fields throughout the microstructure. Thus, it makes sense to account for this distortion and to attempt to limit it during the design process.

Consider the deviation of the stress field from its volumetric average, $\delta\sigma = \sigma - \langle\sigma\rangle_\Omega$, measured in the following norm ($q > 0$)

$$||\delta\sigma||^q_{\hat{L}^q(\Omega)} \stackrel{\text{def}}{=} \frac{1}{|\Omega|}\int_\Omega (\delta\sigma : \delta\sigma)^q \, d\Omega \geq 0. \tag{9.7}$$

The physical meaning of this norm, for $q = 1$, is the standard deviation of the stress field from its volumetric average. Higher q indicate more sensitive measures to the differences between the internal field and its average. A meaningful constraint is

$$w_q P_q(\Lambda) = w_q MAX\left(0, \left(\frac{||\delta\sigma||^q_{\hat{L}^q(\Omega)}}{TOL_q ||\sigma||^q_{\hat{L}^q(\Omega)}} - 1\right)\right). \tag{9.8}$$

To add an entire range of "higher stress moment" constraints, we write

$$\mathbf{w}\mathbb{P}(\Lambda) \stackrel{\text{def}}{=} \sum_{q=1}^{N} w_q P_q(\Lambda), \tag{9.9}$$

where the components of $\mathbf{q} = (q_1, q_2, ...q_N)$ and $\mathbf{w} = (w_1, w_2, ...w_N)$ are nonnegative. The final augmented function is

$$\Pi^P(\Lambda, \mathbf{w}) = \underbrace{\left(\frac{||\mathbb{E}^* - \mathbb{E}^{*,D}||}{||\mathbb{E}^{*,D}||}\right)^{q_0}}_{\text{MACRO MODULI}} + \underbrace{\mathbf{w}\mathbb{P}(\Lambda)}_{\text{MICROFIELD DISTORTION}}. \tag{9.10}$$

For an isotropic objective we write

$$\Pi^P = w_K \underbrace{\left(\frac{\kappa^* - \kappa^{*,D}}{\kappa^{*,D}}\right)^{q_0}}_{\text{MACROSCOPIC BULK}} + w_\mu \underbrace{\left(\frac{\mu^* - \mu^{*,D}}{\mu^{*,D}}\right)^{q_0}}_{\text{MACROSCOPIC SHEAR}} + \underbrace{\sum_{q=1}^{N} w_q P_q(\Lambda)}_{\text{MICROFIELD DISTORTION}} \tag{9.11}$$

Essentially this is an exterior point (penalty) method to enforce constraints. Let us define the following penalized function $\Pi^P(\Lambda_K, \mathbf{w}^K) \stackrel{\text{def}}{=} \Pi(\Lambda_K) + \mathbf{w}^K \mathbb{P}(\Lambda_K)$, where (1) $\mathbb{P}(\Lambda_K) \in C^0(\mathbb{R}^N)$, (2) $\mathbb{P}(\Lambda_K) \geq 0$ and (3) $\mathbb{P}(\Lambda_K) = 0$, if and only if $\Lambda_K \in \Theta_f$, $\Theta_f \stackrel{\text{def}}{=}$ the feasible region. We have the following properties, where Λ^*_K is the minimizer of Π^P for weight $\mathbf{w}^K$ and Λ^* is a true minimizer Π with the constraints:

$$
\begin{aligned}
&(\mathrm{a})\Pi^P(\Lambda_K^*,\mathbf{w}^K) \leq \Pi^P(\Lambda_{K+1}^*,\mathbf{w}^{K+1})\\
&(\mathrm{b})\mathbb{P}(\Lambda_{K+1}^*) \leq \mathbb{P}(\Lambda_K^*)\\
&(\mathrm{c})\Pi(\Lambda_K^*) \leq \Pi(\Lambda_{K+1}^*)\\
&(\mathrm{d})\Pi(\Lambda_K^*) \leq \Pi^P(\Lambda_K^*,\mathbf{w}^K) \leq \Pi(\Lambda^*)\\
&(\mathrm{e})\lim_{\mathbf{w}^K\to\infty}\Pi^P(\Lambda_K^*,\mathbf{w}^K) = \Pi(\Lambda^*).
\end{aligned}
\tag{9.12}
$$

The proofs are constructive and follow. By definition

$$
\begin{aligned}
\Pi^P(\Lambda_{K+1}^*,\mathbf{w}^{K+1}) &= \Pi(\Lambda_{K+1}^*) + \mathbf{w}^{K+1}\mathbb{P}(\Lambda_{K+1}^*)\\
&\geq \Pi(\Lambda_{K+1}^*) + \mathbf{w}^{K}\mathbb{P}(\Lambda_{K+1}^*)\\
&\geq \Pi(\Lambda_{K}^*) + \mathbf{w}^{K}\mathbb{P}(\Lambda_{K}^*) \stackrel{\text{def}}{=} \Pi^P(\Lambda_K^*,\mathbf{w}^K) \Rightarrow (a).
\end{aligned}
\tag{9.13}
$$

By definition

$$
\Pi(\Lambda_K^*) + \mathbf{w}^K\mathbb{P}(\Lambda_K^*) \leq \Pi(\Lambda_{K+1}^*) + \mathbf{w}^K\mathbb{P}(\Lambda_{K+1}^*) \tag{9.14}
$$

and

$$
\Pi(\Lambda_{K+1}^*) + \mathbf{w}^{K+1}\mathbb{P}(\Lambda_{K+1}^*) \leq \Pi(\Lambda_K^*) + \mathbf{w}^{K+1}\mathbb{P}(\Lambda_K^*). \tag{9.15}
$$

Adding the two previous results yields

$$
(\mathbf{w}^{K+1}-\mathbf{w}^K)\mathbb{P}(\Lambda_{K+1}^*) \leq (\mathbf{w}^{K+1}-\mathbf{w}^K)\mathbb{P}(\Lambda_K^*) \Rightarrow (b). \tag{9.16}
$$

Since Λ_K is a minimizer of the augmented form

$$
\Pi(\Lambda_{K+1}^*) + \mathbf{w}^K\mathbb{P}(\Lambda_{K+1}^*) \geq \Pi(\Lambda_K^*) + \mathbf{w}^K\mathbb{P}(\Lambda_K^*). \tag{9.17}
$$

From part (b) we have $\mathbb{P}(\Lambda_{K+1}^*) \leq \mathbb{P}(\Lambda_K^*)$, which, when combined with the above yields

$$
\begin{aligned}
\Pi(\Lambda_{K+1}^*) + \mathbf{w}^K\mathbb{P}(\Lambda_K^*) &\geq \Pi(\Lambda_{K+1}^*) + \mathbf{w}^K\mathbb{P}(\Lambda_{K+1}^*)\\
&\geq \Pi(\Lambda_K^*) + \mathbf{w}^K\mathbb{P}(\Lambda_K^*) \Rightarrow (c).
\end{aligned}
\tag{9.18}
$$

For each K,

$$
\begin{aligned}
\Pi(\Lambda^*) \stackrel{\text{def}}{=} \Pi(\Lambda^*) + \underbrace{\mathbf{w}^K\mathbb{P}(\Lambda^*)}_{=0} &\geq \Pi(\Lambda_K^*) + \mathbf{w}^K\mathbb{P}(\Lambda_K^*) \stackrel{\text{def}}{=} \Pi^P(\Lambda_K^*,\mathbf{w}^K)\\
&\geq \Pi(\Lambda_K^*) \Rightarrow (d).
\end{aligned}
\tag{9.19}
$$

Utilizing the previous results, one can show that the method converges. Let $\{\Lambda_K^*\}$, $K \in \mathcal{Q}$, be a convergent subsequence of $\{\Lambda_K^*\}$, having limit $\hat{\Lambda}$. Therefore, assuming continuity of Π we have $\lim_{K\in\mathcal{Q}} \Pi(\Lambda_K^*) = \Pi(\hat{\Lambda})$. From the previous results

$$\Pi^P(\Lambda_{K+1}^*, \mathbf{w}^{K+1}) \geq \Pi^P(\Lambda_K^*, \mathbf{w}^K), \tag{9.20}$$

which implies

$$\lim_{K\in\mathcal{Q}} \Pi^P(\Lambda_K^*, \mathbf{w}^K) \stackrel{\text{def}}{=} \Pi^{P*} \leq \Pi(\Lambda^*) \quad (By\, d). \tag{9.21}$$

Therefore,

$$\Pi^{P*}(\Lambda^*) - \Pi(\hat{\Lambda}) = \lim_{K\in\mathcal{Q}} \mathbf{w}^K \mathbf{I\!P}(\Lambda_K^*) \leq \infty, \tag{9.22}$$

which implies $\lim_{K\in\mathcal{Q}} \mathbf{I\!P}(\Lambda_K^*) = 0$. Since $\mathbf{I\!P}(\Lambda_K^*)$ is continuous, $\mathbf{I\!P}(\hat{\Lambda}) = 0$, and therefore $\hat{\Lambda}$ is feasible. $\hat{\Lambda}$ must be optimal since, $\Lambda_K^* \to \hat{\Lambda}$ and

$$\Pi(\hat{\Lambda}) = \lim_{K\in\mathcal{Q}} \Pi(\Lambda_K^*) \leq \Pi(\Lambda^*) \Rightarrow (e). \tag{9.23}$$

Remark. By increasing the penalty weights, we force the augmented form's sequence to approach the unaugmented optimum from below. For general properties of exterior point methods see Fiacco and McCormick [38] or Luenberger [131].

9.4 Fatigue-Type Constraints

As mentioned during the construction of the distortion constraints, a drawback of adding particulate material to a homogeneous base matrix is that the presence of a second phase particles will perturb the otherwise smooth stress fields in the matrix. This locally amplifies the stress field intensity throughout the microstructure. Under cyclic loading, this can lead to fatigue-induced damage. Therefore, when designing new solids with heterogeneous microstructure, estimates of the aggregate amount of fatigue damage are valuable in characterizing the long term performance of the material. Accordingly, we construct constraints, representing tolerable damage limits, in a similar way as for the distortion measures.

9.4.1 Classical Fatigue Relations

At a material point, the classical Basquin relation for fatigue life estimation are as follows: (Basquin [13]):

$$||\sigma_a|| = (||\sigma_f|| - ||\sigma_m||)(2N_f)^b \Rightarrow N_f = \frac{1}{2}\left(\frac{||\sigma_a||}{||\sigma_f|| - ||\sigma_m||}\right)^{\frac{1}{b}}, \tag{9.24}$$

which has been extended to a multiaxial stress state by use of Euclidean norms ($||\sigma|| \stackrel{\text{def}}{=} \sqrt{\sigma : \sigma}$). Here the norm of the mean stress is $||\sigma_m|| = \frac{||\sigma_{max}+\sigma_{min}||}{2}$ while the norm of the fluctuating stress is $||\sigma_a|| = \frac{||\sigma_{max}-\sigma_{min}||}{2}$, where we denote static failure stress by σ_f and typically, $-0.12 \leq b \leq -0.05$. Classical relations of this type, which hold at a material point, are discussed in Suresh et al. [195].

9.4.2 Construction of a Constraint

Clearly, from direct numerical computations, say with the finite element method, one can post process the stresses, by using (10.3), directly to compute the estimated fatigue life of every point $\mathbf{x}_k$, denoted by N_{fk}. Once the N_{fk}'s are computed, we employ classical Palmgren [166]– Miner [132] accumulated damage relations, $\kappa_k^n = MAX\left(0, 1-\frac{n}{N_{fk}}\right)\kappa_k^o$ and $\mu_k^n = MAX\left(0, 1-\frac{n}{N_{fk}}\right)\mu_k^o$, where κ_k^o and μ_k^o are the the undamaged values of the bulk and shear moduli at $\mathbf{x}_k$, respectively, and n is the number of applied load cycles. In the isotropic case, the damage is simply

$$\alpha_k^n \stackrel{\text{def}}{=} \frac{\kappa_k^n}{\kappa_k^o} = \frac{\mu_k^n}{\mu_k^o} = MAX\left(0, \left(1-\frac{n}{N_{fk}}\right)\right). \tag{9.25}$$

The overall fatigue damage is characterized volumetric average, $\langle\alpha^n\rangle_\Omega$. A fatigue constraint is constructed first by setting a tolerance, where ideally,

$$\langle\alpha^n\rangle_\Omega > TOL_\alpha, \tag{9.26}$$

which is then incorporated as an active-inactive unilateral condition into a design cost function, in order to minimize ($q_0, q_F > 0$)

$$\Pi^P = \underbrace{\left(\frac{||\mathbb{E}^* - \mathbb{E}^{*,D}||}{||\mathbb{E}^{*,D}||}\right)^{q_0}}_{\text{MACRO OBJECTIVE}} + \underbrace{w_F P_F}_{\text{FATIGUE DAMAGE}}, \tag{9.27}$$

where

$$w_F P_F \stackrel{\text{def}}{=} w_F \left(MAX\left(0, \left(\frac{TOL_\alpha - \langle\alpha^n\rangle_\Omega}{TOL_\alpha}\right)\right)\right)^{q_F}. \tag{9.28}$$

In the isotropic case, we write

$$\Pi^P = w_\kappa \left(\frac{\kappa^* - \kappa^{*,D}}{\kappa^{*,D}}\right)^{q_0} + w_\mu \left(\frac{\mu^* - \mu^{*,D}}{\mu^{*,D}}\right)^{q_0} + w_F P_F. \tag{9.29}$$

Remark I. As in the case of distortion constraints, the formulation in (9.11) is an exterior point (penalty) method to enforce constraints. Therefore, by increasing the penalty weight, we force the augmented form approach the unaugmented optimium from below. In other words, as the penalty parameter is increased, the exterior point formulation more accurately approximates the original constrained problem. However, the exterior point formulation becomes harder to minimize with gradient based methods, if they are applicable, due to ill-conditioning. However, this appears to be a nonissue for nonderivative methods.

Remark II. One can restrict the constraint activation to specific "zones of interest" by simply adding a weighting function during the computation of the average damage, $\phi(\langle\alpha^n\rangle_\Omega)$, where $\phi = 0$ outside of the zones. For example, such zones could be interfacial regions enveloping the material interfaces between the matrix and particles.

9.4.3 Qualitative Behavior of the Fatigue Constraints

One way to qualitatively characterize the aggregate fatigue damage is via concentration tensors, which provide a measure of the deviation away from the mean fields throughout the material. One may write $\langle\sigma\rangle_\Omega = \frac{1}{|\Omega|}(\int_{\Omega_1}\sigma\, d\Omega + \int_{\Omega_2}\sigma\, d\Omega) = v_1\langle\sigma\rangle_{\Omega_1} + v_2\langle\sigma\rangle_{\Omega_2}$, where Ω_1 is the volume of the matrix phase and Ω_2 is the volume fraction of the second phase, and consequently

$$\langle\sigma\rangle_\Omega = v_1\langle\sigma\rangle_{\Omega_1} + v_2\langle\sigma\rangle_{\Omega_2} = ((\mathbb{E}_1 + v_2(\mathbb{E}_2 - \mathbb{E}_1)) : \mathbf{C}) : \langle\epsilon\rangle_\Omega, \quad (9.30)$$

where $\mathbf{C} : \langle\epsilon\rangle_\Omega = \langle\epsilon\rangle_{\Omega_2}$, $\mathbf{C} \stackrel{\text{def}}{=} \left(\frac{1}{v_2}(\mathbb{E}_2 - \mathbb{E}_1)^{-1} : (\mathbb{E}^* - \mathbb{E}_1)\right)$. Thereafter, we may write, for the variation in the stress, $\mathbf{C} : \mathbb{E}^{*-1} : \langle\sigma\rangle_\Omega = \mathbb{E}_2^{-1} : \langle\sigma\rangle_{\Omega_2}$, which reduces to $\mathbb{E}_2 : \mathbf{C} : \mathbb{E}^{*-1} : \langle\sigma\rangle_\Omega \stackrel{\text{def}}{=} \overline{\mathbf{C}} : \langle\sigma\rangle_\Omega = \langle\sigma\rangle_{\Omega_2}$, where $\overline{\mathbf{C}}$ is known as the stress concentration tensor. Therefore, once either $\overline{\mathbf{C}}$ or $\mathbb{E}^*$ are known, the other can be determined. Clearly, the microstress fields are minimally distorted when $\overline{\mathbf{C}} = \mathbf{I}$. For the matrix, we have

$$\langle\sigma\rangle_{\Omega_1} = \frac{\langle\sigma\rangle_\Omega - v_2\langle\sigma\rangle_{\Omega_2}}{v_1} = \frac{\langle\sigma\rangle_\Omega - v_2\overline{\mathbf{C}} : \langle\sigma\rangle_\Omega}{v_1} = \frac{(\mathbf{I} - v_2\overline{\mathbf{C}}) : \langle\sigma\rangle_\Omega}{v_1} = \overline{\overline{\mathbf{C}}} : \langle\sigma\rangle_\Omega. \quad (9.31)$$

We have for the alternating stresses, $\langle\sigma_a\rangle_{\Omega_2} = \overline{\mathbf{C}} : \langle\sigma_a\rangle_\Omega$ and $\langle\sigma_a\rangle_{\Omega_1} = \overline{\overline{\mathbf{C}}} : \langle\sigma_a\rangle_\Omega$, as well as for the mean stresses $\langle\sigma_m\rangle_{\Omega_2} = \overline{\mathbf{C}} : \langle\sigma_m\rangle_\Omega$ and $\langle\sigma_m\rangle_{\Omega_1} = \overline{\overline{\mathbf{C}}} : \langle\sigma_m\rangle_\Omega$. We remark that the failure stresses, are independent of the applied loads, since they are considered material parameters. We remark that the relative decrease in $\overline{\mathbf{C}}$ is accompanied by a corresponding increase $\overline{\overline{\mathbf{C}}}$, since

$$\frac{(\mathbf{I} - v_2\overline{\mathbf{C}})}{1 - v_2} = \overline{\overline{\mathbf{C}}}. \quad (9.32)$$

In other words, $\frac{\partial\overline{\overline{\mathbf{C}}}}{\partial\overline{\mathbf{C}}}$ is a negative-definite tensor function. These previous relations simply indicate that when there is amplification of the stresses somewhere in the microheterogeneous solid there is also simultaneous shielding ("de-amplification") somewhere in the solid. For the second phase

$$N_{f2} = \frac{1}{2}\left(\frac{||\langle\sigma_a\rangle_{\Omega_2}||}{||\langle\sigma_{f2}\rangle_{\Omega_2}||-||\langle\sigma_m\rangle_{\Omega_2}||}\right)^{\frac{1}{b_2}} = \frac{1}{2}\left(\frac{||\overline{\mathbf{C}}:\langle\sigma_a\rangle_{\Omega}||}{||\langle\sigma_{f2}\rangle_{\Omega_2}||-||\overline{\mathbf{C}}:\langle\sigma_m\rangle_{\Omega}||}\right)^{\frac{1}{b_2}}, \tag{9.33}$$

which is a monotonically increasing function of $\overline{\mathbf{C}}$. For the first phase,

$$N_{f1} = \frac{1}{2}\left(\frac{||\langle\sigma_a\rangle_{\Omega_1}||}{||\langle\sigma_{f1}\rangle_{\Omega_1}||-||\langle\sigma_m\rangle_{\Omega_1}||}\right)^{\frac{1}{b_1}} = \frac{1}{2}\left(\frac{||\overline{\overline{\mathbf{C}}}:\langle\sigma_a\rangle_{\Omega}||}{||\langle\sigma_{f1}\rangle_{\Omega_1}||-||\overline{\overline{\mathbf{C}}}:\langle\sigma_m\rangle_{\Omega}||}\right)^{\frac{1}{b_1}}, \tag{9.34}$$

which is a monotonically decreasing function of $\overline{\mathbf{C}}$. The rates of increase and decrease of the fatigue lives are controlled by the magnitudes of the b's (material parameters).

9.5 Nonconvex–Nonderivative Genetic Search

Due to difficulties with objective function nonconvexity and nondifferentiability, we employ a certain class of nonderivative search methods, usually termed "genetic" algorithms (GA), which stem from the pioneering work of John Holland and his colleagues starting the late 1960s [84]. For reviews of such methods, the interested reader is referred to Goldberg [63], Davis [32] or Kennedy and Eberhart [107]. A recent overview of the state of the art of the field can be found in a collection of recent articles, edited by Goldberg and Deb [64]. In Zohdi [249] a genetic algorithm was developed by combining the basic ideas used in the Genetic Algorithm community. Presently, we build upon this algorithm further. The central idea is that the microscale parameters form a genetic string and a survival of the fittest algorithm is applied to a population of such strings. The overall process is (I) A population (S) of different designs (strings) are generated at random with the design space, represented by a ("genetic") string of the design (N) parameters, (II) The performance of each design is tested, (III) The designs are ranked from top to bottom according to their performance, (IV) The best designs are mated pairwise producing two offspring, i.e. each best pair exchanges information by taking random convex combinations of the design components of the parents' genetic strings and (V) The worst performing genetic strings are eliminated and new replacement designs (strings) are introduced into the remaining population of best performing genetic strings and (VI) steps (I–V) are then repeated. An implementation of such ideas is as follows:

- **STEP 1:** Randomly select a population of N starting genetic strings, Λ^i, $(i = 1,...,N)$, where

$$\Lambda^i \stackrel{\text{def}}{=} \{\Lambda_1^i,\Lambda_2^i,\Lambda_3^i,\Lambda_4^i,\Lambda_5^i,\Lambda_6^i,\Lambda_7^i,...,\} = \{\kappa_2^i,\mu_2^i,\nu_2^i,AR^i,s_1^i,s_2^i,s_3^i...\}.$$

- **STEP 2:** Compute the fitness of each string, $\Pi(\Lambda^i)$, (i = 1, ..., N).
- **STEP 3:** Rank the genetic strings: Λ^i, (i = 1, ..., N).
- **STEP 4:** Mate the nearest pairs and produce two offspring, (i = 1, ..., N)

$$\lambda^i \stackrel{\text{def}}{=} \Phi^{(I)}\Lambda^i + (1-\Phi^{(I)})\Lambda^{i+1}$$

and

$$\lambda^{i+1} \stackrel{\text{def}}{=} \Phi^{(II)}\Lambda^i + (1-\Phi^{(II)})\Lambda^{i+1},$$

where $0 \leq \Phi^{(I)}, \Phi^{(II)} \leq 1$, with $\Phi^{(I)}$ and $\Phi^{(II)}$ being different random values for each component.
- **STEP 5:** Kill off the bottom $M < N$ strings and keep the top $K < N$ parents (K offspring + K parents + $M = N$).
- **STEP 6:** Repeat STEPS 1–6 with the top gene pool (K offspring and K parents), plus M new, randomly generated, genetic strings.

Remark I. Previous numerical studies by Zohdi [249] have found that the retention of the top old fit genetic strings is critical. For sufficiently large populations, the benefits of parent retention outweigh any disadvantages of "inbreeding", i.e. a stagnant population, provided sufficient new genes are introduced after each generation. For more details on this so-called "inheritance property" see Davis [32], Onwubiko [160]. This stems from the fact that the objective functions are highly nonconvex and there exists a strong possibility that the inferior offspring will replace superior parents. Retaining the top parents is not only less computationally expensive, since these designs do not have to be reevaluated in future generations, it is theoretically superior.[2] With parent retention, the minimization of the cost function is guaranteed to be monotone with increasing generations (Zohdi [249]).

Remark II. The overall genetic minimization strategy can be enhanced several ways. For example, every few generations, the search domain can be restricted and rescaled to be centered around the best current design.

Remark III. Typically, for samples with a finite number of particles, there will be slight variations in the performance for different random realizations. In order to stabilize the objective function's value with respect to the randomness of the realization, for a fixed design (Λ^I), a regularization procedure is applied, whereby the performances of a series of different random realizations are averaged until the (ensemble) average converges, i.e. until the following condition is met ($i = 1, 2, \ldots, E$):

$$\left|\frac{1}{E+1}\sum_{i=1}^{E+1}\Pi^{(i)}(\Lambda^I) - \frac{1}{E}\sum_{i=1}^{E}\Pi^{(i)}(\Lambda^I)\right| \leq TOL\left|\frac{1}{E+1}\sum_{i=1}^{E+1}\Pi^{(i)}(\Lambda^I)\right|. \quad (9.35)$$

[2] The monotonicity is obvious since the top design will not be replaced until a better design is found.

The index i indicates a random realization that has been generated and E indicates the total number of realizations tested. In order to implement this in STEP 2 of the algorithm, one simply replaces "COMPUTE" with "ENSEMBLE-COMPUTE", which requires a further inner loop to test the performance of multiple realizations. Similar ideas have been applied to randomly dispersed particulate solids in Zohdi [249]. The number needed to stabilize the objective function is far less than that needed to stabilize the gradients, and higher order derivatives that would be needed *if a gradient-type approach were even possible.*

9.6 Numerical Examples

Throughout the numerical examples, we considered the following objective function

$$\Pi^P = \underbrace{w_\kappa \left(\frac{\kappa^* - \kappa^{*,D}}{\kappa^{*,D}}\right)^2}_{\text{MACRO-BULK}} + \underbrace{w_\mu \left(\frac{\mu^* - \mu^{*,D}}{\mu^{*,D}}\right)^2}_{\text{MACRO-SHEAR}} + \underbrace{\sum_{q=1}^{10} w_q P_q(\Lambda)}_{\text{MICRO-DISTORTION}} + \underbrace{w_F P_F}_{\text{MICRO-FATIGUE}} . \tag{9.36}$$

A meaningful measure to track during the minimization process is the percentage error of the moduli from the desired target, $\sqrt{\Pi^P}$. To illustrate the algorithm, we considered a cube of matrix material, with normalized dimensions $1 \times 1 \times 1$, containing randomly distributed inhomogeneities. We considered the following boundary conditions on the exterior of the cube: $\mathbf{u}|_{\partial\Omega} = \mathscr{E} \cdot \mathbf{x}$, $\mathscr{E}_{ij}$=0.001, $i, j = 1, 2, 3$, where $\mathbf{x}$ is a position vector to the boundary of the cube. During the upcoming numerical experiments we tested larger and larger samples of material, keeping the volume fraction fixed and found that results stabilized when approximately 20 (nonintersecting) particles were used in a sample, i.e. the same final designs occurred using larger samples. Also, over the course of such tests the finite element meshes were repeatedly refined, and a mesh density of approximately $9 \times 9 \times 9$ trilinear hexahedra (approximately between 2200 and 3000 DOF for the vector-valued balance of momentum) *per particle* was found to deliver mesh independent results. For 20 particles, this resulted in 46875 DOF, see Fig. 9.5. This mesh density delivered mesh independent results over the course of the numerical experiments. In other words, the same final designs occurred using finer meshes. During the computations, a "2/5" Gauss rule was used, whereby elements containing material discontinuities had increased Gauss rules ($5 \times 5 \times 5$) to enhance the resolution of the internal geometry, while elements with no material discontinuities had the nominal $2 \times 2 \times 2$ rule. For details of such meshing procedures, see Zohdi and Wriggers [233, 235], Zohdi et al. [236] and Zohdi [241]. To illustrate the search process, continuing with 20 particle samples, the effective response produced by a sample containing a particulate stiffener, 22% boron ($\mu_1 = 230\,\text{GPa}$ $\kappa_1 = 172\,\text{GPa}$) spheres in an aluminum matrix ($\mu_1 = 25.9\,\text{GPa}$ $\kappa_1 = 77.9\,\text{GPa}$) was first computed.[3] The effective response was approximately $\kappa^* = 96\,GPa$ and $\mu^* = 42\,GPa$. Our objective was to find alternative microstructures which could deliver the same effective response ($\kappa^{*D} = 96\,GPa$ and $\mu^{*D} = 42\,GPa$), while obeying the constraints, to a specified tolerance. The constraint tolerance for fatigue was $TOL_\alpha = 0.9$ (10 percent damage) and for the distortion constraints, $TOL_q = 0.1, \forall q$. The matrix material was fixed to be aluminum, however, all other design parameters, with the exception of the particle orientations, since isotropic objectives were sought, were allowed to vary. The macroscopic weights were fixed at $w_\kappa = w_\mu = 1$, while the penalty weights were increased. The volume fraction was controlled via a particle/sample size ratio (one variable), defined by a subvolume size $V \stackrel{\text{def}}{=} \frac{L\times L\times L}{N}$, where N is the number of particles in the entire sample and where L is the length of the (cubical) sample, $L \times L \times L$. A generalized diameter (and radius) is defined, $d = 2r$, which is the diameter of the smallest sphere that can encompass a single particle of possibly non-spherical shape. The ratio between the generalized diameter and the subvolume is one design parameter defined by $\xi \stackrel{\text{def}}{=} \frac{r}{V^{\frac{1}{3}}}$. The a priori constraints on the design search space were:

$$
\begin{aligned}
&0.1\kappa_1 = \kappa_2^- \le \kappa_2^{(i)} \le \kappa_2^+ = 10\kappa_1,\\
&0.1\mu_1 = \mu_2^- \le \mu_2^{(i)} \le \mu_2^+ = 10\mu_1,\\
&0.1 = AR^- \le AR^{(i)} \le AR^+ = 10,\\
&1 = s^- \le s^{(i)} \le s^+ = 10,\\
&0.2 = \xi^- \le \xi^{(i)} \le \xi^+ = 0.4.
\end{aligned} \tag{9.37}
$$

For the fatigue constraints, we characterized the maximum and minimum applied boundary loading in the form[4] $\mathbf{u}^{\max}|_{\partial\Omega} = \mathscr{E}_{\max} \cdot \mathbf{x} = a^+ \mathscr{E}_{ij}$, and $\mathbf{u}^{\min}|_{\partial\Omega} = \mathscr{E}_{\min} \cdot \mathbf{x} = a^- \mathscr{E}_{ij}, i,j = 1,2,3$. For purposes of numerical experiment, the cyclic loading amplitudes were set to $a^- = 0.9$ and $a^+ = 1.25$. The Basquin exponents were chosen to be the same for both materials, $b_1 = -0.06$ and $b_2 = -0.06$, in order to isolate the effects of the seven micromechanical variables introduced earlier on the fatigue behavior. The number of applied cycles was set to $n = 10^5$. The failure stresses for the particles, σ_{f2}, were set to $\sigma_{f2,ij} = 1$ GPa, $i,j, = 1,2,3$, while for the matrix σ_{f1}, we selected $\sigma_{f1,11} = \sigma_{f1,22} = \sigma_{f1,33} = 8\times 10^7$ and $\sigma_{f1,12} = \sigma_{f1,23} = \sigma_{f1,31} = 4\times 10^7$.

The number of genetic strings was set to 20, keeping the offspring of the top 6 parents after each generation. This resulted in eight new strings being introduced after each generation. After every four generations, the search domain was restricted

[3] Material combinations such as aluminum/boron are relatively common due to the ease of forming the aluminum matrix and the lightweight of the boron.

[4] By superposition, one only needs to compute a single loading.

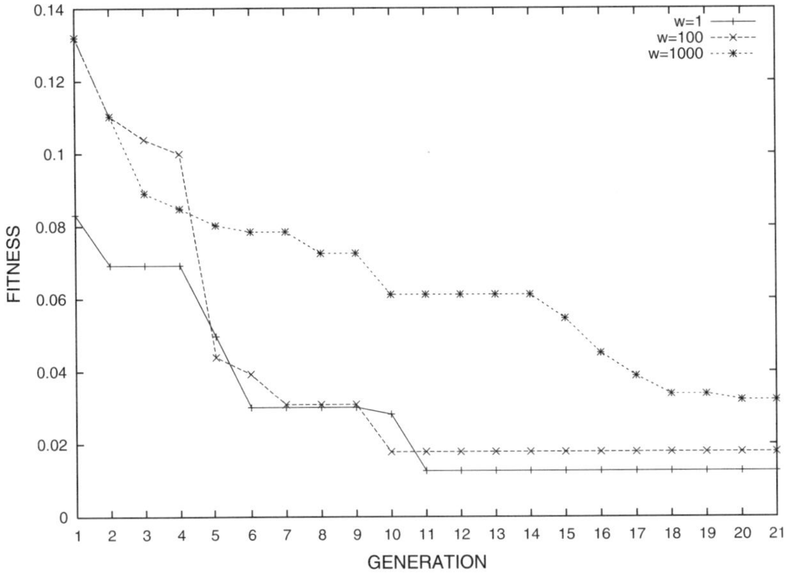

Fig. 9.4 The top design behavior for various weights after 21 generations. Here FITNESS $\stackrel{\text{def}}{=} \sqrt{\Pi^P}$. After every four generations, the search domain was restricted and rescaled to be centered around the best current design

and rescaled to be centered around the best current design. For various penalty weights, the procedure converged to a stable best design in no more than 21 generations (see Fig. 9.4 and 9.5). The total number of global evaluations was $T+(G-1)\times(T-Q)=300$, where $G=21$ was the number of generations, $T=20$ was the total number of genetic strings in the population, and $Q=6$ is the number of parents kept after each generation. As the theory predicted (Box 9.12), at the

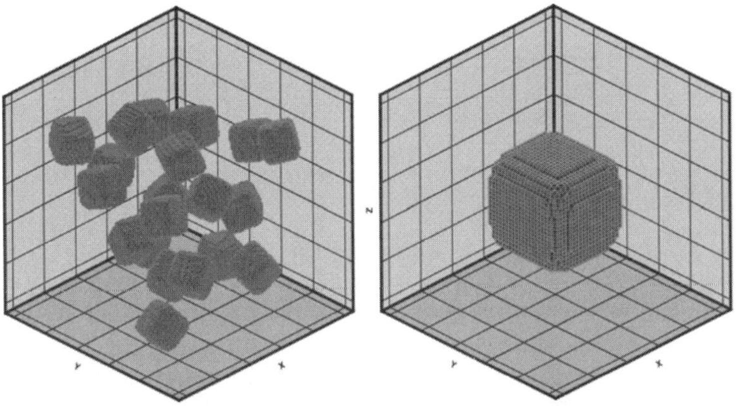

Fig. 9.5 **Left:** A realization of the optimal design. **Right:** A zoom on an individual particle

Table 9.1 Top design after 21 generations. Note $\langle\alpha\rangle_\Omega = 1$ means *no estimated damage due to fatigue*

Penalty	$\sqrt{\Pi^P}$	κ^* (GPa)	μ^* (GPa)	$\langle\alpha\rangle_\Omega$
$w_q = w_F = 1$	0.01260	93.464	40.290	0.92155
$w_q = w_F = 100$	0.01780	92.498	37.716	0.92121
$w_q = w_F = 1000$	0.03208	94.195	38.728	0.90430

converged state (generation 21), with increasing penalty weights $\Pi^P(\Lambda_K^*, \mathbf{w}^K)$ converges from below. There were no changes in the results for further increases in the penalty weights beyond $w_q = w_F = 1000$. Clearly, the constraints restrict the objective from ever attaining the original spherical microstructure, and thus effective properties, $\kappa^* = 96\,GPa$ and $\mu^* = 42\,GPa$, that deliver a zero value of the unconstrained objective function. Various statistics pertaining to these test are tabulated in Tables 9.1, 9.2, and 9.3.

Table 9.2 Top design after 21 generations

Penalty	$\frac{\kappa_2}{\kappa_1}$	$\frac{\mu_2}{\mu_1}$	ξ	v_2	AR	s_1	s_2	s_3	$\sqrt{\Pi^P}$
$w_q = w_F = 1$	4.9755	6.0648	0.22680	0.07524	0.9392	4.9036	5.3449	5.7754	0.01260
$w_q = w_F = 100$	4.1183	6.0271	0.23042	0.07898	1.0065	5.7951	5.0898	3.0565	0.01780
$w_q = w_F = 1000$	5.2445	4.9966	0.23714	0.08668	0.9513	5.6470	6.9010	3.7959	0.03208

Table 9.3 The number of samples needed for ensemble average stabilization for various penalty weights

Penalty	Total Samples	$\frac{Total\ Samples}{Genetic\ String}$
$w_q = w_F = 1$	5616	18.720
$w_q = w_F = 100$	5251	17.503
$w_q = w_F = 1000$	4837	16.123

9.7 Scope of Use

The genetic algorithm that was developed can handle difficulties due to objective function nonconvexity and lack of regularity characterizing the constrained design of random particulate media. The computational approach was constructed in such a way that it can be used in conjunction with a variety of methods designed for large-scale micro–macro simulations, such as

- *Multiscale methods*: Fish and Wagiman [39], Fish et al. [40], Fish and Belsky [41, 42, 43, 44, 45], Fish and Shek [46], Fish and Ghouli [47], Fish and Yu [48], Fish and Chen [49], Belsky et al. [16], Chen and Fish [25] and Wentorf et al. [213];
- *Voronoi cell methods*: Ghosh and Mukhopadhyay [54], Ghosh and Moorthy [55], Ghosh et al. [56], Ghosh and Moorthy [57, 58, 59, 60], Lee et al. [116], Li et al. [123], Moorthy and Ghosh [146] and Raghavan et al. [171];
- *Transformation methods*: Moulinec and Suquet [147], Michel et al. [144],
- *Multipole methods*: adapted to such problems by Fu et al. [51];
- *Partitioning methods*: Huet [88, 89, 92], Hazanov and Huet [77], Hazanov and Amieur [78] and Huet [99, 100];
- *Adaptive hierarchical modeling methods*: Zohdi et al. [223], Oden and Zohdi [156], Moes et al. [145], Oden and Vemaganti [157], Oden et al. [158] and Vemaganti and Oden [209]; and
- *Micro–macro domain decomposition*: Ladeveze and Leguillon [112], Zohdi and Wriggers [225, 233, 235], Zohdi [249] and Zohdi et al. [236].

Chapter 10
Modeling Coupled Multifield Processes

10.1 Introduction

In many applications, multifield models arise from a description of thermo-chemical reactions occurring in deformable multiphase solids (Fig. 10.1). Relevant examples include certain problems of environmental mechanics concerned with the detrimental chemical attack on solids by volatile gases or liquid solutes which come in contact with structural surfaces and then diffuse into the subsurface. The subsequent reactions lead to loss of structural integrity.[1] In this chapter, a relatively general multifield model is developed which describes the diffusion of a dilute solute into a heterogeneous solid material, the subsequent reactions, the production of heat, the changes in the stress fields and the evolution of material changes and inelastic strains within the solid. The modeling and solution algorithms are general enough to be applicable to a wide range of long-term thermo-chemo-mechanical phenomena associated with damage in materials with heterogeneous microstructure. For experimental and theoretical overviews, we refer the reader to the series of works by Huet and coworkers [90, 91, 92, 93, 94, 95, 96, 97, 98, 99, 100]. For general overviews in the area of heterogeneous materials see Nemat-Nasser and Hori [151]. Although the emphasis will be on the simulation of long-term multifield phenomena, the analysis in this chapter is also applicable to shorter time-scale problems involving thermo-chemical processing such as self-propagating high-temperature synthesis (SHS), whereby chemical reactions are initiated on the surface of a material to be processed. In the method, the substantial heat evolved by surface chemical reactions sustain and propagate thermo-chemical processes. This approach is a relatively new economical way of manufacturing advanced materials. For example by initiating a highly exothermic surface thermal reaction between bonded titanium powder in a nitrogen gas atmosphere, an initial amount of titanium-nitride, $Ti(s) + 1/2N_2(g) = TiN(s)$, which is a desirable product, is produced. This process releases a substantial amount of heat, which produces a combustion front which propagates throughout the solid. A wide variety of materials, ceramics, intermetallic compounds and composites can

[1] This chapter follows from work found in Zohdi [250, 251].

Fig. 10.1 Diffusion of a small species into a heterogeneous medium

be produced by SHS. For an introduction, see the texts of Ashby and Jones [8] or Shackelford [178]. Generally speaking, such processes are related to controlled combustion methods (see Schmidt [174]), whereby a chemical species acts as a catalyst to promote reactions, for example in chemical separation and polymerization. A related processing method is Shock Induced Chemical Reactions (SICR), whereby a shock wave is passed through chemical reacting powders, which sinters them together. Some relevant work in the modeling and simulation of such processes can be found in Thadhani [200], Nesterenko et al. [153], Vecchio and Meyers [208] and Meyers et al. [141]. For a review of such methods see Meyers [142] or Nesterenko [152].

10.2 A Model Problem Involving Multifield Processes in Multiphase Solids

A structure which occupies an open bounded domain in $\Omega \in I\!R^3$, with boundary $\partial\Omega$, is considered. The boundary consists of (1) Γ_c and Γ_g, where the solute concentrations (c) and solute fluxes are respectively specified, (2) Γ_u on which the displacements $(\mathbf{u})$ are prescribed and a part Γ_t on which tractions are prescribed and (3) Γ_θ on which the temperature (θ) is prescribed, and a part Γ_q on which thermal fluxes are prescribed. The primary (familiar) mechanical, thermal, and diffusive properties of the heterogeneous material are characterized by a spatially varying elasticity tensor $I\!E \in I\!R^{3^2\times 3^2}$, a spatially varying conductivity tensor $I\!K \in I\!R^{3\times 3}$, and a spatially varying diffusivity tensor $I\!D_0 \in I\!R^{3\times 3}$ (at a reference temperature), all of which are assumed to be symmetric, bounded, positive-definite, tensor-valued functions. Other material properties will be introduced during the development of the overall model. For reasons of clarity, strong forms are used to derive the governing equations, possibly assuming more regularity than warranted. Afterwards, only the weak forms, which produce solutions which coincide with strong forms when the solutions are smooth enough, are employed. The use of weak forms are important for this class of problems due to the heterogeneous microstructure, which leads to particularly rough solution fields.

10.3 Constitutive Assumptions

We consider the case of moderate finite deformations involving elastic and inelastic strains. Later in the chapter, a staggering-type scheme, formulated directly in the (deformed) current configuration, will be developed, and thus Eulerian-based material laws are advantageous. A relatively straightforward extension to classical *isotropic* infinitesimal deformation constitutive models is to employ the Eulerian (or Almansi) strain tensor[2]

$$\sigma = \alpha \mathbb{E}_0 : (\mathbf{e} - \mathbf{B}), \tag{10.1}$$

where σ is the Cauchy stress, $\mathbf{e} = \frac{1}{2}(\nabla_x \mathbf{u} + (\nabla_x \mathbf{u})^T - (\nabla_x \mathbf{u})^T \cdot \nabla_x \mathbf{u})$, $\mathbf{B} = \mathbf{e}_\theta + \mathbf{e}_\lambda + \mathbf{e}_\varphi$, $\mathbf{e}_\theta \stackrel{\text{def}}{=} \gamma \cdot (\theta - \theta_0)\mathbf{1}$, $\mathbf{e}_\lambda$ represents deviatoric-like inelastic strains, for example plastic-like strains, $\mathbf{e}_\varphi$ represents dilatational-like inelastic strains, for example representing gas that occurs as a reaction byproduct, and $\mathbf{e}_\theta$ represents thermal strains. Here, the current value of the elasticity tensor is $\mathbb{E} = \alpha \mathbb{E}_0$, where $\mathbb{E}_0$ represents the virgin isotropic undamaged material, $0 \leq \alpha \leq 1$ is the scalar continuity (isotropic damage) parameter (Kachanov [104]), $\alpha(t=0) = 1$ indicates the initial undamaged state and $\alpha \rightarrow 0$ indicates a completely damaged state. The introduced quantities are modeled as being governed by evolution over-stress functions of the form

$$\dot{\alpha} = \underbrace{\left(a_1 c + a_2 \left(\frac{||\sigma'||}{||\sigma'_{crit}||} - 1 \right) + a_3 \left(\frac{|tr\sigma|}{|tr\sigma_{crit}|} - 1 \right) \right)}_{g_\alpha} \alpha \qquad (0 < \alpha \leq 1),$$

$$\dot{\lambda} = \underbrace{\left(a_4 \left(\frac{||\mathbf{S}'||}{||\mathbf{S}'_{crit}||} - 1 \right) \right)}_{g_\lambda} \lambda \Rightarrow \mathbf{e}_\lambda \stackrel{\text{def}}{=} \mathbf{F} \cdot \left(\int_0^t \dot{\lambda} \mathbf{h}\, dt \right) \cdot \mathbf{F}^T \qquad (\mathbf{h} \stackrel{\text{def}}{=} \frac{\mathbf{S}'}{||\mathbf{S}'||}), \tag{10.2}$$

$$\dot{\varphi} = \underbrace{(a_5 c)}_{g_\varphi} \varphi \Rightarrow \mathbf{e}_\varphi \stackrel{\text{def}}{=} \int_0^t \dot{\varphi} \mathbf{1}\, dt,$$

where $\mathbf{S}$ is the second Piola-Kirchhoff stress, where the normalized concentration of the solute is c, given in molecules per unit volume, where $\sigma'_{crit} \stackrel{\text{def}}{=} k_1 \frac{\sigma'}{||\sigma'||}$,

[2] Such a law is frame indifferent under rigid body rotations and translations for isotropic $\mathbb{E}$.

$\frac{tr\sigma_{crit}}{3} \stackrel{\text{def}}{=} k_2$, k_1 and k_2 being material constants, $\sigma' = \sigma - \frac{tr\sigma}{3}\mathbf{1}$, $\mathbf{S}' = J\mathbf{F}^{-1}\cdot\sigma'\cdot\mathbf{F}^{-T}$, $\mathbf{S}'_{crit} = J\mathbf{F}^{-1}\cdot\sigma'_{crit}\cdot\mathbf{F}^{-T}$, $\mathbf{F} = \nabla_X\mathbf{x}$ is the deformation gradient and $J = det\mathbf{F}$ is the Jacobian of the deformation. The displacement is given by $\mathbf{u} = \mathbf{x} - \mathbf{X}$, where $\mathbf{X}$ are referential coordinates, $\mathbf{x}$ are current coordinates and a_1 through a_5 are spatially variable material parameters governed by the following activation conditions:

$$\text{If } c < c_{crit} \text{ then } a_1 = a_5 = 0,$$

$$\text{If } c \geq c_{crit} \text{ then } a_1 = a_1^*,\ a_5 = a_5^*,$$

$$\text{If } ||\sigma'|| \stackrel{\text{def}}{=} \sqrt{\sigma' : \sigma'} < \sigma'_{crit} \text{ then } a_2 = a_4 = 0,$$

$$\text{If } ||\sigma'|| \stackrel{\text{def}}{=} \sqrt{\sigma' : \sigma'} \geq \sigma'_{crit} \text{ then } a_2 = a_2^*,\ a_4 = a_4^*,$$

$$\text{If } |\frac{tr\sigma}{3}| < |\frac{tr\sigma_{crit}}{3}| \text{ then } a_3 = 0,$$

$$\text{If } |\frac{tr\sigma}{3}| \geq |\frac{tr\sigma_{crit}}{3}| \text{ then } a_3 = a_3^*, \tag{10.3}$$

where c_{crit} is a spatially variable critical (threshold) concentration parameter. The parameters a_1^* through a_5^* are given material parameters that are specified later in the analysis. For further details on these types of phenomenological (damage) formulations, the interested reader is referred to the seminal work of Kachanov [104]. Clearly, further evolution laws can be written for other material property changes, such as $\mathbb{K}$ or $\mathbb{D}_0$, although only changes in $\mathbb{E}$ are considered during the simulations to follow.

10.3.1 An Energy Balance Including Growth

As discussed in earlier chapters, the interconversions of mechanical, thermal and chemical energy are governed by the first law of thermodynamics, which states that the time rate of change of the total energy, $\mathscr{K} + \mathscr{I}$, is equal to the work rate, $\mathscr{P}$, and the net heat supplied, $\mathscr{H} + \mathscr{T}$, i.e.

$$\frac{d}{dt}(\mathscr{K} + \mathscr{I}) = \mathscr{P} + \mathscr{H} + \mathscr{T}. \tag{10.4}$$

Consider an arbitrary subvolume of material contained within Ω, referred to as ω, and

- the kinetic energy given by $\mathscr{K} \stackrel{\text{def}}{=} \int_\omega \frac{1}{2}\rho\dot{\mathbf{u}}\cdot\dot{\mathbf{u}}\,d\omega$,
- the stored energy is $\mathscr{I} \stackrel{\text{def}}{=} \int_\omega \rho w\,d\omega$,

- the rate of work or power of external volumetric ($\rho\mathbf{b}$) and surface ($\sigma\cdot\mathbf{n}$) forces acting on ω is given by $\mathscr{P}\stackrel{\text{def}}{=}\int_\omega \rho\mathbf{b}\cdot\dot{\mathbf{u}}\,d\omega+\int_{\partial\omega}\sigma\cdot\mathbf{n}\cdot\dot{\mathbf{u}}\,da$,
- the heat flow into the volume by conduction is $\mathscr{T}\stackrel{\text{def}}{=}-\int_{\partial\omega}\mathbf{q}\cdot\mathbf{n}\,da=-\int_\omega\nabla_x\cdot\mathbf{q}\,d\omega$,
- the heat generated due to sources, *such as chemical reactions*, is $\mathscr{H}\stackrel{\text{def}}{=}\int_\omega\rho z\,d\omega$.

Combining the expressions leads to

$$\frac{d}{dt}\int_\omega\rho\left(\frac{1}{2}\dot{\mathbf{u}}\cdot\dot{\mathbf{u}}+w\right)d\omega=\int_\omega\rho\mathbf{b}\cdot\dot{\mathbf{u}}\,d\omega+\int_{\partial\omega}\sigma\cdot\mathbf{n}\cdot\dot{\mathbf{u}}\,da-\int_\omega\nabla_x\cdot\mathbf{q}\,d\omega+\int_\omega\rho z\,d\omega. \tag{10.5}$$

Converting the first integral to a reference configuration leads to

$$\begin{aligned}\frac{d}{dt}\int_{\omega_0}\rho\left(\frac{1}{2}\dot{\mathbf{u}}\cdot\dot{\mathbf{u}}+w\right)J\,d\omega_0=&\int_{\omega_0}\frac{d}{dt}(\rho J)\left(\frac{1}{2}\dot{\mathbf{u}}\cdot\dot{\mathbf{u}}+w\right)d\omega_0\\&+\int_{\omega_0}(\rho J)\left(\ddot{\mathbf{u}}\cdot\dot{\mathbf{u}}+\dot{w}\right)d\omega_0,\end{aligned} \tag{10.6}$$

where J is the Jacobian of the deformation gradient. The third integral in (10.5) can be converted via Gauss's divergence theorem:

$$\int_{\partial\omega}\sigma\cdot\mathbf{n}\cdot\dot{\mathbf{u}}\,da=\int_\omega\nabla_x\cdot(\sigma\cdot\dot{\mathbf{u}})\,d\omega=\int_\omega(\nabla_x\cdot\sigma\cdot\dot{\mathbf{u}}+\sigma:\nabla_x\dot{\mathbf{u}})\,d\omega. \tag{10.7}$$

Thus, writing all relations in the current configuration, and since the subvolume ω is arbitrary, yields the following local form yields

$$\begin{aligned}\frac{1}{J}\frac{d}{dt}(\rho J)\left(\frac{1}{2}\dot{\mathbf{u}}\cdot\dot{\mathbf{u}}+w\right)=&-\rho\left(\ddot{\mathbf{u}}\cdot\dot{\mathbf{u}}+\dot{w}\right)+\left(\nabla_x\cdot\sigma\cdot\dot{\mathbf{u}}+\rho\mathbf{b}\cdot\dot{\mathbf{u}}+\sigma:\nabla_x\dot{\mathbf{u}}\right)\\&-\nabla_x\cdot\mathbf{q}+\rho z.\end{aligned} \tag{10.8}$$

Substituting a balance of linear momentum with the extra term arising from the fact that the mass is not conserved (due to the diffusion of the small species)[3]

$$\nabla_x\cdot\sigma+\rho\mathbf{b}=\rho\ddot{\mathbf{u}}+\frac{\dot{\mathbf{u}}}{J}\frac{d}{dt}(\rho J), \tag{10.9}$$

leads to

$$\frac{1}{J}\frac{d}{dt}(\rho J)\left(w-\frac{1}{2}\dot{\mathbf{u}}\cdot\dot{\mathbf{u}}\right)+\rho\dot{w}-\sigma:\nabla_x\dot{\mathbf{u}}+\nabla_x\cdot\mathbf{q}-\rho z=0. \tag{10.10}$$

The mass of the solid in the current configuration is given by

$$\int_\omega\rho\,d\omega=\int_{\omega_0}\rho J\,d\omega_0\approx\int_{\omega_0}(\rho_0+\mathscr{Y}(c))\,d\omega_0, \tag{10.11}$$

[3] Normally, if the mass was conserved, then $\frac{d}{dt}(\rho J)=0$.

where $\mathscr{Y}(c)$ represents the density changes due to uptake of c. Since the subvolume ω_0 is arbitrary, this implies

$$\rho J \approx \rho_0 + \mathscr{Y}(c) \Rightarrow \frac{d}{dt}(\rho J) = \dot{\mathscr{Y}}(c). \tag{10.12}$$

Finally, we have the following local form,

$$\tfrac{1}{J}\dot{\mathscr{Y}}(c)\left(w - \tfrac{1}{2}\dot{\mathbf{u}}\cdot\dot{\mathbf{u}}\right) + \rho\dot{w} - \sigma : \nabla_x\dot{\mathbf{u}} + \nabla_x\cdot\mathbf{q} - \rho z = 0. \tag{10.13}$$

We make the following (non-hyperelastic) moderate finite strain approximation for the stored energy

$$\rho w = W \approx \tfrac{1}{2}(\mathbf{e}-\mathbf{B}) : \mathbb{E} : (\mathbf{e}-\mathbf{B}) + \rho\mathscr{C}\theta, \tag{10.14}$$

where $\mathbf{B}$ are all of the eigenstrains, which implies, since $\dot{w} = \dot{\overline{(\frac{W}{\rho})}} = \frac{\dot{W}}{\rho} - \frac{W}{\rho^2}\dot{\rho}$,

$$\begin{aligned}\rho\dot{w} = \dot{W} - W\frac{\dot{\rho}}{\rho} &= \frac{1}{2}(\mathbf{e}-\mathbf{B}) : \dot{\mathbb{E}} : (\mathbf{e}-\mathbf{B}) + (\dot{\mathbf{e}}-\dot{\mathbf{B}}) : \mathbb{E} : (\mathbf{e}-\mathbf{B}) \\ &+\rho\mathscr{C}\dot{\theta} + \rho\dot{\mathscr{C}}\theta + \dot{\rho}\mathscr{C}\theta - \frac{\dot{\rho}}{\rho}\left(\frac{1}{2}(\mathbf{e}-\mathbf{B}) : \mathbb{E} : (\mathbf{e}-\mathbf{B}) + \rho\mathscr{C}\theta\right),\end{aligned} \tag{10.15}$$

and thus the first law becomes

$$\begin{aligned}\nabla_x\cdot(\mathbb{K}\cdot\nabla_x\theta) &= \frac{1}{J}\dot{\mathscr{Y}}(c)\left(\frac{1}{2\rho}(\mathbf{e}-\mathbf{B}) : \mathbb{E} : (\mathbf{e}-\mathbf{B}) + \mathscr{C}\theta - \frac{1}{2}\dot{\mathbf{u}}\cdot\dot{\mathbf{u}}\right) \\ &+(\dot{\mathbf{e}}-\dot{\mathbf{B}}) : \mathbb{E} : (\mathbf{e}-\mathbf{B}) + \frac{1}{2}(\mathbf{e}-\mathbf{B}) : \dot{\mathbb{E}} : (\mathbf{e}-\mathbf{B}) + \rho(\dot{\mathscr{C}}\theta + \mathscr{C}\dot{\theta}) \\ &-\rho z - \nabla_x\dot{\mathbf{u}} : \mathbb{E} : (\mathbf{e}-\mathbf{B}) - \frac{1}{2}(\frac{\dot{\rho}}{\rho})(\mathbf{e}-\mathbf{B}) : \mathbb{E} : (\mathbf{e}-\mathbf{B}),\end{aligned} \tag{10.16}$$

where Fourier's law, $\mathbf{q} = -\mathbb{K}\cdot\nabla_x\theta$, has been employed.

Remark. The chemical production of energy at a point is modeled as being related to the change in the rate of damage, $\rho z = \rho\zeta|\dot{\alpha}|$, where ζ is a spatially variable material parameter. The parameter ζ is negative for exothermic reactions and positive for endothermic reactions.

10.3.2 Mass Transfer and Reaction-Diffusion Models

The mass balance for a small diffusing species consists of a storage term ($\dot{c}$), a reaction term ($\dot{s}$), and an inward normal flux term ($-\mathbf{G}\cdot\mathbf{n}$). Since the domain is undergoing simultaneous finite deformations, consider a control volume for the dilute mass (m) written in the current (deformed) configuration

$$\frac{dm}{dt} = \frac{d}{dt}\int_{\omega} \mathscr{M}\, d\omega = \int_{\omega_0} \frac{d(\mathscr{M}J)}{dt}\, d\omega_0 = \int_{\omega_0} (\dot{\mathscr{M}}J + \mathscr{M}\dot{J})\, d\omega_0$$
$$= \int_{\omega} (\dot{\mathscr{M}} + \mathscr{M}\frac{\dot{J}}{J})\, d\omega = -\int_{\partial\omega} \mathbf{G}\cdot\mathbf{n}\, da. \qquad (10.17)$$

After using the divergence theorem and since the volume ω is arbitrary, one has the local form

$$\dot{\mathscr{M}} + \mathscr{M}\frac{\dot{J}}{J} = -\nabla_x \cdot \mathbf{G}. \qquad (10.18)$$

We decompose the total dilute mass into two parts, the concentration c and the products of reaction s, $\mathscr{M} = c + s$. It is a classical *stoichiometrically inexact* approximation to assume that the diffusing species reacts (is created or destroyed) in a manner such that the production of the reactant (s) is directly proportional to the concentration (c) of the diffusing species itself (Crank [30]),

$$s = \tau c. \qquad (10.19)$$

Upon substitution of this relation into the conservation law for the diffusing species, one has a diffusion-reaction model in strong form

$$\dot{c}(1+\tau) + c(1+\tau)\tfrac{\dot{J}}{J} = \nabla_x \cdot (\mathbb{D}_0 e^{-\frac{U}{R\theta}} \cdot \nabla_x c). \qquad (10.20)$$

When $\tau > 0$, the diffusing species is destroyed as it reacts, while $\tau < 0$ means that the diffusing species is created as it reacts, i.e. an autocatalytic reaction occurs. In (10.20), the familiar Arrhenius form $\mathbb{D} = \mathbb{D}_0 e^{-\frac{U}{R\theta}}$ has been used, where $\mathbb{D}_0$ is the diffusivity tensor at a reference temperature, U is the activation energy per mole of diffusive species, R is the universal gas constant and θ is the temperature.

Remark. It is sometimes observed that, in regions of relatively high positive triaxial stress, the diffusion is accelerated, while in regions of high negative triaxial stress, diffusion is decelerated. Diffusion models with explicit pressure dependency will not be considered in here, however, we remark that a particularly simple constitutive model to incorporate stress-dependency phenomena is given by a pseudo-Fickian/Arrhenius law, $\mathbf{G} = -\mathbb{D}_0 e^{\frac{-U(\sigma)}{R\theta}} \cdot \nabla c$, motivated by thermodynamical arguments found in the classical works of Flynn [50] or Crank [30]. [4]

Remark. It is important to note that instabilities can be induced by diffusion, i.e. a coupled mechano-chemical system can be stable when no diffusion is present and unstable in the presence of diffusion. An indepth mathematical analysis of such effects has been conducted by Markenscoff [134, 135, 136].

---

[4] An additive split for stress dependency of the form $U(\sigma) = U_0 + \tilde{U}(P)$, where U_0 is a stress-independent reference activation energy and $p = -\frac{tr\sigma}{3}$ is the pressure, has been given in Zohdi [231] and Zohdi et al. [226] for certain applications.

10.4 Staggered Multifield Weak Formulations

Staggering schemes are a natural choice for the solution of the multifield system developed thus far. Generally, such approaches proceed, within a discretized time step, by solving each field equation individually, allowing only the corresponding primary field variable to be active. This effectively decouples the system of differential equations. After the solution of each field equation, the primary field variable is updated, and the next field equation is solved in a similar manner, with only the corresponding primary variable being active. Usually, after this process has been applied only once to all of the field equations, the time step is incremented and the procedure is repeated. This nonrecursive process can be highly sensitive to the order in which the field variables are determined. For accurate numerical solutions, the approach requires small time steps, primarily because the staggering error accumulates with each passing increment. For details, see Park and Felippa [167], Zienkiewicz [221], Schrefler [175], Lewis et al. [121], Doltsinis [34], [35], Piperno [169], Lewis and Schrefler [122] and Le Tallec and Mouro [120]. In an attempt to improve such approaches, a recursive staggering strategy which allowed the adaptive control of time step sizes was developed in Zohdi and Wriggers [233] for a restricted class of coupled geometrically-linear/materially-nonlinear problems. In that approach, in order to reduce the error within a time step, the staggering methodology was formulated as a recursive fixed-point iteration, whereby the system was repeatedly re-solved until fixed-point type convergence was achieved. A sufficient condition for the convergence of such a fixed-point scheme is that the spectral radius of the coupled operator, which depends on the time step size, must be less than unity. This observation was used to adaptively maximize the time step sizes, while simultaneously controlling the coupled operator's spectral radius, in order to deliver solutions below an error tolerance within a prespecified number of iterations. This recursive staggering error control allowed substantial reduction of computational effort by the adaptive use of large time steps. Furthermore, the recursive process is insensitive to the order in which the individual equations are solved, since it is self correcting. In the next section, the approach is extended to the wider class of more complicated problems presently considered. For the sake of completeness the presentation is self-contained.

10.4.1 A Recursive Algorithm

Within a staggering scheme, we employ a relatively straightforward approach, which is amenable to time step adaptivity is

$$\ddot{\mathbf{u}}^{L+1} \approx \frac{\dot{\mathbf{u}}^{L+1}-\dot{\mathbf{u}}^{L}}{\Delta t} = \frac{\frac{\mathbf{u}^{L+1}-\mathbf{u}^{L}}{\Delta t}-\dot{\mathbf{u}}^{L}}{\Delta t} = \frac{\mathbf{u}^{L+1}-\mathbf{u}^{L}}{(\Delta t)^2} - \frac{\dot{\mathbf{u}}^{L}}{\Delta t} = \frac{\mathbf{u}^{L+1}}{(\Delta t)^2} - \frac{\mathbf{u}^{L}}{(\Delta t)^2} - \frac{\dot{\mathbf{u}}^{L}}{\Delta t}, \tag{10.21}$$

where the rate of change of thermal and concentration values are approximated with the same time step size, i.e. $\dot{\theta}^{L+1} \approx \frac{\theta^{L+1}-\theta^{L}}{\Delta t}$ and $\dot{c}^{L+1} \approx \frac{c^{L+1}-c^{L}}{\Delta t}$. *During the staggering process, the geometric configuration of the system is frozen during each iteration, and is updated only at the end of a system recursion.* In this particular case, the last calculated configuration is the previously computed one within the staggering scheme. Algorithmically, employing weak formulations, the staggering scheme is as follows:

$$(\star)\mathrm{AT\ A\ TIME\ STEP\,(L):\ START\ AN\ INTERNAL\ ITERATION\ I=1}$$

$$(\star\star)\underline{\mathrm{UPDATE\ GEOMETRICAL\ CONFIGURATION:}\ \Omega^{L+1,I}=\Omega^{L+1,I-1}}$$

$$\underline{\mathrm{MASS\ BALANCE\ OF\ DIFFUSING\ SPECIES:}}$$

Find $c^{L+1,I+1} \in U_c(\Omega^{L+1,I}), c^{L+1,I+1}|_{\Gamma_c} = C,$ such that $\forall v \in V_c(\Omega^{L+1,I}), v|_{\Gamma_c} = 0$

$$\int_{\Omega^{L+1,I}} \nabla_x v \cdot \mathbb{D}_0^{L+1,I} e^{-\frac{U}{R\theta^{L+1,I}}} \cdot \nabla_x c^{L+1,I+1}\, d\Omega$$
$$+\int_{\Omega^{L+1,I}} v(1+\tau)c^{L+1,I+1}\left(\frac{1}{\Delta t}+\frac{\dot{J}^{L+1,I}}{J^{L+1,I}}\right) d\Omega = \int_{\Omega^{L+1,I}} v\frac{(1+\tau)}{\Delta t}c^{L}\, d\Omega$$
$$+\int_{\Gamma_G} v\mathbf{G}^{L+1}\cdot \mathbf{n}\, dA.$$

$$\underline{\mathrm{COMPUTE\ REACTIONS:}}\,(INTEGRATE\ EVOLUTION\ EQUATIONS)$$

$$\dot{\alpha}^{L+1,I+1} = g_\alpha^{L+1,I+\frac{1}{3}}\alpha^{L+1,I+1} \Rightarrow \alpha^{L+1,I+1} = \alpha^{L}e^{g_\alpha^{L+1,I+\frac{1}{3}}\Delta t^{L+1}},$$
$$\dot{\lambda}^{L+1,I+1} = g_\lambda^{L+1,I+\frac{1}{3}}\lambda^{L+1,I+1} \Rightarrow \lambda^{L+1,I+1} = \lambda^{L}e^{g_\lambda^{L+1,I+\frac{1}{3}}\Delta t^{L+1}} \Rightarrow \mathbf{e}_\lambda^{L+1,I+1}$$
$$\dot{\varphi}^{L+1,I+1} = g_\varphi^{L+1,I+\frac{1}{3}}\varphi^{L+1,I+1} \Rightarrow \varphi^{L+1,I+1} = \varphi^{L}e^{g_\varphi^{L+1,I+\frac{1}{3}}\Delta t^{L+1}} \Rightarrow \mathbf{e}_\varphi^{L+1,I+1}$$

WHERE

$$g^{L+1,I+\frac{1}{3}} \stackrel{\text{def}}{=} g(c^{L+1,I+1},\theta^{L+1,I},\mathbf{u}^{L+1,I}),$$
$$g^{L+1,I+\frac{2}{3}} \stackrel{\text{def}}{=} g(c^{L+1,I+1},\theta^{L+1,I+1},\mathbf{u}^{L+1,I}),$$
$$g^{L+1,I+1} \stackrel{\text{def}}{=} g(c^{L+1,I+1},\theta^{L+1,I+1},\mathbf{u}^{L+1,I+1}).$$

$$\underline{\mathrm{COMPUTE\ HEAT\ GENERATION}}:\ (\rho z)^{L+1,I+1} = \zeta\rho^{L+1,I}\dot{\alpha}^{L+1,I+1}, \tag{10.22}$$

with the algorithm proceeding with

ENERGY EQUATION :

Find $\theta^{L+1,I+1} \in U_\theta(\Omega^{L+1,I}), \theta^{L+1,I+1}|_{\Gamma_\theta^{L+1,I}} = \Theta^{L+1}$ such that $\forall v \in V_\theta(\Omega^{L+1,I}), v|_{\Gamma_\theta^{L+1,I}} = 0$

$$
\begin{aligned}
&\int_{\Omega^{L+1,I}} \nabla_x v \cdot \mathbb{K}^{L+1,I+1} \cdot \nabla_x \theta^{L+1,I+1} \, d\Omega + \frac{1}{\Delta t} \int_{\Omega^{L+1,I}} v \rho^{L+1,I} \mathscr{C}^{L+1,I+1} \theta^{L+1,I+1} \, d\Omega = \\
&\frac{1}{\Delta t} \int_{\Omega^{L+1,I}} v \rho^{L+1,I} \mathscr{C}^{L+1,I+1} \theta^{L} \, d\Omega \\
&- \int_{\Omega^{L+1,I}} v \left(\rho^{L+1,I} \dot{\mathscr{C}}^{L+1,I+1} + \frac{1}{J^{L+1,I}} \dot{\mathscr{Y}}(c^{L+1,I+1}) \mathscr{C}^{L+1,I+1} \right) \theta^{L+1,I+1} \, d\Omega \\
&+ \int_{\Omega^{L+1,I}} v (\rho z)^{L+1,I+1} \, d\Omega + \int_{\Gamma_q^{L+1,I}} v \mathbf{q}^{L+1} \cdot \mathbf{n} \, dA \\
&- \int_{\Omega^{L+1,I}} v \left((\dot{\mathbf{e}}^{L+1,I} - \dot{\mathbf{B}}^{L+1,I+\frac{1}{3}}) : \mathbb{E}^{L+1,I+1} : (\mathbf{e}^{L+1,I} - \mathbf{B}^{L+1,I+\frac{1}{3}}) \right) d\Omega \\
&- \int_{\Omega^{L+1,I}} v \left((\frac{1}{2} (\mathbf{e}^{L+1,I} - \mathbf{B}^{L+1,I+\frac{1}{3}}) : \dot{\mathbb{E}}^{L+1,I+1} : (\mathbf{e}^{L+1,I} - \mathbf{B}^{L+1,I+\frac{1}{3}}) \right) d\Omega \\
&+ \int_{\Omega^{L+1,I}} v \left(\nabla_x \dot{\mathbf{u}}^{L+1,I} : (\mathbb{E}^{L+1,I+1} : (\mathbf{e}^{L+1,I} - \mathbf{B}^{L+1,I+\frac{1}{3}})) \right) d\Omega \\
&+ \int_{\Omega^{L+1,I}} v \left(\frac{\dot{\rho}^{L+1,I}}{\rho^{L+1,I}} \frac{1}{2} (\mathbf{e}^{L+1,I} - \mathbf{B}^{L+1,I+\frac{1}{3}}) : \mathbb{E}^{L+1,I+1} : (\mathbf{e}^{L+1,I} - \mathbf{B}^{L+1,I+\frac{1}{3}}) \right) d\Omega \\
&+ \int_{\Omega^{L+1,I}} v \frac{1}{J^{L+1,I}} \dot{\mathscr{Y}}(c^{L+1,I+1}) \\
&\qquad \times \left(\frac{1}{2\rho^{L+1,I}} (\mathbf{e}^{L+1,I} - \mathbf{B}^{L+1,I}) : \mathbb{E}^{L+1,I+1} : (\mathbf{e}^{L+1,I} - \mathbf{B}^{L+1,I}) - \frac{1}{2} \dot{\mathbf{u}}^{L+1,I} \cdot \dot{\mathbf{u}}^{L+1,I} \right) d\Omega
\end{aligned}
$$

BALANCE OF MOMENTUM :

Find $\mathbf{u}^{L+1,I+1} \in \mathbf{U}_u(\Omega^{L+1,I}), \mathbf{u}^{L+1,I+1}|_{\Gamma_u^{L+1,I}} = \mathbf{d}^{L+1}$ such that $\forall \mathbf{v} \in \mathbf{V}_u(\Omega^{L+1,I}), \mathbf{v}|_{\Gamma_u^{L+1,I}} = \mathbf{0}$

$$
\begin{aligned}
&\int_{\Omega^{L+1,I}} \nabla_x \mathbf{v} : \alpha^{L+1,I+1} \mathbb{E}_0 : \left(\nabla_x \mathbf{u}^{L+1,I+1} - \frac{1}{2} (\nabla_x \mathbf{u}^{L+1,I})^T \cdot \nabla_x \mathbf{u}^{L+1,I} - \mathbf{B}^{L+1,I+\frac{2}{3}} \right) d\Omega \\
&+ \int_{\Omega^{L+1,I}} \rho^{L+1,I} \frac{\mathbf{u}^{L+1,I+1}}{(\Delta t)^2} \cdot \mathbf{v} \, d\Omega - \int_{\Omega^{L+1,I}} \rho^{L+1,I} \left(\frac{\mathbf{u}^{L}}{(\Delta t)^2} + \frac{\dot{\mathbf{u}}^{L}}{\Delta t} \right) \cdot \mathbf{v} \, d\Omega \\
&+ \int_{\Omega^{L+1,I}} \frac{\dot{\mathscr{Y}}(c^{L+1,I+1})}{J^{L+1,I}} \frac{\mathbf{u}^{L+1,I+1}}{\Delta t} \cdot \mathbf{v} \, d\Omega - \int_{\Omega^{L+1,I}} \frac{\dot{\mathscr{Y}}(c^{L+1,I+1})}{J^{L+1,I}} \frac{\mathbf{u}^{L,I+1}}{\Delta t} \cdot \mathbf{v} \, d\Omega \\
&- \int_{\Omega^{L+1,I}} \mathbf{f}^{L+1,I} \cdot \mathbf{v} \, d\Omega - \int_{\Gamma_t^{L+1,I}} \mathbf{t}^{L+1,I} \cdot \mathbf{v} \, dA = 0
\end{aligned}
$$

CHECK FOR CONVERGENCE :

$$
\frac{||\mathbf{u}^{L+1,I+1} - \mathbf{u}^{L+1,I}||_{\mathbf{L}^1(\Omega^{L+1,I})}}{||\mathbf{u}^{L+1,I+1}||_{\mathbf{L}^1(\Omega^{L+1,I})}} \le TOL_u,
$$

$$
\frac{||c^{L+1,I+1} - c^{L+1,I}||_{\mathbf{L}^1(\Omega^{L+1,I})}}{||c^{L+1,I+1}||_{\mathbf{L}^1(\Omega^{L+1,I})}} \le TOL_c,
$$

$$
\frac{||\theta^{L+1,I+1} - \theta^{L+1,I}||_{\mathbf{L}^1(\Omega^{L+1,I})}}{||\theta^{L+1,I+1}||_{\mathbf{L}^1(\Omega^{L+1,I})}} \le TOL_\theta.
$$

IF TOLERANCES NOT MET THEN $\mathrm{I} = \mathrm{I}+1$, GO TO $(\star\star)$

IF TOLERANCES MET THEN INCREMENT TIME : $\mathrm{L} = \mathrm{L}+1$,
UPDATE ALL VARIABLES, GO TO $(\star)$.

(10.23)

Here $U_c(\Omega)$, $U_\theta(\Omega)$ and $\mathbf{U}_u(\Omega)$ are spaces of admissible trial functions, while $V_c(\Omega)$, $V_\theta(\Omega)$ and $\mathbf{V}_u(\Omega)$ are spaces of admissible test functions. For most loading cases and data these spaces will correspond to $H^1(\Omega)$ and $\mathbf{H}^1(\Omega)$. In an abstract setting, one can consider the following staggering solution strategy where only the underlined variable is allowed to be active, the corresponding field equations solved, and active variable updated, and the process repeated for the next field equation:

$$
\begin{aligned}
&\mathscr{A}_1(\underline{c^{I+1}},\alpha^I,\mathbf{e}^I_\lambda,\mathbf{e}^I_\varphi,\theta^I,\mathbf{u}^I)=\mathscr{F}_1(c^I,\alpha^I,\mathbf{e}^I_\lambda,\mathbf{e}^I_\varphi,\theta^I,\mathbf{u}^I) \quad (\mathit{MASS\ TRANSFER}),\\
&\mathscr{A}_2(c^{I+1},\underline{\alpha^{I+1}},\mathbf{e}^I_\lambda,\mathbf{e}^I_\varphi,\theta^I,\mathbf{u}^I)=\mathscr{F}_2(c^{I+1},\alpha^I,\mathbf{e}^I_\lambda,\mathbf{e}^I_\varphi,\theta^I,\mathbf{u}^I) \quad (\mathit{DEGRADATION}),\\
&\mathscr{A}_3(c^{I+1},\alpha^{I+1},\underline{\mathbf{e}^{I+1}_\lambda},\mathbf{e}^I_\varphi,\theta^I,\mathbf{u}^I)=\mathscr{F}_3(c^{I+1},\alpha^{I+1},\mathbf{e}^I_\lambda,\mathbf{e}^I_\varphi,\theta^I,\mathbf{u}^I) \quad (\mathit{DEV.\ EIGENSTRAINS}),\\
&\mathscr{A}_4(c^{I+1},\alpha^{I+1},\mathbf{e}^{I+1}_\lambda,\underline{\mathbf{e}^{I+1}_\varphi},\theta^I,\mathbf{u}^I)=\mathscr{F}_4(c^{I+1},\alpha^{I+1},\mathbf{e}^{I+1}_\lambda,\mathbf{e}^I_\varphi,\theta^I,\mathbf{u}^I) \quad (\mathit{DIL.\ EIGENSTRAINS}),\\
&\mathscr{A}_5(c^{I+1},\alpha^{I+1},\mathbf{e}^{I+1}_\lambda,\mathbf{e}^{I+1}_\varphi,\underline{\theta^{I+1}},\mathbf{u}^I)=\mathscr{F}_5(c^{I+1},\alpha^{I+1},\mathbf{e}^{I+1}_\lambda,\mathbf{e}^{I+1}_\varphi,\theta^I,\mathbf{u}^I) \quad (\mathit{ENERGY}),\\
&\mathscr{A}_6(c^{I+1},\alpha^{I+1},\mathbf{e}^{I+1}_\lambda,\mathbf{e}^{I+1}_\varphi,\theta^{I+1},\underline{\mathbf{u}^{I+1}})=\mathscr{F}_6(c^{I+1},\alpha^{I+1},\mathbf{e}^{I+1}_\lambda,\mathbf{e}^{I+1}_\varphi,\theta^{I+1},\mathbf{u}^I) \quad (\mathit{MOMENTUM}).
\end{aligned}
\tag{10.24}
$$

Remark. Consistent with the fully staggered solution approach introduced in (10.22), we freeze the g_α, g_λ and g_φ, using the most current state variable values, and integrate analytically.

Remark. Writing the system in the form presented leads to algebraic systems which are symmetric and positive definite. Therefore, somewhat standard iterative solvers such as the preconditioned Conjugate Gradient Method, can be used. Such solvers are highly advantageous since any starting solution, from a previous time step or staggered iteration can be used as the first guess in the solution procedure, thus providing a "head start" in the process.

10.4.2 Convergence and Contraction-Mapping Time Stepping Control

Consider the general abstract equation

$$\mathscr{A}(\mathbf{w})=\mathscr{F}, \tag{10.25}$$

where $\mathbf{w}$ represents all of the fields in the analysis:

$$\mathbf{w}\stackrel{\mathrm{def}}{=}(c,\alpha,\mathbf{e}_\lambda,\mathbf{e}_\varphi,\theta,\mathbf{u}), \tag{10.26}$$

It is advantageous to write this in the form

$$\Pi(\mathbf{w})=\mathscr{A}(\mathbf{w})-\mathscr{F}=\mathscr{G}(\mathbf{w})-\mathbf{w}+\mathbf{r}=\mathbf{0}. \tag{10.27}$$

A straightforward fixed point iterative scheme is

$$\mathscr{G}(\mathbf{w}^{I-1})+\mathbf{r}=\mathbf{w}^I. \tag{10.28}$$

The convergence of such a scheme is dependent on the behavior of $\mathcal{G}$. Namely, a sufficient condition for convergence is that $\mathcal{G}$ is a contraction mapping for all $\mathbf{w}^I$, $I = 1,2,3...$ Convergence of the iteration can be studied by defining the error vector $\varepsilon_I = \mathbf{w}^I - \mathbf{w}$. A necessary condition for convergence is iterative self consistency, i.e. the exact solution must be represented by the scheme $\mathcal{G}(\mathbf{w}) + \mathbf{r} = \mathbf{w}$. Enforcing this condition, a sufficient condition for convergence is the existence of a contraction mapping

$$||\varepsilon_I|| = ||\mathbf{w}^I - \mathbf{w}|| = ||\mathcal{G}(\mathbf{w}^{I-1}) - \mathcal{G}(\mathbf{w})|| \leq \eta ||\mathbf{w}^{I-1} - \mathbf{w}||, \tag{10.29}$$

where, if $\eta < 1$ for each iteration I, then $\varepsilon_I \to \mathbf{0}$ for any arbitrary starting solution $\mathbf{w}^{I=0}$ as $I \to \infty$. Therefore, unconditional convergence is attained if, for $\mathbf{w}^I$, $I = 1,2,3...,\ \eta^I < 1$. This type of convergence criteria is common for linear iterative (Gauss-Jacobi-Seidel) solution methods of relaxation type (Southwell [193, 194], Frankel [53], Young [217] and Axelsson [9]). For reviews of nonlinear techniques, we refer the reader to Perron, [168], Ostrowski [164, 165], Ortega and Rockoff [162], Kitchen [109] or Ames [4]. The algorithm outlined in (10.22) and (10.23) can be considered as fixed-point scheme, whose convergence within each time step is dependent on the time step size itself. The step size can be manipulated, enlarged or reduced, to induce the desired rates of convergence within a time step, in order to achieve an error tolerance within a prespecified number of iterations. Following the approach in Zohdi [243], [250, 251] one approximates the contraction constant $\eta \approx S\Delta t$, where one expects the error within an iteration to behave according to $(S\Delta t)^I \varepsilon_0 = \varepsilon_I$, $I = 1,2,...$, where ε_0 is the initial error and S is a function intrinsic to the system. *Our target or ideal condition is to meet an error tolerance in a given number of iterations, not more, and not less.* One writes this in the following approximate form, $(S\Delta t_{\text{tol}})^{I_d} \varepsilon_0 = \text{TOL}$, where I_d is the number of desired iterations. Therefore, if the error tolerance is not met in a desired number of iterations, the contraction constant η is too large. Accordingly, one can solve for a new smaller step size, under the assumption that S is constant[5]

$$\Delta t_{\text{tol}} = \Delta t \left(\frac{\left(\frac{\text{TOL}}{\varepsilon_0}\right)^{\frac{1}{I_d}}}{\left(\frac{\varepsilon_I}{\varepsilon_0}\right)^{\frac{1}{I}}} \right). \tag{10.30}$$

Clearly, the expression in (10.30) can be used for time step enlargement, if convergence is met in less than I_d iterations. One sees that if $\varepsilon_{I_d} > \varepsilon_{\text{tol}}$ and $I = I_d$, then the expression in (10.30) collapses to a ratio of the error tolerance to the achieved level of iterative error after I_d iterations, $\Delta t_{\text{tol}} = \Delta t(\frac{\text{TOL}}{\varepsilon_{I_d}})^{I_d}$, and thus the step size will be scaled by the ratio of the error to the tolerance. For the multifield system, we define the normalized errors within each time step (L), for the three primary fields,[6]

[5] The assumption that S is constant is not overly severe, since the time steps are to be recursively refined and unrefined.

[6] The other quantities, α, $\mathbf{e}_\lambda$, and $\mathbf{e}_\varphi$ are controlled by the three primary fields $(c, \theta, \mathbf{u})$, thus we monitor only their convergence.

$$
\begin{aligned}
\varepsilon_{uI} &\stackrel{\text{def}}{=} \frac{||\mathbf{u}^{L+1,I} - \mathbf{u}^{L+1,I-1}||_{\mathbf{L}^1(\Omega)}}{||\mathbf{u}^{L+1,I}||_{\mathbf{L}^1(\Omega)}},\\
\varepsilon_{cI} &\stackrel{\text{def}}{=} \frac{||c^{L+1,I} - c^{L+1,I-1}||_{\mathbf{L}^1(\Omega)}}{||c^{L+1,I}||_{\mathbf{L}^1(\Omega)}},\\
\varepsilon_{\theta I} &\stackrel{\text{def}}{=} \frac{||\theta^{L+1,I} - \theta^{L+1,I-1}||_{\mathbf{L}^1(\Omega)}}{||\theta^{L+1,I}||_{\mathbf{L}^1(\Omega)}},
\end{aligned}
\tag{10.31}
$$

and their corresponding violation ratios

$$
\psi_{uI} \stackrel{\text{def}}{=} \frac{\varepsilon_{uI}}{TOL_u}, \qquad \psi_{\theta I} \stackrel{\text{def}}{=} \frac{\varepsilon_{\theta I}}{TOL_\theta}, \qquad \psi_{cI} \stackrel{\text{def}}{=} \frac{\varepsilon_{cI}}{TOL_c}. \tag{10.32}
$$

One then determines the maximum violation $\Psi_I \stackrel{\text{def}}{=} \max(\psi_{uI}, \psi_{cI}, \psi_{\theta I})$ and a minimum scaling factor $\Phi_I \stackrel{\text{def}}{=} \min(\phi_{uI}, \phi_{cI}, \phi_{\theta I})$ from

$$
\phi_{uI} \stackrel{\text{def}}{=} \left(\frac{(\frac{TOL_u}{\varepsilon_{u0}})^{\frac{1}{I_d}}}{(\frac{\varepsilon_{uI}}{e_u^0})^{\frac{1}{I}}} \right), \quad \phi_{cI} \stackrel{\text{def}}{=} \left(\frac{(\frac{TOL_c}{\varepsilon_{c0}})^{\frac{1}{I_d}}}{(\frac{\varepsilon_{cI}}{\varepsilon_{c0}})^{\frac{1}{I}}} \right), \quad \phi_{\theta I} \stackrel{\text{def}}{=} \left(\frac{(\frac{TOL_\theta}{\varepsilon_{\theta 0}})^{\frac{1}{I_d}}}{(\frac{\varepsilon_{\theta I}}{\varepsilon_{\theta 0}})^{\frac{1}{I}}} \right). \tag{10.33}
$$

Thereafter, the following criteria for temporal adaptivity is adopted:

IF TOLERANCES MET (ψ_{uI} *AND* $\psi_{\theta I}$ *AND* $\psi_{cI} \leq 1$) AND $I < I_d$ THEN :

(a) STEP TIME FORWARD : $t = t + \Delta t$

(b) CONSTRUCT NEW TIME STEP : $\Delta t = \Phi_I \Delta t$

(c) SELECT MINIMUM : $\Delta t = MIN(\Delta t^{lim}, \Delta t)$; START NEXT TIME STEP

IF ANY TOLERANCE NOT MET (ψ_{uI} *OR* $\psi_{\theta I}$ *OR* $\psi_{cI} > 1$) OR $I = I_d$ THEN :

(a) CONSTRUCT NEW TIME STEP : $\Delta t = \Phi_{I_d} \Delta t$

(b) RESTART AT OLD TIME

(10.34)

The overall goal is to deliver solutions where staggering (incomplete coupling) error is controlled and the temporal discretization accuracy dictates the upper limits on the time step size (Δt^{lim}).

10.5 Numerical Experiments

To illustrate the algorithm, we considered a microscale cube of matrix material, with dimensions $10^{-4}m \times 10^{-4}m \times 10^{-4}m$, embedded with randomly distributed inhomogeneities. The various physical properties of the two materials are shown in

Table 10.1 Material properties used in the computational examples

Material Property	Matrix	Particles
Mechanical		
κ(GPa)	77.9	230.0
μ(GPa)	25.9	172.0
$\gamma(\frac{1}{{}^oK})$	9.71×10^{-6}	8.92×10^{-6}
Thermal		
$K(\frac{J}{s-m-{}^oK})$	237	148
$\rho(\frac{kg}{m^3})$	2700.84	2330.28
$\mathscr{C}(\frac{J}{kg-{}^oK})$	903	712
Diffusive		
$D_0(\frac{m}{sec^2})$	1.0×10^{-6}	1.0×10^{-7}
$U\frac{kN-m}{mole}$	142	300
$\tau(\frac{1}{sec})$	5.0×10^{-2}	1.0×10^{-2}
Damage Evolution		
$a_1^*(\frac{m^3}{molecules-sec})$	-7.3×10^{-8}	-3.0×10^{-9}
$a_2^*(\frac{m^3}{molecules-sec})$	-7.3×10^{-8}	-3.0×10^{-9}
$a_3^*(\frac{m^3}{molecules-sec})$	-7.3×10^{-8}	-3.0×10^{-9}
$a_4^*(\frac{1}{sec})$	6.27×10^{-10}	3.15×10^{-10}
$a_5^*(\frac{1}{sec})$	6.27×10^{-11}	3.15×10^{-11}
k_1(MPa)	120	3000
k_2(MPa)	120	3000
$c_{crit}(\frac{molecules}{m^3})$	0.0	0.0
$\zeta(N-m)$	-1×10^{11}	-5×10^{11}

Table 10.1, and correspond to a commonly used lightweight aluminum matrix-boron particle industrial composite.

We consider a set of topological microstructural variables which can be conveniently parametrized by a generalized "ellipsoid"

$$\left(\frac{|x-x_o|}{r_1}\right)^{s_1} + \left(\frac{|y-y_o|}{r_2}\right)^{s_2} + \left(\frac{|z-z_o|}{r_3}\right)^{s_3} = 1. \tag{10.35}$$

The types of suspensions to be introduced in the matrix binder can be controlled by (1) the polynomial order, s_1, s_2 and s_3, where values of $s < 1$ produce nonconvex shapes, $s = 1$ produce convex eight-sided diamond shapes, $s = 2$ standard ellipsoids and $s > 2$ values produce "blocklike" shapes, (2) the aspect ratios defined by $AR \stackrel{\text{def}}{=} \frac{r_1}{r_2} = \frac{r_1}{r_3}$, where $r_2 = r_3$, $AR > 1$ for prolate geometries and $AR < 1$ for oblate shapes and (3) the volume fractions, via particle/sample size ratio, which is defined via a subvolume size $V \stackrel{\text{def}}{=} \frac{L\times L\times L}{N}$, where N is the number of particles in the entire sample and where $L = 10^{-4}m$ is the length of the (cubical) sample, $L \times L \times L$. A generalized diameter is defined, $d = 2r$, which is the diameter of the smallest sphere that can enclose a single particle of possibly non-spherical shape. The ratio between the generalized diameter and the subvolume are related by $\xi \stackrel{\text{def}}{=} \frac{r}{V^{\frac{1}{3}}}$. *In order to illustrate the behavior we used the disk-like microstructure shown in Fig.* 10.2. For any microstructural combination, volume fractions, phase contrasts, etc..., the samples must be tested and enlarged, holding the volume fraction constant, but increasing the number of particles, for example from 2, 4, ... etc ...until the macroscopic results stabilize. Approximately a 20 particle sample gave relatively stable results. A more detailed and rigorous analysis of size effects for such systems is beyond the scope of this presentation. The reader is referred to the following series of works Huet [87, 88, 89, 90, 91, 92, 93], Huet et al. [94], Guidoum and Navi [71], Amieur et al. [5], Guidoum [72], Amieur [6] Hazanov and Huet [77], Hazanov and Amieur [78], Amieur et al. [7] and Huet [98, 99, 100], as well as some recent work of the first author (Zohdi [233, 246, 247, 248, 249]).

Over the course of such tests the finite element meshes were repeatedly refined, and a mesh density of approximately $9 \times 9 \times 9$ trilinear hexahedra (approximately 800–1000 degrees of freedom (DOF) for the diffusion-reaction and energy balance equations, and between 2200–3000 DOF for the vector-valued balance of

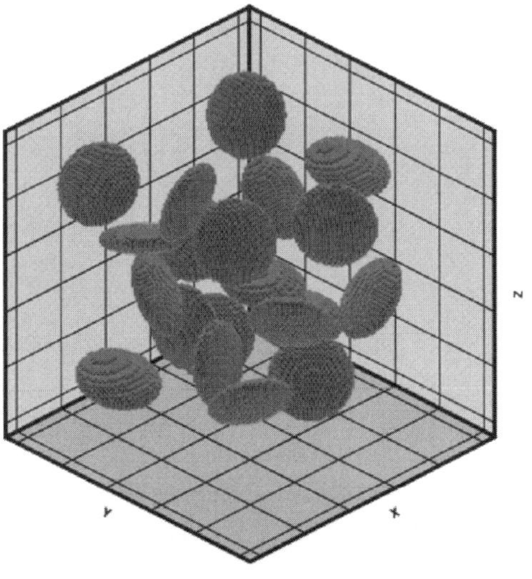

Fig. 10.2 The numerical resolution of $s_1 = s_2 = s_3 = 2$ particles, with $\xi = 0.375$, and an aspect ratio of $\frac{1}{3}$, i.e. $r_1 = \frac{1}{3}r_2 = \frac{1}{3}r_3$ resulting in a volume fraction of $v_2 \approx 0.076$ (20 particles shown)

momentum) *per particle* was found to deliver mesh independent results. Therefore, for example for a 10 particle test, 8000 DOF were needed for the diffusion-reaction and energy balance equation, and 24000 DOF for the balance of momentum equation, for 20 particles, 15625 DOF/46875 DOF, etc. During the computations, a "2/5" Gauss rule was used, whereby elements containing material discontinuities had increased Gauss rules ($5 \times 5 \times 5$) to enhance the resolution of the internal geometry, while elements with no material discontinuities had the nominal $2 \times 2 \times 2$ rule. The numerical resolution of the microstructure is shown in Fig. 10.2, for a topological exponent of $s_1 = s_2 = s_3 = 2$ for oblate spheroids. For related work in the optimization of the material microstructure, controlling the parameters such as s, see Zohdi [246, 248, 249].

The boundary conditions for the cubical domain were: (1) $c|_{\partial\Omega} = C = 1$, $c(\mathbf{x}, t = 0) = 0$, (2) $\theta|_{\partial\Omega} = \Theta = 30°\,Celcius = 303.13°\,Kelvin$, $\theta(\mathbf{x}, t = 0) = 0°\,Celcius$ and (3) $\mathbf{u}|_{\partial\Omega} = \frac{t}{T}\mathscr{E} \cdot \mathbf{X}$, $\mathscr{E}_{ij}$=0.02, $i, j = 1, 2, 3$, where $\mathbf{X}$ is a referential position vector to the boundary of the cube, t is the time, and T is the total simulation time. The material parameters, *selected only for the purposes of numerical experiment*, are shown in Table 10.1. From a practical engineering point of view, macroscopic quantities, which are volumetrically averaged outcomes of the simulated microstructural events, are of interest. Such quantities include, (1) the mechanical response, $\langle\sigma\rangle_\Omega$, (2) the average change in the material, for example damage, $\langle\alpha\rangle_\Omega$, (3) the average temperature, $\langle\theta\rangle_\Omega$, and (4) the average concentration $\langle c\rangle_\Omega$. Clearly, in this external pure displacement controlled regime, the stresses will relax over time, since the material stiffness is deteriorating in the interior.

The algorithmic staggering tolerance was set to $\max(\varepsilon_u, \varepsilon_\theta, \varepsilon_c) \leq 0.0001$ for the normalized/global error control. The designated maximum number of internal iterations, I_d, was set to five. The starting time step size was $\Delta t = 10^3$ seconds. In order to smoothly refine and unrefine the time steps, the adjustments were bounded *between successive time steps* (L) to be $0.9 < \frac{\Delta t^{L+1}}{\Delta t^L} \leq 1.1$. The total simulation time was set to one year. *For the purposes of numerical experiment only*, the damage rate parameters were chosen such that for a material point undergoing constant damage at unit concentration, with no stress, after one year $\alpha(t = T) = 0.1 = e^{a_1^* t}$,

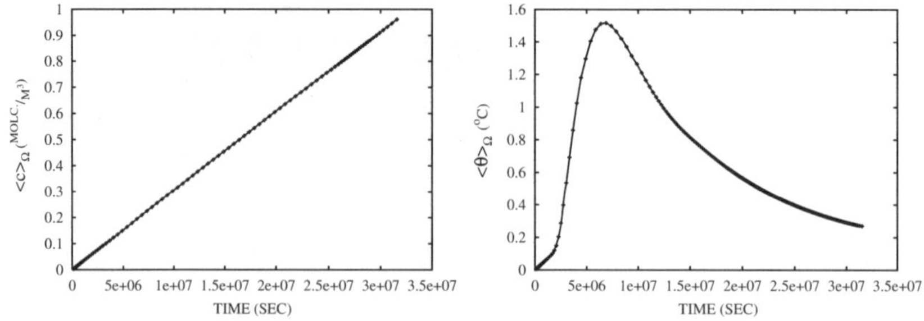

Fig. 10.3 Left: volumetric average of the concentration over time. **Right:** volumetric average of the temperature over time

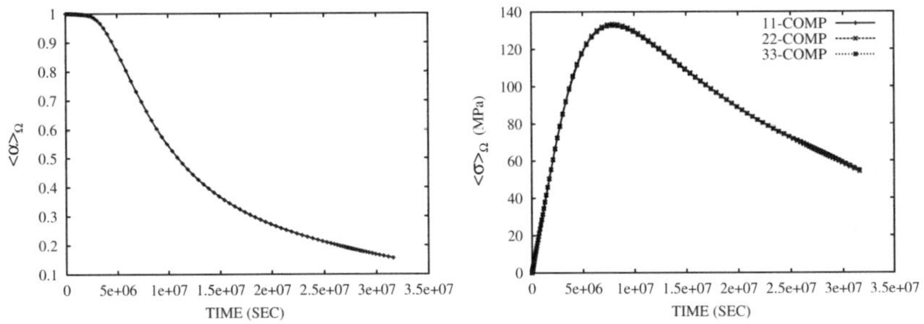

Fig. 10.4 Left: volumetric average of the degradation over time. **Right:** volumetric average of the normal stresses over time

$T = 31536000$ seconds, which led to $a_1^* = -7.3 \times 10^{-8} (\frac{\mathrm{m}^3}{\mathrm{molecules-sec}})$. This rate was used for the matrix. For the particulate material, we set the rate parameter to be significantly smaller, $a_1^* = -3 \times 10^{-9} (\frac{\mathrm{m}^3}{\mathrm{molecules-sec}})$. These values were also selected for the a_2^* and a_3^* rates as well. The deterioration rates of all material parameters other than $\mathbb{E}$, such as $\mathscr{C}$, $\mathbb{K}$, $\mathbb{D}_0$, etc., were set to zero. For the deviatoric eigenstrain, the rate coefficient a_4^* was calibrated in such a way that $\lambda = 0.02$ in the matrix and $\lambda = 0.01$ in the particle, for constant $\dot{\lambda}$'s, at $t = T$. A value of one-tenth was used for the dilatational eigenstrain. Finally, the function for the changes in density due to the uptake of the diffusing species was modeled as $\mathscr{Y}(c) \approx ac$, with $a = 0.1 \frac{\mathrm{kg}}{\mathrm{molecule}}$. The time step limit size was set to 5×10^5 seconds, which was set as the upper bound for reasons of truncation error control. The total number of system solves needed was 462, as opposed to 31,536, had the system been solved nonrecursively with no time step adaptation. Figs. 10.3, 10.4, and 10.5 illustrate the results for the major quantities of interest. The volumetric average of the concentration (Fig. 10.3) exhibits a quasilinear growth over time. The volumetric average of the temperature (Fig. 10.3) exhibits a nonlinear, nonmonotone, behavior over time, representing initial heating then cooling off. The volumetric average of the damage is also highly nonlinear (Fig. 10.4), and the stresses growth, then

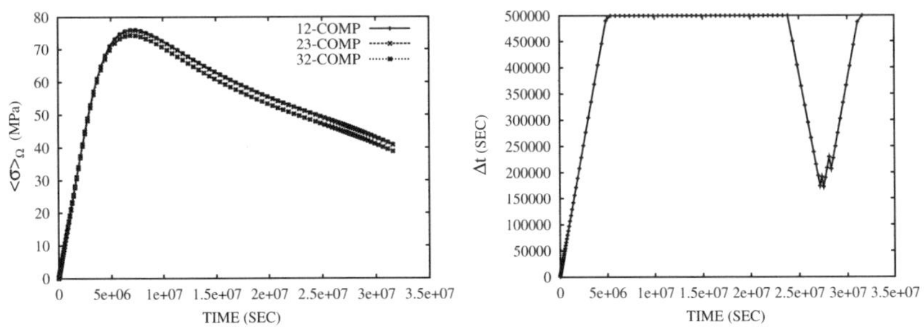

Fig. 10.5 Left: volumetric average of the shear stresses over time. **Right:** adapted time step sizes over time

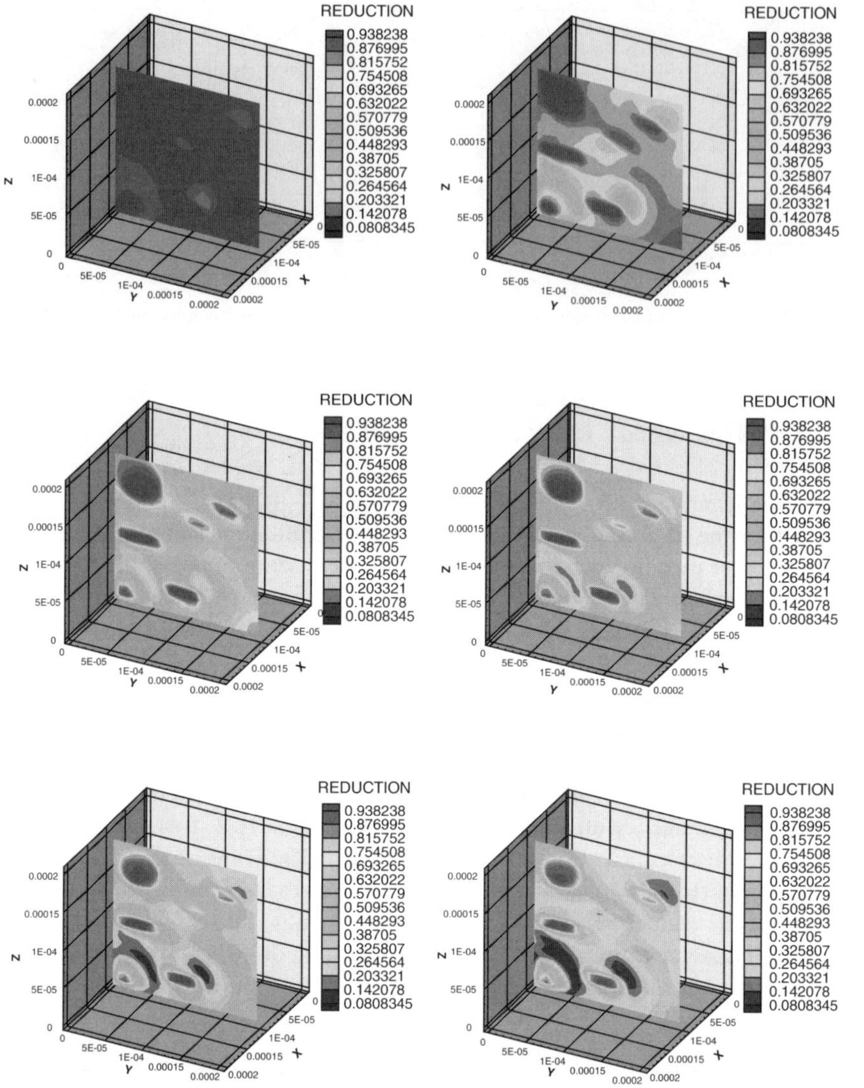

Fig. 10.6 Starting from the top, left to right: The time history of the degradation (α) throughout the solid, in increments of 0.1 years

relax in a manner that is characteristic of solids undergoing softening (Figs. 10.4 and 10.5). In Fig. 10.5 minor anisotropic texture is exhibited, indicating that the sample is slightly non-statistically representative. Finally, Fig. 10.5 shows the variation in the time step sizes to meet the staggering and discretization error requirements. A spatial time history of the degradation throughout the solid is shown in Figs. 10.6 and 10.7.

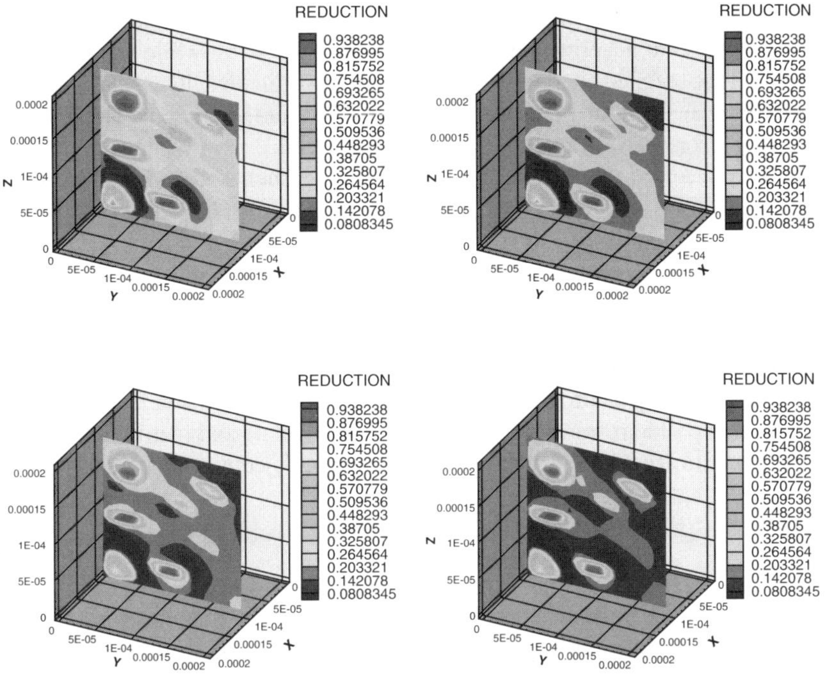

Fig. 10.7 Starting from the top, left to right: The time history, continued from Fig. 10.6, of the degradation (α) throughout the solid, in increments of 0.1 years

10.6 Concluding Remarks

In this chapter, the (staggering) error due to incompletely resolving the coupling between multifield equations, describing time-dependent thermo-chemo-mechanical processes in solids possessing irregular heterogeneous microstructure, was characterized in such a way to be amenable to a relatively simple method of adaptive control. A solution strategy was developed, whereby the time step size was manipulated, enlarged or reduced, to control the contraction mapping constant of the system operator in order to induce desired staggering rates of convergence within each time step. The overall goal was to deliver accurate solutions where the staggering error was controlled while simultaneously obeying time-step size limits dictated by discretization error concerns.

Generally speaking, the staggering error, which is a function of the time step size, is temporally variable and can become stronger, weaker, or possibly oscillatory, is extremely difficult to ascertain a-priori as a function of the time step size. Therefore, to circumvent this problem, the adaptive strategy presented in this chapter was developed to provide accurate solutions by iteratively adjusting the time steps. Specifically, a sufficient condition for the convergence of the presented fixed-point scheme was that the spectral radius or contraction constant of the coupled operator, which depends on the time step size, must be less than unity. This observation was used to adaptively maximize the time step sizes, while simultaneously controlling the coupled operator's spectral radius, in order to deliver solutions below an error tolerance within a prespecified number of desired iterations. This recursive staggering error control can allow for substantial reduction of computational effort by the adaptive use of large time steps. Furthermore, such a recursive process has a reduced sensitivity, relative to an explicit staggering approach, to the order in which the individual equations are solved, since it is self-correcting.

In the ideal case, one would like to make predictions of whether a certain type of matrix-particulate combination will have poor multifield and inelastic behavior via numerical simulations, in order to minimize expensive laboratory tests. As an extension to this chapter, emphasis should be placed on microstructural optimization to resist severe-loading environments. For example, this could involve the construction of an inverse problem, where combinations of particulate and matrix materials are sought to minimize the following history-dependent objective function:

$$\Pi = \int_0^T \left(\frac{||\langle \sigma(\mathbf{x},t) \rangle_\Omega - \sigma^D(t)||}{||\sigma^D(t)||} \right) dt + \text{CONSTRAINTS}, \tag{10.36}$$

where $\sigma(\mathbf{x},t)$ is the microscopic stress response of the material, $\sigma^D(t)$ is a prespecified desired effective macroscopic stress response, T is the total time interval of interest, and where $||\cdot||$ is an appropriate norm. Such an objective function depends in a nonconvex and nondifferentiable manner on the design parameters, denoted by the vector Λ. One approach is to employ algorithms which combine gradient methods for local searches with genetic algorithms for global searches. A recent overview of the state of the art of genetic algorithms can be found in a collection of articles edited by Goldberg and Deb [64]. There are a variety of such methods, which employ concepts of species evolution, such as reproduction, mutation and crossover. The application of such ideas to non-multifield material design can be found in Zohdi [248] and [249] as well as in previous chapters.

Chapter 11
Closing Comments

It is frequently asked of a theoretical or computational analyst: "Have the theoretical and computational results been compared to experiments?" Computational and theoretical micro–macro mechanics open the reverse question: "Have the experimental results been compared to computations and theory?" At the risk of closing on a controversial note, it is not unfair to think that, in many situations, computational methods are perhaps more predictive than experiments. Furthermore, the continual rise in computational power will only push research more towards reliance on numerical simulations, employing non-phenomenological micro–macro models. *For successful modern research, computations, theory and experiments should be symbiotic and interactive.*

References

1. Aboudi, J. (1992). Mechanics of composite materials-a unified micromechanical approach. Elsevier, 29.
2. Ainsworth, M. and Oden, J. T. (2000). A posterori error estimation in finite element analysis. John-Wiley.
3. Ames, W. F. (1965). Nonlinear Partial differential equations in engineering. Academic Press.
4. Ames, W. F. (1977). Numerical methods for partial differential equations. 2nd edition. Academic Press.
5. Amieur, M., Hazanov, S. and Huet, C. (1993). Numerical and experimental study of size and boundary conditions effects on the apparent properties of specimens not having the representative volume. In: C. Huet, (Ed.). Micromechanics of Concrete and Cementitious Composite.
6. Amieur, M. (1994). Etude numérique et expérimentale des effets d'échelle et de conditions aux limites sur des éprouvettes de béton n'ayant pas le volume représentatif. Doctoral dissertation No 1256, Ecole Polytechnique Fédérale de Lausanne, Lausanne, Switzerland.
7. Amieur, M., Hazanov, S. and Huet, C. (1995). Numerical and experimental assessment of the size and boundary conditions effects for the overall properties of granular composite bodies smaller than the representative volume. In: D. F. Parker and A. H. England, (Eds.). IUTAM Symposium on Anisotropy, Inhomogeneity and Nonlinearity in Solid Mechanics, Kluwer Academic Publishers, The Netherlands. 149–154.
8. Ashby, M. A. and Jones, D. R. H. (1992). Engineering materials 2. An introduction to microstructures, processing and design. Pergamon Press.
9. Axelsson, O. (1994). Iterative solution methods. Cambridge University Press.
10. Babúska, I. and Rheinbolt, W. C. (1978). A posteriori error estimates for the finite element method. The International Journal for Numerical Methods in Engineering. **12**, 1597–1615.
11. Babúska, I. and Miller, A. D. (1987). A feedback finite element method with a-posteriori error estimation. Part I. Computer Methods in Applied Mechanics and Engineering. **61**, 1–40.
12. Babúska, I., Andersson, B., Smith, P. J. and Levin, K. (1999). Hierarchical modeling of heterogeneous solids. Damage analysis of fiber composites Part I: Statistical analysis on fiber scale. Computer Methods in Applied Mechanics and Engineering. **172**, 27–77.
13. Basquin, O. H. (1910). The exponential law of endurance tests. Proceedings of the American Society for Testing and Materials. **10**, 625–630.
14. Bathe, K. J. (1996). Finite element procedures. Prentice-Hall.
15. Becker, E. B., Carey, G. F. and Oden J. T. (1980). Finite elements: an introduction. Prentice-Hall.
16. Belsky, V., Beall, M. W., Fish, J., Shephard, M. S. and Gomaa, S. (1995). Computer-aided multiscale modeling tools for composite materials and structures. International Journal of Computing Systems in Engineering. **6**(3), 213–223.

17. Bonet, J. and Wood, R. D. (1997). Nonlinear continuum mechanics for finite element analysis. Cambridge University Press.
18. Briggs, W. (1987). A multigrid tutorial. Siam Issue.
19. Budiansky, B. (1965). On the elastic moduli of some heterogeneous materials. Journal of the Mechanics and Physics of Solids. **13**, 223–227.
20. Carol, I., Rizzi, E. and Willam, K. (1994). A unified theory of elastic degradation and damage based on a loading surface. The International Journal of Solids and Structures. **31**(20), 2835–2865.
21. Carey, G. F. and Oden J. T. (1983). Finite elements: a second course. Prentice-hall.
22. Casey, J. (1985). Approximate kinematical relations in plasticity. The International Journal of Solids and Structures. **21**, 671–682.
23. Champaney, L., Cognard, J., Dureisseix D. and Ladeveze, P. (1997). Large scale applications on parallel computers of a mixed domain decomposition method. Computational Mechanics. **19**(4), 253–263.
24. Chandrasekharaiah, D. S. and Debnath, L. (1994). Continuum mechanics. Academic press.
25. Chen, W. and Fish, J. (2001). A dispersive model for wave propagation in periodic heterogeneous media based on homogenization with multiple spatial and temporal scales. Journal of Applied Mechanics. **68**(2), 153–161.
26. Cherkaev, A. V. and Gibiansky, L. V. (1992). The exact coupled bounds for effective tensors of electrical and magnetic properties of two-component two-dimensional composites. Proceedings of the Royal Society of Edinburgh. **122A**, 93–125.
27. Christensen, R. (1990). A critical evaluation for a class of micromechanics models. Journal of the Mechanics and Physics of Solids. **38**(3), 379–404.
28. Ciarlet, P. G. (1993). Mathematical elasticity. Elsevier.
29. Courant, R. (1943). Variational methods for the solution of problems of equilibrium and vibration. Bulletin of the American Mathematical Society. **49**, 1–23.
30. Crank, J. (1975). The mathematics of diffusion. 2nd. edition. Oxford Science Publications,
31. Davidon, W. C. (1959). Variable metric method for minimization. Research and development report. ANL-5990 (Ref.). U.S. Atomic Energy Commision, Argonne National Laboratories.
32. Davis, L. (1991). Handbook of genetic algorithms. Thompson Computer Press.
33. Do, I. P. H. and Benson, D. (2001). Micromechanical modeling of shock-induced chemical reactions in multi-material powder mixtures. International Journal of Plasticity. **17**, 641–668.
34. Doltsinis, I. St. (1993). Coupled field problems–solution techniques for sequential and parallel processing. In: M. Papadrakakis (Ed.). Solving Large–scale Problems in Mechanics.
35. Doltsinis, I. St. (1997). Solution of coupled systems by distinct operators. Engineering Computations. **14**, 829–868.
36. Eshelby, J. D. (1957). The elastic field of an ellipsoidal inclusion, and related problems. Proceedings of the Royal Society **A241**, 376–396.
37. Eustathopoulos, N. and Mortensen, A. (1993). Capillary phenomena, interfacial bonding and reactivity. In: S. Suresh, A. Mortensen and A. Needleman (Eds.). Fundamentals of Metal Matrix Composites. Butterworth-Heineman publishers.
38. Fiacco, A. V. and McCormick G. P. (1990). Nonlinear programming: sequential unconstrained minimization techniques. Siam reissue. John Wiley and Sons, New York and Toronto.
39. Fish, J. and Wagiman, A. (1993). Multiscale finite element method for a locally nonperiodic heterogeneous medium. Computational Machanics. **12**, 164–180.
40. Fish, J., Pandheeradi, M. and Belsky, V. (1995). An efficient multilevel solution scheme for large scale nonlinear systems. International Journal for Numerical Methods in Engineering. **38**, 1597–1610.
41. Fish, J. and Belsky, V. (1995). Multigrid method for periodic heterogeneous media Part I: Convergence studies for one dimensional case. Computer Methods in Applied Mechanics and Engineering. **126**, 1–16.
42. Fish, J. and Belsky, V. (1995). Multigrid method for periodic heterogeneous media Part II: Multiscale modeling and quality control in multidimensional case. Computer Methods in Applied Mechanics and Engineering. **126**, 17–38.

43. Fish, J. and Belsky, V. (1997). Generalized aggregation multilevel solver. International Journal for Numerical Methods in Engineering. **40**, 4341–4361.
44. Fish, J., Shek, K., Pandheeradi, M. and Shephard, M. S. (1997). Computational plasticity for composite structures based on mathematical homogenization: Theory and practice. Computer Methods in Applied Mechanics and Engineering. **148**(1–2), 53–73.
45. Fish, J., Yu, Q. and Shek, K. L. (1999). Computational damage mechanics for composite materials based on mathematical homogenization. International Journal for Numerical Methods in Engineering. **45**, 1657–1679.
46. Fish, J. and Shek, K. (1999). Finite deformation plasticity for composite structures: Computational models and adaptive strategies. Computer Methods in Applied Mechanics and Engineering. **172**, 145–174.
47. Fish, J. and Ghouli, A. (2001). Multiscale analytical sensitivity analysis for composite materials. International Journal for Numerical Methods in Engineering. **50**, 1501–1520.
48. Fish, J. and Yu, Q. (2001). Multiscale damage modeling for composite materials: theory and computational framework. International Journal for Numerical Methods in Engineering. **52**(1–2), 161–192.
49. Fish, J. and Chen, W. (2001). Uniformly valid multiple spatial-temporal scale modeling for wave propagation in heterogeneous media. Mechanics of Composite Materials and Structures. **8**, 81–99.
50. Flynn, C. P. (1972). Point defects and diffusion. Clarendon Press, Oxford.
51. Fu, Y., Klimkowski, K., Rodin, G. J., Berger, E., Browne, J. C., Singer, J. K., Van de Geijn, R. A. and Vemaganti, K. (1998). Fast solution method for three-dimensional many-particle problems of linear elasticity. The International Journal of Numerical Methods in Engineering. **42**, 1215–1229.
52. Fletcher, R. and Powell, M. J. D. (1963). A rapidly convergent descent method for minimization. Computer Journal. **7**, 149–154.
53. Frankel, S. P. (1950). Convergence rates of iterative treatments of partial differential equations. Mathematical Tables and Other Aids to Computations. **4**, 65–75.
54. Ghosh, S. and Mukhopadhyay, S. N. (1993). A material based finite element analysis of heterogeneous media involving Dirichlet tessellations. Computer Methods in Applied Mechanics and Engineering. **104**, 211–247.
55. Ghosh, S. and Moorthy, S. (1995). Elastic-plastic analysis of arbitrary heterogeneous materials with the Voronoi Cell finite element method. Computer Methods in Applied Mechanics and Engineering. **121**(1–4), 373–409.
56. Ghosh, S., Kyunghoon, L. and Moorthy, S. (1996). Two scale analysis of heterogeneous elastic-plastic materials with asymptotic homogenization and Voronoi cell finite element model. Computer Methods in Applied Mechanics and Engineering. **132**(1–2), 63–116
57. Ghosh, S. and Moorthy, S. (1998). Particle fracture simulation in non-uniform microstructures of metal-matrix composites. Acta mater. **46**(3), 965–982.
58. Ghosh, S., Lee, K. and Raghavan, P. (2001). A multi-level computational model for multiscale damage analysis in composite and porous materials. International Journal of Solids and Structures. **38**, 2335–2385.
59. Ghosh, S., Ling, Y., Majumdar, B. and Kim, R. (2001). Interfacial debonding analysis in multiple fiber reinforced composites. Mechanics of Materials. **32**, 562–591.
60. Ghosh, S., Kyunghoon, L. and Moorthy, S. (1996). Two scale analysis of heterogeneous elastic-plastic materials with asymptotic homogenization and Voronoi cell finite element model. Computer Methods in Applied Mechanics and Engineering. **132**(1–2), 63–116.
61. Gill, P. Murray, W. and Wright, M. (1995). Practical optimization. Academic Press.
62. Golanski, D., Terada, K. and Kikuchi, N. (1997). Macro and micro scale modeling of thermal residual stresses in metal matrix composite surface layers by the homogenization method. Computational Mechanics. **19**(3), 188–202.
63. Goldberg, D. E. (1989). Genetic algorithms in search, optimization and machine learning. Addison-Wesley.
64. Goldberg, D. E. and Deb, K. (2000). Special issue on Genetic Algorithms. Computer Methods in Applied Mechanics and Engineering. **186**(2–4), 121–124.

65. González, C. and Llorca, J. (2003). An analysis of the effect of hydrostatic pressure on the tensile deformation of aluminum-matrix composites. Materials Science and Engineering: A, Elsevier. **341**(1), 256–263(8).
66. González, C. and Llorca, J. (2001). Micromechanical modelling of deformation and failure in Ti-6Al-4V/SiC composites. Acta Materialia. **49**, 3505–3519.
67. González, C. and Llorca, J. (2000). A self-consistent approach to the elasto-plastic behaviour of two-phase materials including damage. Journal of the Mechanics and Physics of Solids. **48**, 675–692.
68. Green, A. E. and Naghdi, P. M. (1965). A general theory of an elasto-plastic continuum. Archive for Rational Mechanics and Analysis. **18**, 251–281.
69. Green, A. E. and Naghdi, P. M. (1971). Some remarks on elastoplastic deformation at finite strain. International Journal of Engineering Science. **9**, 1219–1229.
70. Guedes, J. M. and Kikuchi, N. (1990). Preprocessing and postprocessing for materials based on the homogenization method with adaptive finite element methods. Computer Methods in Applied Mechanics and Engineering. **83**, 143–198.
71. Guidoum, A. and Navi, P. (1993). Numerical simulation of thermo-mechanical behaviour of concrete through a 3-D granular cohesive model. In: C. Huet, (Ed.) Micromechanics of Concrete and Cementitious Composites. Presses Polytechniques et Universitaires Romandes, Lausanne. pp. 213–228.
72. Guidoum, A. (1994). Simulation numérique 3D des comportements des bétons en tant que composites granulaires. Doctoral Dissertation No 1310. Ecole Polytechnique de Lausanne, Switzerland.
73. Gurtin, M. (1981). An introduction to continuum mechanics. Academic Press.
74. Hashin, Z. and Shtrikman, S. (1962). On some variational principles in anisotropic and nonhomogeneous elasticity. Journal of the Mechanics and Physics of Solids. **10**, 335–342.
75. Hashin, Z. and Shtrikman, S. (1963). A variational approach to the theory of the elastic behaviour of multiphase materials. Journal of the Mechanics and Physics of Solids. **11**, 127–140.
76. Hashin, Z. (1983) Analysis of composite materials: a survey. ASME Journal of Applied Mechanics. **50**, 481–505.
77. Hazanov, S. and Huet, C. (1994). Order relationships for boundary conditions effect in heterogeneous bodies smaller than the representative volume. Journal of the Mechanics and Physics of Solids. **42**, 1995–2011.
78. Hazanov, S. and Amieur, M. (1995). On overall properties of elastic heterogeneous bodies smaller than the representative volume. International Journal of Engineering Science. **33**(9), 1289–1301.
79. Hill, R. (1952). The elastic behaviour of a crystalline aggregate. Proceedings of the Physical Society (London) **A65**, 349–354.
80. Hill, R. (1965). A self consistent mechanics of composite materials. Journal of the Mechanics and Physics of Solids. **13**, 213–222
81. Hill, R. and Rice, J. R. (1973). Elastic potentials and the structure of inelastic constitutive laws. SIAM Journal of Applied Mathematics. **25**, 448–461.
82. Hill, R. (1963). New derivations of some elastic extremum principles. In Progress in Applied Mechanics, Prager Anniversary Volume New York. pp. 91–106.
83. Hill, R. (1963). Elastic Properties of reinforced solids: some theoretical principles. Journal of the Mechanics and Physics of Solids. **11**, 357–372.
84. Holland, J. H. (1975). Adaptation in natural and artificial systems. University of Michigan Press, Ann Arbor, Michigan.
85. Horst, R. and Tuy, H. (1996). Global optimization-deterministic approaches. Springer–Verlag.
86. Hsiao, S. W. and Kikuchi, N. (1997). Numerical analysis of deep drawing process for thermoplastic composite laminates. Journal of Engineering Materials and Technology, ASME Transactions. **119**, 314–318.

87. Huet, C. (1981). Remarques sur l'assimilation d'un matériau hétérogéne á un milieu continu équivalent. In: C. Huet and A. Zaoui, (Eds.). Rheological behaviour and Structure of Materials. Presses ENPC, Paris. pp. 231–245.
88. Huet, C. (1982) Universal conditions for assimilation of a heterogeneous material to an effective medium. Mechanics Research Communications. 9(3), 165–170.
89. Huet, C. (1984). On the definition and experimental determination of effective constitutive equations for heterogeneous materials. Mechanics Research Communications. **11**(3), 195–200.
90. Huet, C. (1987). Thermo-hygro-mechanical couplings in wood technology and Rheological Behaviours. In: H. D. Bui and Q. S. Nguyen, (Eds.). Thermomechanical couplings in solids, IUTAM symposium, Paris, 1986. North-Holland. pp. 163–182.
91. Huet, C. (1988). Modelizing the kinetics of the thermo-hygro-viscoelastic behavior of wood under constant climatic conditions. In: R. Itani, (Ed.). Proceedings of the 1988 International Conference on Timber Engineering, Seattle. pp. 395–401
92. Huet, C. (1990). Application of variational concepts to size effects in elastic heterogeneous bodies. Journal of the Mechanics and Physics of Solids. **38**, 813–841.
93. Huet, C. (1991). Hierarchies and bounds for size effects in heterogeneous bodies. In: G. A. Maugin, (Ed). Continuum Models and Discrete Systems, Vol. 2, pp. 127–134.
94. Huet, C., Navi, P. and Roelfstra, P. E. (1991). A homogenization technique based on Hill's modification theorem. In Maugin, G.A. (Eds), Continuum Models and Discrete Systems, Longman, Harlow, Vol. 2 , pp. 135–43.
95. Huet, C. (1995). Hybrid continuum thermodynamics framework and numerical simulations examples for the delayed micromechanical behavior of heterogeneous materials with chemical, climatic and defect sensitivity. In: Q. S. Nguyen and V. D. Nguyen, (Eds.). Engineering Mechanics Today. Proceedings of the International Conference Hanoi, University of Hanoi, pp. 170–184.
96. Huet, C., Guidoum, A. and Navi, P. (1995). A 3D micromechanical model for numerical analysis and prediction of long term deterioration in concrete. In: K. Sakai, N. Banthia, and O. E. Gjorv, (Eds.). Concrete Under Severe Conditions: Environment and Loading. Vol. 2, Spon, London, pp. 1458–1467.
97. Huet, C. (1996). Recent advances in the long term deformation and deterioration behavior of structural materials and components through the integrated micromechanics and Thermodynamics of Solids Approach. In: A. Gerdes, (Ed.). Advances in Building Materials. Aedificatio, Freiburg. pp. 161–196.
98. Huet, C. (1997). Activities 1989–1996. Laboratory for Building Materials. P. Navi and A. Tolou (Eds.). Department of Materials Report.
99. Huet C. (1997). An integrated micromechanics and statistical continuum thermodynamics approach for studying the fracture behaviour of microcracked heterogeneous materials with delayed response. Engineering Fracture Mechanics, Special Issue. **58**(5–6), 459–556.
100. Huet, C. (1999). Coupled size and boundary condition effects in viscoelastic heterogeneous bodies. Mechanics of Materials. **31**(12), 787–829.
101. Hughes, T. J. R. (1989). The finite element method. Prentice Hall.
102. Hyun, S. and Torquato, S. (2001). Designing material microstructures with target properties. Journal of Materials Research. **16**, 280.
103. Jikov, V. V., Kozlov, S. M. and Olenik, O. A. (1994). Homogenization of differential operators and integral functionals. Springer-Verlag.
104. Kachanov, L. M. (1986). Introduction to continuum damage mechanics. Martinus Nijoff, Dordricht.
105. Kee, A., Matic, P. and Everett, R. K. (1998). A mesoscale computer simulation of multiaxial yield in gasar porous copper. Material Science and Engineering **A249**, 30–39.
106. Kelly, D. W., Gago, J. R., Zienkiewicz, O. C. and Babùska, I. (1983). A posteriori error analysis and adaptive processes in the finite element method. Part I-error analysis. The International Journal for Numerical Methods in Engineering. **19**, 1593–1619.
107. Kennedy, J. and Eberhart, R. (2001). Swarm Intelligence. Morgan Kaufmann Publishers.

108. Kikuchi, N., Nishiwaki, S., Fonseca, J.O. and Silva, E. C. N. (1998). Design optimization method for compliant mechanisms and material microstructure. Computer Methods in Applied Mechanics and Engineering. **151**, 401–417.
109. Kitchen, J. (1966). Concerning the convergence of iterates to fixed points. Studia Mathematica. **27**, 247–249.
110. Krajcinovic, D. (1996). Damage mechanics. North Holland.
111. Kröner, E. (1972). Statistical continuum mechanics. CISM Lecture Notes. **92**, Springer-Verlag.
112. Ladeveze, P. and Leguillon, D. (1983). Error estimate procedure in the finite element method and applications. SIAM Journal of Numerical Analysis **20**, 485–509.
113. Ladeveze, P. and Dureisseix, D. (2000). A micro/macro approach for parallel computing of heterogeneous structures. International Journal for Computational Civil and Structural Engineering. **1**(1), 18–28.
114. Ladeveze, P. (1998). A modelling error estimator for dynamic structural model updating. Advances in Adaptive Computational Methods in Mechanics. P. Ladeveze and J. T. Oden, Elsevier, pp. 135–151. Proceedings of the Workshop on New Advances in Adaptive Computational Methods in Mechanics, Cachan, 1997.
115. Ladeveze, P. and Dureisseix, D. (1999). Une nouvelle stratégie de calcul micro/macro en mécanique des strutures. C.R.A.S. Série IIb **327**, 1237–1244.
116. Lee, K., Moorthy, S. and Ghosh, S. (1999). Multiple scale computational model for damage in composite materials, Effective properties of composite materials with periodic microstructure: a computational approach. Computer Methods in Applied Mechanics and Engineering. **172**, 175–201.
117. Lemaitre, J. (1985). Coupled elasto-plasticity and damage constitutive equations. Computer Methods in Applied Mechanics and Engineering. **51**, 31–49.
118. Lemaitre, J. and Chaboche, J.-L. (1990) Mechanics of solid materials. Cambridge University Press, Cambridge.
119. Le Tallec, P. (1994). Domain decomposition methods in computational mechanics. Computational Mechanics Advances. **1**, 121–220.
120. Le Tallec, P. and Mouro, J. (2001). Fluid structure interaction with large structural displacements. Computer Methods in Applied Mechanics and Engineering. **190**(24–25), 3039–3067.
121. Lewis, R. W., Schrefler, B. A. and Simoni, L. (1992). Coupling versus uncoupling in soil consolidation. International Journal for Numerical and Analytical Methods in Geomechanics. **15**, 533–548.
122. Lewis, R. W. and Schrefler, B. A. (1998). The finite element method in the static and dynamic deformation and consolidation of porous media. 2nd edition. Wiley press.
123. Li, M., Ghosh, S. and Richmond, O. (1999). An experimental-computational approach to the investigation of damage evolution in discontinuously reinforced aluminum matrix composite. Acta mater. **47**(12) 3515–3532.
124. Liu, H. W. (1970). Stress corrosion cracking and the interaction between crack-tip stress-field and solute atoms. Journal of Basic Engineering ASME. **92D**, 633–638.
125. Llorca, J. (2000). Void formation in Metal Matrix Composites. Comprehensive Composite Materials. Vol. 3, Metal Matrix Composites. (T. W. Clyne, Ed.), Pergamon, Amsterdam, pp. 91–115159.
126. Llorca, J., Elices, M., and Termonia, Y. (2000). Elastic properties of sphere-reinforced composites with a mesophase. Acta Materialia. **48**, 4589–4597.
127. Llorca, J. and González, C. (1998). Microstructural factors controlling the strength and ductility of particle-reinforced metal-matrix composites. Journal of the Mechanics and Physics of Solids. **46**, 1–28.
128. Llorca, J. (1994). A numerical stydy of the mechanisms of cyclic strain hardening in metal-ceramic composites. Acta Metallurgica et Materialia. **42**, 151–162.
129. Llorca, J., Needleman, A. and Suresh, S. (1991). An analysis of the effects of matrix void growth on deformation and ductility in metal-ceramic composites. Acta Metallurgica et Materialia. **39**, 2317–2335.

130. Lubliner, J. (1990). Plasticity theory. Macmillan Publishers.
131. Luenberger, D. (1974). Introduction to linear and nonlinear programming. Addison-Wesley, Menlo Park.
132. Miner, M. A. (1945). Cumulative damage in fatigue. Journal of Applied Mechanics. **12**, 159–164.
133. Malvern, L. (1968). Introduction to the mechanics of a continuous medium. Prentice Hall.
134. Markenscoff, X. (2001). Diffusion induced instability. Quarterly of Applied Mechanics. **59**(1), 147–151.
135. Markenscoff, X. (2001). Instabilities of a thermo-mechano-chemical system. Quarterly of Applied Mechanics. **59**(3), 471–477.
136. Markenscoff, X. (2003). On conditions of "negative creep" in amorphous solids. Mechanics of Materials. **35**(3–6), 553–557.
137. Marsden, J. E. and Hughes, T. J. R. (1983). Mathematical foundations of elasticity. Prentice Hall.
138. Maxwell, J. C. (1867). On the dynamical theory of gases. Philosophical Transactions of the Royal Society of London. **157**, 49.
139. Maxwell, J. C. (1873). A treatise on electricity and magnetism. 3rd. edition. Clarendon Press, Oxford.
140. Metals Handbook. (1975). Failure Analysis and Prevention. **8**, The American Society for Metals.
141. Meyers, M. A., Yu, L. H. and Vecchio, K. S. (1994). Shock synthesis in silicides-ii. Thermodynamics and kinetics. Acta Metallurgica Materialia. **42**(3), 715–729.
142. Meyers, M. A. (1994). Dynamic behavior of materials. John-Wiley.
143. Michaud, V. (1992). Liquid state processing. In: S. Suresh, A. Mortensen and A. Needleman, (Eds.). Fundamentals of Metal Matrix Composites.
144. Michel, J. C., Moulinec, H. and Suquet, P. (1999). Effective properties of composite materials with periodic microstructure: a computational approach. Computer Methods in Applied Mechanics and Engineering. **172**, 109–143.
145. Moes, N., Oden, J. T. and Zohdi, T. I. (1998). Investigation of the interaction of numerical error and modeling error in the Homogenized Dirichlet Projection Method. Computer Methods in Applied Mechanics and Engineering. **159**, 79–101.
146. Moorthy, S. and Ghosh, S. (2000). Adaptivity and convergence in the Voronoi cell finite element model for analyzing heterogeneous materials. Computer Methods in Applied Mechanics and Engineering. **185**, 37–74.
147. Moulinec, H. and Suquet, P. (1998). A numerical method for computing the overall response of nonlinear composites with complex microstructure. Computer Methods in Applied Mechanics and Engineering. **157**, 69–94.
148. Mori, T. and Tanaka, K. (1973) Average stress in matrix and average energy of materials with misfitting inclusions. Acta. Metall. **21**, 571–574.
149. Moorthy, S. and Ghosh, S. (2000). Adaptivity and convergence in the Voronoi cell finite element model for analyzing heterogeneous materials. Computer Methods in Applied Mechanics and Engineering. **185**, 37–74.
150. Mura, T. (1993). Micromechanics of defects in solids, 2nd edition. Kluwer Academic Publishers.
151. Nemat-Nasser, S. and Hori, M. (1999). Micromechanics: overall properties of heterogeneous solids. 2nd edition. Elsevier, Amsterdam.
152. Nesterenko, V. F. (2001). Dynamics of heterogeneous materials. Springer-Verlag.
153. Nesterenko, V. F., Meyers, M. A., Chen, H. C. and LaSalvia, J. C. (1994). Controlled high rate localized shear in porous reactive media. Applied Physics Letters. **65**(24), 3069–3071.
154. Nye, J. (1984). Physical properties of crystals. Oxford science publications.
155. Oden J. T. and Carey, G. F. (1984). Finite elements: mathematical aspects. Prentice-hall.
156. Oden, J. T. and Zohdi, T. I. (1997). Analysis and adaptive modeling of highly heterogeneous elastic structures. Computer Methods in Applied Mechanics and Engineering. **148**, 367–391.
157. Oden, J. T. and Vemaganti, K. (1999). Adaptive hierarchical modeling of heterogeneous structures. Physica D. **133**, 404–415.

158. Oden, J. T., Vemaganti, K. and Moes, N. (1999). Hierarchical modeling of heterogeneous solids. Computer Methods in Applied Mechanics and Engineering. **172**, (1–27).
159. Ogden, R. W. (1999). Nonlinear elasticity. Dover reissue.
160. Onwubiko, C. (2000). Introduction to engineering design optimization. Prentice Hall.
161. Orowan, E. (1944). Proceedings if the Institution of Mechanical Engineers. **151**, 133 (discussion of paper by H. O'Neill).
162. Ortega, J. and Rockoff, M. (1966). Nonlinear difference equations and Gauss-Seidel type iterative methods. SIAM. Journal on Numerical Analysis **3**, 497–513.
163. Ortiz, M. and Pandolfi, A. (1999). Finite deformation irreversible cohesive elements for three-dimensional crack-propagation analysis. The International Journal of Numerical Methods in Engineering. **44**, 1267–1282.
164. Ostrowski, A. (1957). Les points d' attraction et de r'epulsion pour l'it'eration dans l'espace a' n dimensions. C. R. Acad. Sci. Paris. **244**, 288–289.
165. Ostrowski, A. (1966). Solution of equations and systems of equations. Academic Press, New York.
166. Palmgren, A. (1924). Die Lebensdauer von Kugellagern. Zeitschrift des Vereins Deutscher Ingenieure. **68**, 339–341.
167. Park, K. C. and Felippa, C. A. (1983). Partitioned analysis of coupled systems. In: T. Belytschko and T. J. R. Hughes, (Eds.). Computational Methods for Transient Analysis.
168. Perron, O. (1929). Über Stabilität und asyptotisches Verhalten der Lösungen eines Systems endlicher Differenzengleichungen. J. Reine Angew. Math. **161**, 41–64.
169. Piperno, S. (1997). Explicit/implicit fluid/structure staggered procedures with a structural predictor and fluid subcycling for 2D inviscid aeroelastic simulations. International Journal for Numerical Methods in Fluids. **25**, 1207–1226.
170. Poza, P. and Llorca, J. (1999). Mechanical Behaviour of Al-Li/SiC composites. Part III: Micromechanical modeling. Metallurgical and Materials Transactions **30A**, 869–878.
171. Raghavan, P., Moorthy, S., Ghosh, S. and Pagano, N. J. (2001). Revisiting the composite laminate problem with an adaptive multi-level computational model. Composites Science and Technology. **61**, 1017–1040.
172. Rayleigh, J. W. (1892). On the influence of obstacles arranged in rectangular order upon properties of a medium. Philosophical Magazine **32**, 481–491.
173. Reuss, A. (1929). Berechnung der Fliessgrenze von Mischkristallen auf Grund der Plastizitätsbedingung für Einkristalle. Z. angew. Math. Mech. **9**, 49–58.
174. Schmidt, L.(1998). The engineering of chemical reactions, Oxford University Press.
175. Schrefler, B. A. (1985). A partitioned solution procedure for geothermal reservoir analysis. Communications in Applied Numerical Methods **1**, 53–56.
176. Segurado, J. and Llorca, J. (2002). A numerical approximation to the elastic properties of sphere-reinforced composites. Journal of the Mechanics and Physics of Solids, **50**.
177. Segurado, J., Llorca, J. and González, C. (2002). On the accuracy of mean-field approaches to simulate the plastic deformation of composites. Scripta Materialia. **46**, 525–529.
178. Shackelford, J. (1996). Introduction to materials science for engineers, 4th edition, Macmillan Publishers.
179. Silva, E. C. N., Fonseca, J. S. O. and Kikuchi, N. (1997). Optimal design of piezoelectric microstructures. Computational Mechanics. **19**, 397–410.
180. Silva, E. C. N., Fonseca, J. S. O. and Kikuchi, N. (1998). Optimal design of periodic piezocomposites. Computer Methods in Applied Mechanics and Engineering. **159**, 49–77.
181. Sigmund, O. (2000). Topology optimization: A tool for the tailoring of structures and materials. A special issue of the philosophical. Transactions of the Royal Society: Science into the next Millennium (Issue III, Mathematics, Physics and Engineering). **358**(1765), 211–228.
182. Sigmund, O. (2000). A new Class of Extremal Composites. Journal of the Mechanics and Physics of Solids. **48**(2), 397–428.
183. Sigmund, O. and Torquato, S. (1999). Design of smart composite materials using topology optimization. Smart Materials and Structures. 8, pp. 365–379.
184. Sigmund, O. and Torquato, S. (1996). Composites with extremal thermal expansion coefficients. Applied Physics Letters. **69**(21): 3203–3205.

185. Sigmund, O. (1995). Tailoring materials with prescribed elastic properties. Mechanics of Materials. **20**, 351–368.
186. Sigmund, O. and Torquato, S. (1999). Design of smart composite materials using topology optimization. Smart Materials and Structures. **8**, 365.
187. Sigmund, O., Torquato, S. and Aksay, I. A. (1998). On the optimization of 1–3 piezocomposites using topology optimization. Journal of Materials Research. **14**, 1038.
188. Sigmund, O. (2001). Design of multiphysics actuators using topology optimization – Part I: One-material structures. Computer Methods in Applied Mechanics and Engineering. 190(49–50), 6577–6604.
189. Sigmund, O. (2001). Design of multiphysics actuators using topology optimization – Part II: Two-material structures. Computer Methods in Applied Mechanics and Engineering **190**(49–50), 6605–6627.
190. Simo, J. C. and Hughes, T. J. R (1998). Computational inelasticity. Springer-Verlag.
191. Simo, J. C. and Ortiz, M. (1985). A unified approach to finite deformation elastoplastic analysis based on the use of hyperelastic constitutive laws. Computer Methods in Applied Mechanics and Engineering. **49**, 221–245.
192. Simo, J. C. (1988). A framework for finite strain elastoplasticity based on maximum plastic dissipation and the multiplicative decomposition: Part 1. Continuum formulation. Computer Methods in Applied Mechanics and Engineering. **66**, 199–219.
193. Southwell, R. V. (1940). Relaxation methods in engineering science. Oxford University Press.
194. Southwell, R. V. (1946). Relaxation methods in theoretical physics. Oxford University Press.
195. Suresh, S., Mortensen, A. and Needleman, A. (1993). Fundamentals of metal matrix composites. Butterworth-Heineman publishers.
196. Szabo, B. and Babúska, I. (1991). Finite element analysis. Wiley Interscience.
197. Terada, K. and Kikuchi, N. (1996). Microstructural design of composites using the homogenization method and digital images. Materials Science Research International. 2(2), 65–72.
198. Terada, K., and Kikuchi, N. (1996). Global-local constitutive modeling of composite materials by the homogenization method. Materials Science Research International. 2(2), 73–80.
199. Terada, K., Hori, M., Kyoya, T. and Kikuchi, N. (2000). Simulation of the multi-scale convergence in computational homogenization approaches. The International Journal of Solids and Structures. **37**, 2229–2361.
200. Thadhani, N. N. (1993). Shock-induced chemical reactions and synthesis of materials. Progress in material science. **37**, 117–226.
201. Torquato, S. (1991). Random heterogeneous media: microstructure and improved bounds on effective properties. Applied Mechnics Reviews **44**, 37–76.
202. Torquato, S. (1997). Effective stiffness tensor of composite media I. Exact series expansions. Journal of the Mechanics and Physics of Solids. **45**, 1421–1448.
203. Torquato, S. (1998). Effective stiffness tensor of composite media II. Applications to isotropic dispersions. Journal of the Mechanics and Physics of Solids. **46**, 1411–1440.
204. Torquato, S. (2002). Random heterogeneous materials: Microstructure and macroscopic properties. Springer-Verlag, New York.
205. Torquato, S. and Hyun, S. (2001). Effective-medium approximation for composite media: Realizable single-scale dispersions. Journal of Applied Physics **89**, 1725–1729.
206. Truesdell, C. and Noll, W. (1992). The nonlinear field theories. Springer-Verlag.
207. Van Vlack, L. H. (1966). Elements of materials science. Addison-Wesley.
208. Vecchio, K. S. and Meyers, M. A. (1994). Shock synthesis in silicides-i. Experimentation and microstructural evolution, Acta Metallurgica Materialia. **42**(3), 701–714.
209. Vemaganti, K. S. and Oden, J. T. (2001). Estimation of local modeling error and goal-oriented adaptive modeling of heterogeneous materials. Computer Methods in Applied Mechanics and Engineering. **190**, 46–47, 6089–6124.
210. Vigdergauz, S. (2001). Genetic algorithm perpective to identify energy optimizing inclusions in an elastic plate. The International Journal of Solids and Structures. **38**, 6851–6867.

211. Vigdergauz, S. (2001). The effective properties of a perforated elastic plate: numerical optimization by genetic algorithm. The International Journal of Solids and Structures. **38**, 8593–8616.
212. Voigt, W. (1889). Über die Beziehung zwischen den beiden Elastizitätskonstanten isotroper Körper. Wied. Ann. **38**, 573–587.
213. Wentorf, R., Collar, R., Shephard, M. S. and Fish, J. (1999). Automated modeling for complex woven mesostructures. Computer Methods in Applied Mechanics and Engineering. **172**, 273–291.
214. Wriggers, P. (2001). Nichtlineare Finite-Element-Methoden. Springer-Verlag.
215. Wriggers, P., Zavarise, G. and Zohdi, T. I. (1998). A computational study of interfacial debonding damage in fibrous composite materials. Computational Materials Science. **12**, 39–56.
216. Wriggers, P. (2002). Computational contact mechanics. John-Wiley.
217. Young. D. M. (1950). Iterative methods for solving partial difference equations of elliptic type. Doctoral thesis. Harvard University.
218. Zhang, J., Lin, Z., Wong, A., Kikuchi, N., Li, V. C., Yee, A. F. and Nusholtz, G. S. (1997). Constitutive modeling and material characterization of polymetric foams. Journal of Engineering Materials and Technology, ASME Transactions. **119**, 279–283.
219. Zhigljavsky, A. A. (1991). Theory of global random search. Kluwer
220. Zienkiewicz, O. C. and Taylor R. L. (1991). The finite element method. Vols. I and II. McGraw-Hill.
221. Zienkiewicz, O. C. (1984). Coupled problems and their numerical solution. In: R. W. Lewis, P. Bettess and E. Hinton, (Eds.). Numerical methods in coupled systems. Wiley, Chichester. pp. 35–58.
222. Zienkiewicz, O. C. and Zhu, J. Z. (1987). A simple error estimator and adaptive procedure for practical engineering analysis. The International Journal for Numerical Methods in Engineering. **24**, 337–357.
223. Zohdi, T. I., Oden, J. T. and Rodin, G. J. (1996). Hierarchical modeling of heterogeneous bodies. Computer Methods in Applied Mechanics and Engineering. **138**, 273–298.
224. Zohdi, T. I., Feucht, M., Gross, D. and Wriggers, P. (1998). A description of macroscopic damage via microstructural relaxation. The International Journal of Numerical Methods in Engineering. **43**, 493–507.
225. Zohdi, T. I. and Wriggers, P. (1999). A domain decomposition method for bodies with microstructure based upon material regularization. The International Journal of Solids and Structures. **36**(17) 2507–2526.
226. Zohdi, T. I., Hutter, K. and Wriggers, P. (1999). A technique to describe the macroscopic pressure dependence of diffusive properties of solid materials containing heterogeneities. Computational Materials Science. **15**, 69–88.
227. Zohdi, T. I. and Wriggers, P. (1999). On the effects of microstress on macroscopic diffusion processes. Acta Mechanica. **136**(1–2), 91–107.
228. Zohdi, T. I. and Wriggers, P. (2000). A computational model for interfacial damage through microstructural cohesive zone relaxation. The International Journal of Fracture. **101**(3), L9–L14.
229. Zohdi, T. I. and Wriggers, P. (2000). On the sensitivity of homogenized material responses at infinitesimal and finite strains. Communications in Numerical Methods in Engineering. **16**, 657–670.
230. Zohdi, T. I. (2000). Overall solution-difference bounds on the effects of material inhomogeneities. The Journal of Elasticity. **58**(3), 249–255.
231. Zohdi, T. I. (2000). Some remarks on hydrogen trapping, The International Journal of Fracture, **106**(2), L9–L14.
232. Zohdi, T. I. and Wriggers, P. (2001). A model for simulating the deterioration of structural-scale material responses of microheterogeneous solids. Computer Methods in Applied Mechanics and Engineering. **190**(22–23), 2803–2823.

233. Zohdi, T. I. and Wriggers, P. (2001). Aspects of the computational testing of the mechanical properties of microheterogeneous material samples. The International Journal of Numerical Methods in Engineering. **50**, 2573–2599.
234. Zohdi, T. I. and Wriggers, P. (2001). Modeling and simulation of the decohesion of particulate aggregates in a binding matrix. Engineering Computations. **18**(1–2), 79–95.
235. Zohdi, T. I. and Wriggers, P. (2001). Computational micro-macro material testing. Archives of Computational Methods in Engineering. **8**(2), 131–228.
236. Zohdi, T. I., Wriggers, P. and Huet, C. (2001). A method of substructuring large-scale computational micromechanical problems. Computer Methods in Applied Mechanics and Engineering. **190**(43–44), 5639–5656.
237. Zohdi, T. I. (2001). Computational optimization of vortex manufacturing of advanced materials. Computer Methods in Applied Mechanics and Engineering. **190**(46–47), 6231–6256.
238. Zohdi, T. I. (2002). An adaptive-recursive staggering strategy for simulating multifield coupled processes in microheterogeneous solids. The International Journal of Numerical Methods in Engineering. **53**, 1511–1532.
239. Zohdi, T. I. (2001). On the propagation of microscale material uncertainty in a class of hyperelastic finite deformation stored energy functions. The International Journal of Fracture. **112**(2), 13–17.
240. Zohdi, T. I. (2002). Incorporation of microfield distortion into rapid effective property design. Mathematics and Mechanics of Solids. **7**(3), 237–254.
241. Zohdi, T. I. (2002). On the tailoring of microstructures for prescribed effective properties. The International Journal of Fracture. **114**(3), 15–20.
242. Zohdi, T. I., Kachanov, M. and Sevostianov, I. (2002). On perfectly-plastic flow in porous material. The International Journal of Plasticity. **18**, 1649–1659.
243. Zohdi, T. I. (2002). An adaptive-recursive staggering strategy for simulating multifield coupled processes in microheterogeneous solids. The International Journal of Numerical Methods in Engineering. **53**, 1511–1532.
244. Zohdi, T. I., Monteiro, P. J. M. and Lamour, V. (2002). The International Journal of Fracture. Extraction of elastic moduli from granular compacts. **115**(3), 49–54.
245. Zohdi, T. I. (2002). Bounding envelopes in multiphase material design. The Journal of Elasticity. **66**, 47–62.
246. Zohdi, T. I. (2003). Large-scale statistical inverse computation of inelastic accretion in transient granular flows. The International Journal of Nonlinear Mechanics. **8**, 1205–1219.
247. Zohdi, T. I. (2003). Compaction of cohesive hyperelastic granules at finite strains. June Issue of the Proceedings of the Royal Society.
248. Zohdi, T. I. (2003). Constrained inverse formulations in random material design. Computer Methods in Applied Mechanics and Engineering. **192**, 28–30, 18, 3179–3194.
249. Zohdi, T. I. (2003). Genetic optimization of statistically uncertain microheterogeneous solids. Philosophical Transactions of the Royal Society: Mathematical, Physical and Engineering Sciences. **361**(1806), 1021–1043.
250. Zohdi, T. I. (2004). Staggering error control for a class of inelastic processes in random microheterogeneous solids. The International Journal of Nonlinear Mechanics. Issue **39**, 281–297.
251. Zohdi, T. I. (2004). Modeling and simulation of a class of coupled thermo-chemo-mechanical processes in multiphase solids. Computer Methods in Applied Mechanics and Engineering. **193**(6–8), 679–699.
252. Zohdi, T. I. (2005). Statistical ensemble error bounds for homogenized microheterogeneous solids. Journal of Applied Mathematics and Physics. (Zeitschrift für Angewandte Mathematik und Physik). **56**(3), 497–515.